NATURAL HISTORY
UNIVERSAL LIBRARY

西方博物学大系

主编：江晓原

THE NATURAL HISTORY
OF BRITISH BIRDS

不列颠珍稀鸟类志

[英] 爱德华·多诺万 著

华东师范大学出版社

图书在版编目（CIP）数据

不列颠珍稀鸟类志 = The Natural history of British birds：
英文 / (英) 爱德华·多诺万 (Edward Donovan)著. — 上海：
华东师范大学出版社, 2018
（寰宇文献）
ISBN 978-7-5675-8397-9

Ⅰ.①不… Ⅱ.①爱… Ⅲ.①鸟类–动物志–英国–英文
Ⅳ.①Q959.708

中国版本图书馆CIP数据核字(2018)第230003号

不列颠珍稀鸟类志
The Natural history of British birds
(英) 爱德华·多诺万 (Edward Donovan)

特约策划　黄曙辉　徐　辰
责任编辑　庞　坚
特约编辑　许　倩
装帧设计　刘怡霖

出版发行　华东师范大学出版社
社　　　址　上海市中山北路3663号　邮编 200062
网　　　址　www.ecnupress.com.cn
电　　　话　021-60821666　行政传真　021-62572105
客服电话　021-62865537
门市（邮购）电话　021-62869887
地　　　址　上海市中山北路3663号华东师范大学校内先锋路口
网　　　店　http://hdsdcbs.tmall.com/

印　刷　者　虎彩印艺股份有限公司
开　　　本　787×1092　16开
印　　　张　74.25
版　　　次　2018年10月第1版
印　　　次　2018年10月第1次
书　　　号　ISBN 978-7-5675-8397-9
定　　　价　1300.00元（精装全二册）

出　版　人　王　焰

（如发现本版图书有印订质量问题，请寄回本社客服中心调换或电话021-62865537联系）

总　目

《西方博物学大系》总序

江晓原

　　《西方博物学大系》收录博物学著作超过一百种，时间跨度为 15 世纪至 1919 年，作者分布于 16 个国家，写作语种有英语、法语、拉丁语、德语、弗莱芒语等，涉及对象包括植物、昆虫、软体动物、两栖动物、爬行动物、哺乳动物、鸟类和人类等，西方博物学史上的经典著作大备于此编。

中西方"博物"传统及观念之异同

　　今天中文里的"博物学"一词，学者们认为对应的英语词汇是 Natural History，考其本义，在中国传统文化中并无现成对应词汇。在中国传统文化中原有"博物"一词，与"自然史"当然并不精确相同，甚至还有着相当大的区别，但是在"搜集自然界的物品"这种最原始的意义上，两者确实也大有相通之处，故以"博物学"对译 Natural History 一词，大体仍属可取，而且已被广泛接受。

　　已故科学史前辈刘祖慰教授尝言：古代中国人处理知识，如开中药铺，有数十上百小抽屉，将百药分门别类放入其中，即心安矣。刘教授言此，其辞若有憾焉——认为中国人不致力于寻求世界"所以然之理"，故不如西方之分析传统优越。然而古代中国人这种处理知识的风格，正与西方的博物学相通。

　　与此相对，西方的分析传统致力于探求各种现象和物体之间的相互关系，试图以此解释宇宙运行的原因。自古希腊开始，西方哲人即孜孜不倦建构各种几何模型，欲用以说明宇宙如何运行，其中最典型的代表，即为托勒密（Ptolemy）的宇宙体系。

　　比较两者，差别即在于：古代中国人主要关心外部世界"如何"运行，而以希腊为源头的西方知识传统（西方并非没有别的知识传统，只是未能光大而已）更关心世界"为何"如此运行。在线性发展无限进步的科学主义观念体系中，我们习惯于认为"为何"是在解决了"如何"之后的更高境界，故西方的分析传统比中国的传统更高明。

然而考之古代实际情形，如此简单的优劣结论未必能够成立。例如以天文学言之，古代东西方世界天文学的终极问题是共同的：给定任意地点和时刻，计算出太阳、月亮和五大行星（七政）的位置。古代中国人虽不致力于建立几何模型去解释七政"为何"如此运行，但他们用抽象的周期叠加（古代巴比伦也使用类似方法），同样能在足够高的精度上计算并预报任意给定地点和时刻的七政位置。而通过持续观察天象变化以统计、收集各种天象周期，同样可视之为富有博物学色彩的活动。

还有一点需要注意：虽然我们已经接受了用"博物学"来对译 Natural History，但中国的博物传统，确实和西方的博物学有一个重大差别——即中国的博物传统是可以容纳怪力乱神的，而西方的博物学基本上没有怪力乱神的位置。

古代中国人的博物传统不限于"多识于鸟兽草木之名"。体现此种传统的典型著作，首推晋代张华《博物志》一书。书名"博物"，其义尽显。此书从内容到分类，无不充分体现它作为中国博物传统的代表资格。

《博物志》中内容，大致可分为五类：一、山川地理知识；二、奇禽异兽描述；三、古代神话材料；四、历史人物传说；五、神仙方伎故事。这五大类，完全符合中国文化中的博物传统，深合中国古代博物传统之旨。第一类，其中涉及宇宙学说，甚至还有"地动"思想，故为科学史家所重视。第二类，其中甚至出现了中国古代长期流传的"守宫砂"传说的早期文献：相传守宫砂点在处女胳膊上，永不褪色，只有性交之后才会自动消失。第三类，古代神话传说，其中甚至包括可猜想为现代"连体人"的记载。第四类，各种著名历史人物，比如三位著名刺客的传说，此三名刺客及所刺对象，历史上皆实有其人。第五类，包括各种古代方术传说，比如中国古代房中养生学说，房中术史上的传说人物之一"青牛道士封君达"等等。前两类与西方的博物学较为接近，但每一类都会带怪力乱神色彩。

"所有的科学不是物理学就是集邮"

在许多人心目中，画画花草图案，做做昆虫标本，拍拍植物照片，这类博物学活动，和精密的数理科学，比如天文学、物理学等等，那是无法同日而语的。博物学显得那么的初级、简单，甚至幼稚。这种观念，实际上是将"数理程度"作为唯一的标尺，用来衡量一切知识。但凡能够使用数学工具来描述的，或能够进行物理实验的，那就是"硬"科学。使用的数学工具越高深越复杂，似乎就越"硬"；物理实验设备越庞大，花费的金钱越多，似乎就越"高端"、越"先进"……

这样的观念，当然带着浓厚的"物理学沙文主义"色彩，在很多情况下是不正确的。而实际上，即使我们暂且同意上述"物理学沙文主义"的观念，博物学的"科学地位"也仍然可以保住。作为一个学天体物理专业出身，因而经常徜徉在"物理学沙文主义"幻影之下的人，我很乐意指出这样一个事实：现代天文学家们的研究工作中，仍然有绘制星图，编制星表，以及为此进行的巡天观测等等活动，这些活动和博物学家"寻花问柳"，绘制植物或昆虫图谱，本质上是完全一致的。

这里我们不妨重温物理学家卢瑟福（Ernest Rutherford）的金句："所有的科学不是物理学就是集邮（All science is either physics or stamp collecting）。"卢瑟福的这个金句堪称"物理学沙文主义"的极致，连天文学也没被他放在眼里。不过，按照中国传统的"博物"理念，集邮毫无疑问应该是博物学的一部分——尽管古代并没有邮票。卢瑟福的金句也可以从另一个角度来解读：既然在卢瑟福眼里天文学和博物学都只是"集邮"，那岂不就可以将博物学和天文学相提并论了？

如果我们摆脱了科学主义的语境，则西方模式的优越性将进一步被消解。例如，按照霍金（Stephen Hawking）在《大设计》（*The Grand Design*）中的意见，他所认同的是一种"依赖模型的实在论（model-dependent realism）"，即"不存在与图像或理论无关的实在性概念（There is no picture- or theory-independent concept of reality）"。在这样的认识中，我们以前所坚信的外部世界的客观性，已经不复存在。既然几何模型只不过是对外部世界图像的人为建构，则古代中国人干脆放弃这种建构直奔应用（毕竟在实际应用中我们只需要知道七政"如何"运行），又有何不可？

传说中的"神农尝百草"故事，也可以在类似意义下得到新的解读："尝百草"当然是富有博物学色彩的活动，神农通过这一活动，得知哪些草能够治病，哪些不能，然而在这个传说中，神农显然没有致力于解释"为何"某些草能够治病而另一些则不能，更不会去建立"模型"以说明之。

"帝国科学"的原罪

今日学者有倡言"博物学复兴"者，用意可有多种，诸如缓解压力、亲近自然、保护环境、绿色生活、可持续发展、科学主义解毒剂等等，皆属美善。编印《西方博物学大系》也是意欲为"博物学复兴"添一助力。

然而，对于这些博物学著作，有一点似乎从未见学者指出过，而鄙意以为，当我们披阅把玩欣赏这些著作时，意识到这一点是必须的。

这百余种著作的时间跨度为 15 世纪至 1919 年，注意这个时间跨度，正是西方列强"帝国科学"大行其道的时代。遥想当年，帝国的科学家们乘上帝国的军舰——达尔文在皇家海军"小猎犬号"上就是这样的场景之一，前往那些已经成为帝国的殖民地或还未成为殖民地的"未开化"的遥远地方，通常都是踌躇满志、充满优越感的。

作为一个典型的例子，英国学者法拉在（Patricia Fara）《性、植物学与帝国：林奈与班克斯》（*Sex, Botany and Empire, The Story of Carl Linnaeus and Joseph Banks*）一书中讲述了英国植物学家班克斯（Joseph Banks）的故事。1768 年 8 月 15 日，班克斯告别未婚妻，登上了澳大利亚军舰"奋进号"。此次"奋进号"的远航是受英国海军部和皇家学会资助，目的是前往南太平洋的塔希提岛（Tahiti，法属海外自治领，另一个常见的译名是"大溪地"）观测一次比较罕见的金星凌日。舰长库克（James Cook）是西方殖民史上最著名的舰长之一，多次远航探险，开拓海外殖民地。他还被认为是澳大利亚和夏威夷群岛的"发现"者，如今以他命名的群岛、海峡、山峰等不胜枚举。

当"奋进号"停靠塔希提岛时，班克斯一下就被当地美丽的土著女性迷昏了，他在她们的温柔乡里纵情狂欢，连库克舰长都看不下去了，"道德愤怒情绪偷偷溜进了他的日志当中，他发现自己根本不可能不去批评所见到的滥交行为"，而班克斯纵欲到了"连嫖妓都毫无激情"的地步——这是别人讽刺班克斯的说法，因为对于那时常年航行于茫茫大海上的男性来说，上岸嫖妓通常是一项能够唤起"激情"的活动。

而在"帝国科学"的宏大叙事中，科学家的私德是无关紧要的，人们关注的是科学家做出的科学发现。所以，尽管一面是班克斯在塔希提岛纵欲滥交，一面是他留在故乡的未婚妻正泪眼婆娑地"为远去的心上人绣织背心"，这样典型的"渣男"行径要是放在今天，非被互联网上的口水淹死不可，但是"班克斯很快从他们的分离之苦中走了出来，在外近三年，他活得倒十分滋润"。

法拉不无讽刺地指出了"帝国科学"的实质："班克斯接管了当地的女性和植物，而库克则保护了大英帝国在太平洋上的殖民地。"甚至对班克斯的植物学本身也调侃了一番："即使是植物学方面的科学术语也充满了性指涉。……这个体系主要依靠花朵之中雌雄生殖器官的数量来进行分类。"据说"要保护年轻妇女不受植物学教育的浸染，他们严令禁止各种各样的植物采集探险活动。"这简直就是将植物学看成一种"涉黄"的淫秽色情活动了。

在意识形态强烈影响着我们学术话语的时代，上面的故事通常是这样被描述的：库克舰长的"奋进号"军舰对殖民地和尚未成为殖民地的那些地方的所谓"访问"，其实是殖民者耀武扬威的侵略，搭载着达尔文的"小猎犬号"军舰也是同样行径；班克斯和当地女性的纵欲狂欢，当然是殖民者对土著妇女令人发指的践踏；即使是他采集当地植物标本的"科学考察"，也可以视为殖民者"窃取当地经济情报"的罪恶行为。

后来改革开放，上面那种意识形态话语被抛弃了，但似乎又走向了另一个极端，完全忘记或有意回避殖民者和帝国主义这个层面，只歌颂这些军舰上的科学家的伟大发现和成就，例如达尔文随着"小猎犬号"的航行，早已成为一曲祥和优美的科学颂歌。

其实达尔文也未能免俗，他在远航中也乐意与土著女性打打交道，当然他没有像班克斯那样滥情纵欲。在达尔文为"小猎犬号"远航写的《环球游记》中，我们读到："回程途中我们遇到一群黑人姑娘在聚会，……我们笑着看了很久，还给了她们一些钱，这着实令她们欣喜一番，拿着钱尖声大笑起来，很远还能听到那愉悦的笑声。"

有趣的是，在班克斯在塔希提岛纵欲六十多年后，达尔文随着"小猎犬号"也来到了塔希提岛，岛上的土著女性同样引起了达尔文的注意，在《环球游记》中他写道："我对这里妇女的外貌感到有些失望，然而她们却很爱美，把一朵白花或者红花戴在脑后的髮鬐上……"接着他以居高临下的笔调描述了当地女性的几种发饰。

用今天的眼光来看，这些在别的民族土地上采集植物动物标本、测量地质水文数据等等的"科学考察"行为，有没有合法性问题？有没有侵犯主权的问题？这些行为得到当地人的同意了吗？当地人知道这些行为的性质和意义吗？他们有知情权吗？……这些问题，在今天的国际交往中，确实都是存在的。

也许有人会为这些帝国科学家辩解说：那时当地土著尚在未开化或半开化状态中，他们哪有"国家主权"的意识啊？他们也没有制止帝国科学家的考察活动啊？但是，这样的辩解是无法成立的。

姑不论当地土著当时究竟有没有试图制止帝国科学家的"科学考察"行为，现在早已不得而知，只要殖民者没有记录下来，我们通常就无法知道。况且殖民者有军舰有枪炮，土著就是想制止也无能为力。正如法拉所描述的："在几个塔希提人被杀之后，一套行之有效的易货贸易体制建立了起来。"

即使土著因为无知而没有制止帝国科学家的"科学考察"行为，这事也很像一个成年人闯进别人的家，难道因为那家只有不懂事的小孩子，闯入者就可以随便打探那家的隐私、拿走那家的东西、甚至将那家的房屋土地据为己有吗？事实上，很多情况下殖民者就是这样干的。所以，所谓的"帝国科学"，其实是有着原罪的。

如果沿用上述比喻，现在的局面是，家家户户都不会只有不懂事的孩子了，所以任何外来者要想进行"科学探索"，他也得和这家主人达成共识，得到这家主人的允许才能够进行。即使这种共识的达成依赖于利益的交换，至少也不能单方面强加于人。

博物学在今日中国

博物学在今日中国之复兴，北京大学刘华杰教授提倡之功殊不可没。自刘教授大力提倡之后，各界人士纷纷跟进，仿佛昔日蔡锷在云南起兵反袁之"滇黔首义，薄海同钦，一檄遥传，景从恐后"光景，这当然是和博物学本身特点密切相关的。

无论在西方还是在中国，无论在过去还是在当下，为何博物学在它繁荣时尚的阶段，就会应者云集？深究起来，恐怕和博物学本身的特点有关。博物学没有复杂的理论结构，它的专业训练也相对容易，至少没有天文学、物理学那样的数理"门槛"，所以和一些数理学科相比，博物学可以有更多的自学成才者。这次编印的《西方博物学大系》，卷帙浩繁，蔚为大观，同样说明了这一点。

最后，还有一点明显的差别必须在此处强调指出：用刘华杰教授喜欢的术语来说，《西方博物学大系》所收入的百余种著作，绝大部分属于"一阶"性质的工作，即直接对博物学作出了贡献的著作。事实上，这也是它们被收入《西方博物学大系》的主要理由之一。而在中国国内目前已经相当热的博物学时尚潮流中，绝大部分已经出版的书籍，不是属于"二阶"性质（比如介绍西方的博物学成就），就是文学性的吟风咏月野草闲花。

要寻找中国当代学者在博物学方面的"一阶"著作，如果有之，以笔者之孤陋寡闻，唯有刘华杰教授的《檀岛花事——夏威夷植物日记》三卷，可以当之。这是刘教授在夏威夷群岛实地考察当地植物的成果，不仅属于直接对博物学作出贡献之作，而且至少在形式上将昔日"帝国科学"的逻辑反其道而用之，岂不快哉！

2018 年 6 月 5 日
于上海交通大学
科学史与科学文化研究院

爱德华·多诺万（Edward Donovan，1768–1837），英国博物学家。生于爱尔兰的科克。为伦敦林奈学会会员。在那个博物学家和收藏家纷纷开设个人博物馆的时代，他于1807年也在伦敦设立了博物学研究所，展示数百件动物标本与植物标本。他本人并不去海外采集标本，而是运用自己良好的社会关系，委托诸多探险家为自己工作，因而得以聚集起大量罕见标本。对这些标本的研究促使他出版了一批颇负盛名的博物学著作，如《不列颠珍稀鸟类志》、《中国昆虫志》、《印度昆虫志》。他亲自制作铜版并调色，使书中的插画色彩尽可能鲜艳逼真，加上所展示的标本都是珍稀之物，因此这些著作一经出版，便声名鹊起。

《不列颠珍稀鸟类志》于1794年在英国印行第一版，本书即据该版影印。全书10卷，约1200页，附有240余幅根据实体标本测量、描绘的精美彩图，其中可以见到一些今人已经无法亲眼目睹的英伦珍稀鸟类，可谓弥足珍贵。

THE
NATURAL HISTORY
OF
BRITISH BIRDS;

OR, A

SELECTION OF THE MOST RARE, BEAUTIFUL, AND INTERESTING

BIRDS

WHICH INHABIT THIS COUNTRY:

THE DESCRIPTIONS FROM THE

SYSTEMA NATURÆ

OF

LINNÆUS;

WITH

GENERAL OBSERVATIONS,

EITHER ORIGINAL, OR COLLECTED FROM THE LATEST
AND MOST ESTEEMED

ENGLISH ORNITHOLOGISTS;

AND EMBELLISHED WITH

FIGURES,

DRAWN, ENGRAVED, AND COLOURED FROM THE ORIGINAL SPECIMENS.

By E. DONOVAN.

LONDON:

PRINTED FOR THE AUTHOR; AND FOR F. AND C. RIVINGTON,
No. 62, ST. PAUL'S CHURCH-YARD. 1794.

ADVERTISEMENT.

THIS Work being now completed, we conceive our engagement with the Subſcribers, is, in every reſpeꞔt fulfilled, and that it only remains to embrace this opportunity, to repeat our thanks for their favours ; and in particular, for the candour with which the Supplementary Part has been received.

But, in ſubmitting the work to the Public in general, as an illuſtration of an important branch of Britiſh Zoology, it is incumbent on us, to ſtate briefly, the nature, and extent of the undertaking, the information it contains, and its peculiar advantages. In this retroſpeꞔtive ſurvey of the work, we muſt advert to our former obſervations ; ſome ſimilar remarks appeared at the concluſion of the fourth volume, and we retrace thoſe, as they immediately relate to the outline of our deſign.

A   Ornithology,

ADVERTISEMENT.

Ornithology, as a fcience, has undergone various alterations and improvements: different authors have fubmitted
their fyftems to the world, and each has found its admirers
and opponents; nor has the unrivalled genius of Linnæus
devifed an arrangement in which thofe oppofite opinions
may be reconciled. On the importance of Ornithology, in
the great fcale of animated nature, no difference of opinion
can prevail. The beauty and elegance of the feathered race:
their pleafing and various melody; their fagacity, and trac-
table manners, has been admired in every age and country;
and their unerring œconomy and inftinct, has ever engaged
the attention of the moral philofopher. The Birds of this
country are of plainer colours than thofe of warmer climates,
but they are not lefs interefting to the Englifh naturalift.
If, in fome inftances, their beauty has little claim to our notice; in others it excites our admiration; and to the intelligent mind, their beauty, their fingularities, peculiar manners,
and œconomy are equally engaging. They are the fource
of information and improvement to the practical Ornithologift,
and of rational and agreeable amufement to every common
obferver of nature.

At the commencement of this work, it was our intention
to form a complete Hiftory of Britifh Ornithology, and to
include figures of all the known Birds, amounting to more
than two hundred and fifty fpecies; but we have fince conceived it would be advifable to felect only the more beautiful

tiful Birds, in addition to thofe which are interefting to the naturalift: for a confiderable number of the Britifh fpecies are fo well known, that their hiftory would be tedious, and the figures unneceffary, in a work profeffedly defigned to treat of the moft remarkable fpecies only. Not that we have entirely overlooked the common Birds; in feveral inftances fome of thefe are introduced alfo, to point out their fingular habits of life, and other interefting peculiarities; but, in general, we have endeavoured to form an inftructive as well as amufing illuftration of this department of Natural Hiftory.

It will perhaps be contended, that a complete collection of figures and defcriptions of all the Britifh Birds, would be more acceptable than any partial felection, however comprehenfive. We admit the propriety of this objection; but muft obferve, that fuch addition would confiderably increafe the expenfe to the purchafer, and fcarcely contribute to his information. On the other hand, this work, in its prefent limits, may affift the refearches of the uninformed naturalift, and tend to promote a deeper and more extenfive enquiry into this branch of fcience. It embraces in one view the whole of thofe Britifh Birds that are fcarcely known, and of which the inquifitive reader, under many circumftances, may be defirous of information. We have omitted many of thofe Birds which conftantly inhabit this country; but have included all local fpecies, and in particular, thofe,

A 2

whofe

ADVERTISEMENT.

whose haunts and breeding places are difficult of acceſs, and the Birds in conſequence, little known. The extenſive marſhes and lowlands in ſome parts of the kingdom, are the retreats and breeding places of certain ſpecies. Thoſe ſolitary kinds, which retire to the depths, and gloomy receſſes of foreſts, are rarely obſerved; and many of thoſe which ſeek the open plain for ſecurity, elude our vigilance, and are not better known. But the rareſt of the local kinds, are of the rapacious and gallinaceous tribes, which never leave their dreary ſolitudes: their wilds and barren mountains in the north; to viſit the ſouthern parts of Great Britain. The migratory Birds are numerous, and include many well known ſpecies, with others that are uncommonly ſcarce. We have taken an extenſive variety of the beautiful Land Birds, that reſort to this country occaſionally from the ſouth of Europe; and of the aquatic or web-footed tribe, that are driven by the ſeverity of the winter in the Arctic regions to ſeek ſhelter on our ſhores. Hiſtory and tradition inform us of other Birds that formerly inhabited theſe kingdoms, but are now extirpated; and theſe form an intereſting ſequel to this ſelection. We cannot vindicate the propriety of introducing naturalized exotic ſpecies amongſt theſe, though they are arranged by our Naturaliſts in the Britiſh Ornithology; and in ſome inſtances we have followed their authority, for the ſake of embelliſhment and variety. 6

In

ADVERTISEMENT.

In the courſe of publication, we have been fortunate in procuring ſpecimens of many uncommon Birds. Among the moſt remarkable, are *The Roſe-coloured Ouzel, Roller, Little Bittern, Waxen Chatterer, Black Woodpecker*, and in particular that rare and almoſt unknown ſpecies, the *Wood Chat.*—*The Red-necked Grebe, Dartford Warbler, Duſky Lark, Long-legged Plover, Egret* and *Cock of the Wood.* We could enumerate many other ſpecies highly important to the uninformed naturaliſt; but, we refer our readers for the general detail, to the complete Syſtematic Arrangement annexed to this Advertiſement.

In this Syſtematic Arrangement we have followed the *Syſtema Naturæ* of Linnæus, though we totally diſſent from the opinion of that celebrated naturaliſt in his primary diviſions of Ornithology. In every ſyſtem, the Birds which inhabit the land only, are ſeparated from ſuch as frequent the water. This appears to be a natural method of forming two principal diviſions of Ornithology; it was adopted by Ray, and approved by Pennant and Latham. In the Linnæan ſyſtem, thoſe which inhabit the water are ſeparated from the reſt; but the Land Birds are divided into two parts, and the Water Birds are placed between them. To avoid confuſion we have adhered to this arrangement; but we have alſo placed an Index, in the manner of Pennant and Latham, at the concluſion of each volume.

A 3

Having

ADVERTISEMENT.

Having endeavoured to exp'ain the nature and extent of this undertaking, we submit the whole to the candour of the Public; and, though not indifferent to the flattering testimony of approbation, it has received in the course of five years publication, we entreat indulgence for whatever may be thought exceptionable, either in the outline of our undertaking, the selection of species, or the manner in which they are illustrated. To our Subscribers, we shall not presume to address any apology, as their opinion must be already decided. The progressive manner in which the work has appeared, has afforded every opportunity for critical examination, for detecting error, or discovering merit; and, we trust, their continued patronage is some criterion of their approbation, and of the general utility of our design.

SYSTEMATIC

SYSTEMATIC ARRANGEMENT

OF

SELECT BRITISH BIRDS.

ORDER I.

ACCIPITRES

Includes the Falcon, Owl, and Shrike or Butcher Bird.

FALCON.

Falco Offifragus.   Sea Eagle.

Haliætus.   Ofprey.

Apivorus.   Honey Buzzard.

Milvus.   Kite.

* *Peregrinus.*   Peregrine Falcon.

Cyaneus.   Hen Harrier.

Tinnunculus.   Keftril.

Subbuteo.   Hobby.

* *Æfalon.*   Merlin.

---

* Thofe marked with a ftar are not defcribed by Linnæus.

CONTENTS.

OWL.

ORDER II.

· PICÆ ·

Crow, Roller, Oriole, Cuckow, Wryneck, Woodpecker, King's-fifher, Nuthatch, Hoopoe, Creeper.

CROW.

ROLLER.

ORIOLE.

CONTENTS.

ORDER III.

ANSERES

Duck, Merganfer, Auk, Petrel, Pelican, Diver, Gull and Tern.

DUCKS.

CONTENTS.

DUCK.

GULL.

CONTENTS.

GULL.

TERN.

ORDER IV.

GRALLÆ

Heron, Ibis, Snipe, Sandpiper, Plover, Avofet, Oyfter-catcher,
Coot, Rail, Buftard.

HERON.

IBIS.

SNIPE.

SAND-

CONTENTS.

SANDPIPER.

Tringa Pugnax. Ruff.
    Vanellus. Lapwing.
    Cinclus. Purre.

PLOVER.

Charadrius Pluvialis. Golden Plover.
    Himantopus. Long-legged Plover.
    Hiaticula. Ringed Plover.
    Morinellus. Dottrel.

AVOSET.

Recurvirostra Avosetta. Scooping Avoset.

OYSTER-CATCHER.

Hæmatopus Ostralegus. Pied Oyster-catcher.

COOT.

Fulica Atra. Common Coot.
    Chloropus. Water Hen.

RAIL.

Rallus Crex. Land Rail.
    Aquaticus. Water Rail.
    Porzana. Small spotted Water Hen.

ORDER

CONTENTS.

ORDER V.

GALLINÆ.

Pheafant, Grous, Peacock.

PHEASANT.

Phafianus Colchicus. Common Pheafant.

GROUS.

Tetrao Urogallus. Cock of the Wood.
Tetrix. Black Game.
Logopus. White Game or Ptarmigan.

PEACOCK.

Pavo Criftatus. Common Peacock, *(variety.)*

ORDER VI.

PASSERES.

Pigeon, Lark, Stare, Thrufh, Chatterer, Grofbeak, Bunting
Finch, Fly-catcher, Warbler, Wagtail, Titmoufe, Swallow,
Goat Sucker.

PIGEON.

Columba Oenas. Stock Pigeon.

LARK.

Alauda Ofcura. Dufky Lark.

STARE.

CONTENTS.

WAGTAIL,

CONTENTS.

THE
LINNÆAN ARRANGEMENT

OF THE

ORDERS INCLUDED IN THE CLASS

AVES.

ORDER I.

Acciptres.

RAPACIOUS Birds; having the upper mandible of the beak furnished on each fide with an angular procefs; claws arched, ftrong.

ORDER II.

Picæ.

Pies; having the beak a little curved, and rather comprefled on the fides.

ORDER III.

Anferes.

Web-footed; thefe have a beak fomewhat obtufe, and covered with a thin fkin; at the bafe underneath gibbous, and wide at the end; the *faux*, or edges of the bafe, denticulated; the feet palmated, or webbed, and formed for fwimming.

A 2   ORDER

4

ORDER IV.

Grallæ.

Waders. Thefe have the beak fubcylindrical, and fomewhat obtufe; the tongue entire, and flefhy; the thighs naked for fome fpace above the knees; legs very long.

ORDER V.

Gallinæ.

Gallinaceous Birds having the upper mandible convex, or arched, and receiving the edges of the lower noftrils, half covered by means of a convex membrane, rather cartilaginous; the rectrices, or tail-feathers, more than twelve; the feet cloven, but connected by a membrane as far as the firft joint.

ORDER VI.

Paſſeres.

Pafferine. Thefe have a conical acuminated, or pointed, beak; noftrils oval, open, naked.

PLATE

5

PLATE I.

PARUS BIARMICUS.

BEARDED TITMOUSE.

PASSERES.

Bill conic, pointed. Noftrils oval, broad, naked.

GENERIC CHARACTER.

Bill fhort, ftrong, entire, briftles at the bafe. Tongue, blunt, with briftles at the end.

SPECIFIC CHARACTER,

AND

SYNONYMS.

Bill fhort, ftrong, convex, yellow. Head grey. A black tuft, or beard, beneath each Eye. Plumage red yellow. Tail long. Legs black.

> *Lin. Syft. Nat.* 342. *edit.* 12—1766.
> *Scop. ann.* 1. *N*° 241.
> *J. L. Frifch. t.* 8.

PARUS BARBATUS. La Mefange barbue, ou le Mouftache.—
> *Brif. Orn. III. p.* 567. *N*° 12.
> *Buf. Oif. pl.* 18. *v. p.* 418.
> *Pl. enl.* 618. *t.* 1. 2.

Pendulus. *Kram. el. p.* 373.

Beardmanica, *Albin.* 1. *pl.* 48.

Lanius Minimus, *Leaft Butcher Bird.* *Edw. pl.* 55.

Bearded Titmoufe. *Br. Zool.* 1. *N*° 167.
> *Arct. Zool.—Br. Muf.—Lev. Muf.*

A 3

The

The Great Titmoufe, Colemoufe, and Marfh Titmoufe; with the Blue, Long-tailed, and Bearded, Titmice, are the only fpecies of the tribe which inhabit this country: they are all very frequent excepting the latter; which however is not uncommon in certain fituations, though formerly efteemed as rare.

It was defcribed by *Aldrovandus* in his Ornithology publifhed in the years 1610—1613; and appears to be well known at that time in feveral parts of Europe, though unknown in Britain: more than a century after *Aldrovandus*, (1734), it was included in a Hiftory of the Birds of Germany by *J. L. Frifch*; but even at that time, was fo rare with us, that it was fcarcely afcertained to be a native of Britain: and Albin, who feems to have poffeffed fome knowledge of Birds, determines it as a native, only on the authority of the information he received from others; his Hiftory of Birds was publifhed in 1738; therein he gives a figure of the Male Bird, and fays in the Defcriptions annexed, " Thefe two birds (male and female) I bought of Mr. *Bland* on *Tower-hill*, who told me he had them from *Jutland:* I have been fince informed by Sir *Robert Abdy*, that they are found in the *Salt-marfhes* in *Effex*, and by others that they are likewife in the fens in *Lincolnfhire*."

He alfo obferves that it receives the name Beardmanica from the black tuft refembling a picked beard.

Edwards * refers it to the tribe of Butcher-birds, under the title of Lanius *Minimus*, Leaft Butcher-Bird; but Linnæus, in his *Syftema Naturæ*, reduced it to the Parus genus; and late writers have alfo determined it to the fame family.

---

* G. Edwards's Nat. Hift. of Birds, Vol. 7. 4to. London, 1743, &c.

Its

PLATE I. 7

Its length, from the tip of the bill to the end of the tail, is fix inches and a quarter: the bill is thick, and of a bright yellow colour, but fades immediately when the bird dies; in the female it is rather dufky; and the head, which is of a fine grey in the male, is of a brownifh ruft colour, fpotted with black, in the female: indeed, the female is immediately diftinguifhed by the plainnefs of her plumage, it neither poffeffing the beautiful purple colour on the breaft, or the black tufts on the throat, which charaċterizes the male: the vent-feathers of the male are pale black; of the female light brown; as are the other parts of the belly.

Pennant, in the Britifh Zoology, fays, " This fpecies is found in the marfhes near *London:* we have feen it near *Gloucefter:* it is alfo frequent among the great traċts of reeds near *Cowbit* in *Lincolnfhire,* where I fufpeċt they breed.

Latham, in his General Synopfis,—" Thefe birds are found in England, but have hitherto been obferved only in marfhy fituations where *reeds* grow, on the feeds of which it feeds, as well as fmall in-feċts; both of which have been found in their ftomachs. They are pretty frequent, and in not inconfiderable quantities, in the marfhes among the reeds between *Erith* and *London,* and are again met with in fuch-like places near *Gloucefter,* as well as among the great traċts of reeds near *Cowbit,* in *Lincolnfhire.* In all thefe places I make no doubt of their breeding, as I know that they ftay in the firft-named parts the whole year. The neft is not known for certain; but I have feen one, which was compofed of very foft downy materials, fufpended between three reeds drawn together, thought to be the work of that bird."

" They are alfo common in *Denmark*; and *Buffon* fuppofes that a pair of thefe, having efcaped from the cage of the *Countefs* of *Albe-*

A 4 *marle,*

marle, have founded this colony in England. This may have been the cafe in refpect to thofe of *Erith*, being on the borders of the *Thames*, but will not fo well account for their being elfewhere; and I am inclined to think that they are indigenous to us, and have been fo *ab origine*; and that it is merely owing to their frequenting fuch places only where the *reeds* grow, that they have been fo little known; for as thofe birds never go farther than a few yards from the *beds* *, they have ftood a greater chance, which has really happened, of not being earlier obferved."

Kramer fays the neft is built among the *willows*, and is of the fhape of a purfe, made of foft downy materials, fuch as the down of the *Greater Cat's-Tail* †, or that of the *Afp* ‡, hanging the neft on a branch.

What new light Mr. Latham could throw on this fubject, is given in the *Supplement* publifhed in the year 1787.

" I have never yet been able myfelf to afcertain the neft and eggs. In *Sepp's* ‖ plate the neft is placed on the *ground* among the *fedges*. It is of a very loofe texture, compofed of the tops of dry grafs, mixed with the feed-heads of *rufhes* and *reeds*, with narrow leaves intermixed. The eggs four in number, of a reddifh white, marked with fmall brown fpots."

---

* " The *reed-beds* frequently cover many acres of ground; thefe grow in the water, fo as to be overflowed at every tide; and few perfons ever go near them, except in the time of cutting, which they do in boats, as, except at very low tides, one can fcarce fet a footftep within their boundaries."

† *Typha latifolia.* Lin.   ‡ *Populus tremula.* Lin.   ‖ *Sepp.* *Vog.* pl. in p. 83.

" This

PLATE I. 9

" This fpecies is found in *Schonen*, in Sweden; but rarely. Is very common about the *Cafpian Sea* and *Palus Mœotis*, and among the rufhes of the rivers which fall into them; but in no high latitudes in *Afia*. None in Siberia *."

The male fhews much tendernefs and care for its young, and its mate partakes alfo of its affection: it is faid to be ever conftant in its attachments; and at night, when at rooft, the male protects the female, under the concave of his wing.

---

* *Arct. Zool.*

PLATE

- 27 -

PLATE II.

CORVUS GLANDARIUS.

J A Y.

P I C Æ.

Bill compreſſed, convex, or a little curved.

GENERIC CHARACTER.

Bill ſtrong, conic, with briſtles at its baſe reflected downwards. Tongue bifid.

SPECIFIC CHARACTER,

AND

SYNONYMS.

Head covered with long feathers. Forehead white with black ſtrokes. From the angles of the mouth a broad ſtreak of black under each eye. The head, ſides, neck, breaſt, back and ſcapulars, vinaceous buff-colour. Coverts of the wings fine blue, barred with black. Rump white. Tail black-brown.

CORVUS glandarius. *Lin. Syſt.* 1. *p.* 156. *N°* 7.
    Scop. ann. 1. *p.* 36. *N°* 39.
    Mull. p. 12.

PICA glanduria. *Geſn. av.* 700.

Jay. *Aldr. Ornith. l.* 12. *e.* 14.
   Will. Orn. p. 130. *pl.* 19.
   Raii. av. 41. *A.* 2.
   Alb. av. 1. *p.* 16. *t.* 16.

Le

Le Geay, Garrulus. *Brif. Orn.* ii. *p.* 47. *N°* 1.

      Buf. ois. III. p. 107. *pl.* 8.—*Pl. enl.* 481.

      Renel. 481.

Holtz-fchreyer (Wood Cryer), Eichen-heher (Oak Jay), Nus-heher, *Frifch. t.* 55.

Nus-heher, *Kramer El. p.* 335.

Ghiandaia, *Zinan.* 67.

Skoia, Schoga. *Scopoli, N°* 39.

Screch y Coed. (*Antient Britifh.*) *Brit. Zool.*

———————————

The Jay, though very frequent in every part of the country, evidently deferves, for the fingular beauty of its plumage, the firft place in a collection of interefting Britifh Birds: it is a fpecies, however, not merely confined to this kingdom; but generally diffufed over the greater part of the continent of Europe.

Among the foreign birds which are referred to this genus, are feveral fpecies eminently diftinguifhed for their rich and elegant colouring; but the European Jay is certainly fuperior in this particular to many exotics of the fame family.

Its habits are known, in a great degree, to refemble thofe of the Magpie, which alfo belongs to the fame genus. It will devour with avidity cherries, goofeberries, rafpberries, or other fruits; but in autumn and winter it feeds on acorns; whence it long fince derived the name Pica *Glandaria.* It is not deftitute of a carnivorous appetite, as it frequently combats, and devours, fmall birds: it alfo feeds on worms and infects.

                                              Ablin,

PLATE II. 13

Ablin, in Plate 16. Vol. 1. very accurately defcribes the Jay he took his figure from, which nearly correfponds with our fpecimen.

" Its length, from the point of the bill to the end of the tail, was fourteen inches; its breadth, when the wings were extended, twenty-one inches and an half; its weight feven ounces; the bill dark afh colour inclining to black, ftrong, near an inch and an half long; the tongue black, thin, pellucid, and cloven at the tip; the irides of the eyes white. Near on the lower chap of the bill are two black fpots, on each fide one. The chin and lower part of the belly whitifh, elfe the breaft and belly are of a mixt cinereous and red; the rump above is white, the back red, with a mixture of blue; the feathers on the crown of the head variegated with black and white.

" The fails of the wings are in number twenty, of which the firft is fhorter by half than the fecond, the fourth the longeft (being by meafure fix inches and a quarter); the firft or outermoft is black, the bottom or lower part white, which is proper to it alone; the fix next following have their exterior vanes of an afh colour, the three next likewife, but more obfcure, and mingled with blue, being alfo marked toward their bottom with tranfverfe black and white ftrokes; the five fucceeding have their exterior vanes half white, half black, viz. the lower half white, the upper black, but fo that each extremity of the white is terminated with blue; the fixteenth, in place of the white of the four precedent, hath tranfverfe blue, black, and white fpots; the feventeenth is black, having one or two blue fpots; the eighteenth is black with fome little red; the nineteenth red with the tip black; the under fides of all the feathers of the wing are of a dark or dufky colour; the covert-feathers of the fifteen exterior fails are very beautiful, being variegated or chequered with black, white, and lovely fhining blue lines, the reft of the covert-feathers being black.

" The

9

" The tail is fix inches and a quarter long, confifting of twelve feathers, wholly black except toward their roots : under the rump there is fomething of a blue mingled with cinereous.

" The feet and toes are of a ferruginous dufky colour, the middle toe is the longeft; the outmoft is equal to the back toe; the lower joint of the outmoft toe is joined to the middlemoft ; the back claw is the greateft.

" The guts are twenty-four inches long; the blind guts but half an inch; it hath a gall and a long fpleen : the ftomach or gizzard not very flefhy, and having its echinus, wherein were found acorns, &c."

He adds, " The female differs little or nothing from the male, either in bignefs or colour, fo that it is very difficult to know them afunder:" but after the publication of his firft Volume of Birds, he rectifies this error, in a fupplement which is prefixed to the work. " The following obfervations," (fays he) " I received from Dr. Derham, after the defcriptions were printed. He hath obferved the Cock Jay (Plate 16. Vol. 1.) to be fomewhat bigger than the Hen, the feathers on the head to be blacker, the ftripes longer, and the black and blue colours more elegant in the Cock than in the Hen."

Jays generally build in woods, preferring a fituation near the fkirts : the neft is compofed of fibrous roots, and young twigs, erected on a bafe of feveral large fticks, and is placed on the top of a thorn-bufh, or other under-wood, or between the firft branches of low decayed trees : the female lays five or fix eggs, of the fize of a pigeon's, of a cinereous olive colour, marked with very pale brown fpots : the young Jays remain with their parents till the next fpring; and at the pairing time they each choofe his mate, to propagate their future progeny.

8   It

PLATE II. 15

It is a reftlefs and very quarrelfome bird: makes a harfh, chattering and fcreaming noife; and is ever at variance not only with its own fpecies, but with every other inhabitant of the foreft: when deprived of liberty, it may be taught to imitate the human voice; but the original appearance of its plumage is fo altered by confinement, as fcarcely to retain any of the beautiful colours, which are fo eminently confpicuous in the wild ftate.

It is a native of *Denmark*, and of *Ruffia*; of *Scotland*, and of *England*; but does not frequent the iflands adjacent.

Latham fays, " The Jay, I believe, is not fpread fo far as many others of the genus, as we do not hear of its inhabiting further fouth than *Italy* and *Greece*.

" This fpecies is common in the woods both of *Ruffia* and *Siberia*, but none beyond the *Lena* *; *Georgi* mentions it as frequenting the Lake *Baikal*, and *Ruffel* records it as an *Aleppo* fpecies †. I have a fufpicion alfo that it extends to *China*, as it is to be feen in the drawings of birds from that country.

" It is called by the name of *Jay*, about *Arragon* in Spain, as in *England*. In the laft it is not efteemed as food; but in the firft it is expofed to fale along with other birds ‡." It is alfo eaten in *Sweden* ‖. *Supp. Gen. Syn. p.* 79. *N*° 19.

---

* *Arct. Zool.* † *Hift. Alepp. p.* 69. ‡ *Faun. Arag.*
‖ *Mr. Swederus.*

LE

LE GEAY BLANC.

WHITE JAY.

Brif. orn. II. p. 51. A.

Latham defcribes this Jay, as only a variety of the common fort, and fays he had one which was taken in a neft with four other Jays of the ufual colour; the fpecimen he mentions, is at prefent in my collection; it is lefs than the common fize, is wholly white, inclining to a cream colour in the fhades; the legs and bill are white alfo: the irides red.

PLATE

3

PLATE III.

TURRDUS ROSEUS,

ROSE-COLOURED THRUSH

OR OUZEL.

PASSERES.

Bill conic, pointed. Noftrils oval, broad, naked.

GENERIC CHARACTER.

Bill ftrait, fubulate, and fomewhat angular.

SPECIFIC CHARACTER,

AND

SYNONYMS.

Bill at the bafe reddifh, at the point black. A long pendent Creft. Head, Creft, Neck, Wings, and Tail, black, tinged with green-purple. Breaft, Belly, and Back, pale rofe-colour with dark fpots ; Legs dirty orange.

Lin. Syft. 1. *p.* 294. *Nᵒ* 15.

Faun. Suec. 219.

Nov. Com. Ac. Petrop. XV. *p.* 478. *t.* 23. *f.* 1.

STURNUS *Rofeus, Scop. ann.* 1. *p.* 130. *Nᵒ* 191.

TURDUS *Seleucis. Faun. Arab. p.* 6. *a.—p.* 5. *Nᵒ* 16.

Le Merle couleur de Rofe. *Brif. orn.* ii. *p.* 250. *Nᵒ* 20.

Buf. Oif. iii. *p.* 348. *t. pl.* 22.—*pl. enl.* 251.

MERULA Rofea, *Raii. Syn. p.* 67. 9.

Aldr. av. II. 283.

Rofe, or Carnation-coloured Ouzel. *Will. orn. p.* 194.

Edw. pl. 20.

Br. Zool. App. Nᵒ 5. *pl.* 5.

Ar. Zool.—Lev. Muf.

B

PLATE III.

This fpecies is very rare in every part of Europe; and efpecially in this country: we have a figure of it in the Britifh Zoology; but unfortunately, notwithftanding Mr. *Pennant*'s very laudable intentions, he had no opportunity of confulting the original fpecimen; he depended on a drawing done by his friend Mr. G. *Edwards*; and the copy is only a diftant imitation of the bird; it is defective about the body, and exceedingly incorrect as to the circumference of the neck; the account fays, "Mr. *Edwards* difcovered this beautiful bird twice in our ifland, near *London*, at *Norwood*, and another time in *Norfolk*;" the figure of this was copied by permiffion from his beautiful and accurate defign, which we gratefully acknowledge, as well as every other affiftance from our worthy friend; whofe pencil has done fo much honour to his country.

This bird is the fize of a ftarling; eight inches in length: bill three quarters of an inch, of a flefh-colour, with the bafe reddifh: irides pale: the feathers on the head long, they form a creft which impends on the neck: the head, neck, wings, and tail, are black; the two latter are ftrongly gloffed with green; and in fome parts with an inclination to blue and purple: the back, rump, breaft, belly, and leffer wing coverts, pale rofe colour, with a few irregular dark fpots: legs pale red, or orange.

Ruffell in his Hiftory of Aleppo, calls our *rofe* a *flefh*-colour; and in the Peterfburgh Tranfactions the name given to it is *fanguineous*; we may then conclude that the bird varies much in its rofe, or pink colour; the female alfo is paler than the male.

Mr. Latham has added a farther proof of its being found in this country, one was fhot at *Grantham*, in *Lincolnfhire*, and is now in the poffeffion of Sir Jofeph Banks, Bart.; and he fays that he is affured of one, or more, being fhot almoft every feafon about *Ormfkirk* in *Lancafhire*.

It

P L A T E III.

It is more frequent in *France*; and is met with in *Burgundy*, in its paſſage to other parts : *Aldrovandus* mentions it as not uncommon in Italy, where it is called the ſea ſtare, and ſays its uſual haunts are among heaps of dung * : it is alſo found in *Switzerland* and *Lapland*, but it never paſſes the limits of that frozen region †.

In ſome parts of Aſia it is common : " It comes in great numbers about *Aleppo*, in *July* and *Auguſt*, in purſuit of the ſwarms of *locuſts* ‡; whence it is held ſacred by the Turks, as great quantities are deſtroyed by this means : it is alſo ſeen in vaſt flocks, every year in the ſouth of *Ruſſia*; about the river *Don*; and in *Siberia*, about the *Irtiſch*; finding abundance of *locuſts* for food, and convenience for breeding between the rocks : it is alſo common on the borders of the *Caſpian Sea*; about *Aſtrachan*; and from thence all along the *Volga*." *Latham Gen. Syn.* 3. 50. 52.

It extends to India; Sir Joſeph Banks, has one in his collection, which was received from Bombay.

---

* *Aldr. Av.* II. 283.   † *Linnæus.*—Mr. *Ekmarck.*
‡ Hence called the locuſt bird. *Ruſſell. Hiſt. Allep.*

PLATE

4

PLATE IV.

MOTACILLA REGULUS.

GOLDEN-CRESTED WREN.

PASSERES.

Bill conic, pointed. Noftrils oval, broad, naked.

GENERIC CHARACTER.

Bill ftrait, flender. Tongue jagged.

SPECIFIC CHARACTER,

AND

SYNONYMS.

Crown of the head bright yellow, with a longitudinal black margin on each fide, which paffes immediately above the eyes. Back greenifh. Breaft white with a dirty green tinge. Legs yellow-brown.

MOTACILLA REGULUS *Linn. Syft.* 1. *p.* 338. *N°* 48.
    Muller, p. 33, *N°* 280.
    Georgi Reife, p. 175.
    Frifch. t. 24.

GOLDEN-CROWNED WREN *Raii. Syn. p.* 79. *A.* 9.
    Will. Orn. p. 227. *pl.* 42.
    Albin. 1. *pl.* 53. *A.*
    Edw. pl. 254. 1.
    Catefb. Car. App. 36. 37.
    Br. Zool. 1. *N°* 153.
    Arct. Zool.—Br. Muf.—Lev. Muf.
    Latham. Gen. Syn. IV. 508.

             Le

PLATE IV.

Le Poul, ou Souci, ou Roitelet hupé, Calendula, *Brison, av.* III. *p.* 579. *N*° 17. *Pl. enl.* 651. *fig.* 3.

Le Roitelet. *Buff. oif.* V. *p.* 363. *pl.* 16. *f.* 2.

Fior rancio. *Olina. pl. in p.* 6.

La Soulcie. *Belon av.* 345.

Kongs fogel. *Faun. Suec. fp.* 262.

Kratlich. *Scopoli, N*° 240.

Sommer Zaunkoenig (Summer Wren.) *Frisch,* 1. 24.

Goldhannel. *Kram.* 378.

Fugle-Konge. *Brun.* 285.

The golden crested Wren is the fmalleft bird yet difcovered, in either of the Britifh ifles; is common to *France, Auftria, Italy,* and moft other parts of Europe; and in thofe countries, as with us, it appears to be the leaft native fpecies.

But difcoveries in the interior parts of *fouth America,* have verified, that it is not the leaft kind exifting; in that country where the Condor is found, the moft diminutive fpecies of the feathered tribe are alfo taken; and to thofe the leaft European bird bears a gigantic difproportion; for inftance, the length of the golden-crefted Wren is three inches and an half, its weight feventy-fix grains: but the total length of the leaft South-American Humming bird * is not more than an inch and a quarter, and its weight when frefh killed twenty grains †; the female is yet fmaller ‡.

---

* Trochilus minimus *Lin. Syft.* 1. *p.* 193. *N*° 22.

† Sir Hans Sloane. *Jam.* ii. *p.* 307.   ‡ Brown. *Jam. p.* 475.

The

PLATE IV.

The appellations Regulus, and Tyrannus, Little King, or Tyrant, have been given to the golden-crefted Wren by fome authors: it has ability to conceal the orange band on the head; by corrugating the forehead, and drawing together the feathers, which form the black longitudinal band on each fide.

The colour of the plumage of the female is paler, than of the male; the creft or feathers on the crown of the head are yellow, but without the bright orange colour, which foftens into the creft of the male.

It remains with us through the winter *; frequents woods; and builds its neft, either in oak, fir, or yew trees, the neft is of a roundifh form, with an opening on one fide; it is compofed of *mofs*; and lined within with fome downy fubftance, (perhaps cobwebs,) intermixed with fmall filaments.

It lays fix or feven eggs, which are no bigger than large peas †.

Although the fpecies is found in Europe, it is fcattered throughout the other three quarters of the globe, with only fome little variations which mark the influence of climate; Latham mentions a fpecimen received from *Cayenne* with black legs.

It is a native of *Ruffia* ‡, *Sweden*, and *Norway*; and is found as far north as the *Shetland* ifles; but difappears before winter; it bears cold extremely well, and therefore it may be rather the fcarcity of infects, on which it feeds; than merely the approaching feafon, which induces it to take fuch vaft flights.

---

* *Latham.* IV. 509. 145.　　† *Albin Orn.* 1. 51. 53.　　‡ *Georgi.*

Is

PLATE IV.

It is alfo found in the northern parts of *America*, Penfylvania *, and New-York †.

" We have obferved this bird fufpended in the air for a confiderable time over a bufh in flower, whilft it fung very melodioufly. The note does not much differ from that of the common Wren, but is very weak." *Brit. Zool.* 379. 153.

---

* *Edwards.*   † *Colonel Davies.*

PLATE

5

PLATE V.

MOTACILLA ALBA.

WHITE WAGTAIL.

PASSERES.

Bill conic, pointed. Noſtrils oval, broad, naked.

GENERIC CHARACTER.

Bill ſtrait, ſlender. Tongue jagged.

SPÉCIFIC CHARACTER.

AND

SYNONYMS.

Bill dark brown. Head, Tail, and Legs, black. Breaſt, Belly, and ſides of the Tail, white. Upper parts of the Body, and Wing coverts cinereous, Tail and back claw long.

MOTACILLA ALBA. *Linn. Syſt.* I. *p.* 331. *N°* 2.

Geſner, av. 618.

Sepp. Vog. pl. in p. 119.—*Faun. Arag. p.* 88.

Lath. Gen. Syn. IV. *p.* 395. N° I.—*Arct. Zool.* ii. *p.* 396. E.

WHITE WATER-WAGTAIL, *Raii. Syn.* 75. *A.* I.—*Albin.* I.
pl. 49. *Will. Orn. p.* 237.—*Brit. Zool.* I.
N° 142. *pl.* 55. *Er. Muſ.—Lev. Muſ.*

La Lavandiere, *Briſ. Orn.* iii. 461. N° 38.

Buff. Oiſ. v. *p.* 251. *pl.* 14. *f.* 1.—*pl. enl.* 625. *f.* I.—
Variety. f. 2.

Ballarina, Cutrettola. *Olina*, 43.

Monachina. *Zinan.* 51.

Pliſka, Paſtaritra. *Scopoli*, N° 224.

Arla. Sadeſarla. *Faun. Suec. ſp.* 252.

Danis Vip-Stiert, Havre Sæer.

Norvegis Erle, Lin-Erle.

Weiſs und ſchwartze Bachſteltze. *Friſch.* I. 23.
Graue Bachſtelze. Kram 374.

C

This

PLATE V.

This bird is very frequent in England, and is spread throughout the whole of the old Continent; *Latham* says he has more than once met with a representation of it in *Chinese* drawings: it extends as far as *Iceland*, the *Feroe Isles*, and *Drontheim*; it is common in *Russia*, *Siberia*, and *Kamptschatka*; but is not found in the more northern regions.

It also inhabits *India*, a drawing which was done on the spot being in the collection of Lady Impey *.

It frequents the sides of pools and small rivulets; and feeds on insects : it is often seen running on the ground with much celerity, or leaping up after flies : the tail is frequently in motion. Birds of this genus seldom perch; fly in an undulating manner, and have a twittering noise in flight.

Willughby observes, that this species shifts its quarters in the winter; moving from the north to the south of England during that season.

Latham suspects that part of them migrate, as he does not recollect seeing so many in winter, as in the summer season, and says, in *Scotland*, and in the north of England, it is scarce ever seen in hard weather.

This, and others of this class, are called, both by the French and English, *Washer-Women*, or *Dish-Washers*.

It is particularly serviceable to the farmer in Spring and Autumn, by attending the plough to devour the larvæ of insects, worms, &c. which are turned up; hence it is the interest of the farmer to discourage any attempt to destroy those birds, and to this circumstance they may owe much for their preservation.

---

* *Latham Supp.* Gen. Syn. 178.

Tha

PLATE V.

The marks and colours vary very confiderably in different fpecimens. Some have only a crefcent of black on the breaft, the chin and throat being quite white, in others all the white parts are ftrongly tinctured with yellow; in fome the chin, fore part of the neck, and breaft, are black; and in the *Leverian Mufeum* is a fine variety; white, except the hind parts, which are yellowifh.

In the *females*, the top of the head generally inclines to brown. The ufual length of this fpecies is feven inches from the bill to the extremity of the tail.

The neft is built on the ground, is compofed of dry grafs, fine fibres, and mofs; lined with hair, feathers, or foft dry grafs; the eggs are five in number; white, fpotted with brown: for the moft part they have only one brood in a year.

C 2　　　　　　　PLATE

PLATE VI.

PODICEPS RUFICOLLIS.

RED-NECKED GREBE.

ANSERES.

Bill obtufe, covered with a thin membrane, broad, gibbous below the bafe, fwelled at the apex. Tongue flefhy. Legs naked. Feet webbed, or finned.

GENERIC CHARACTER.

Bill ftraight, flender, pointed. Noftrils linear at the bafe of the Bill. Legs placed near the tail. Feet flat, thin, and ferrated behind with a double row of notches.

SPECIFIC CHARACTER,
AND
SYNONYMS.

Bill black, with the bafe of each mandible fine yellow. Irides bright orange-yellow. Crown, and fides of the Head above the Eyes black-brown, with the feathers a little elongated. The hind part of the Neck, Back, and Wings, dark brown; fix of the middle fecondaries white, a little mottled with dufky at the tips: the two or three next outward, are more or lefs white near the tips and inner webs. The Chin, fides under the Eyes, and fore part of the Neck, for above an inch, pale afh-colour; the reft of the Neck ferruginous chefnut, mottled on the Breaft with dufky; thence to the Vent, white, like fattin, mottled on the fides with dufky irregular fpots. Legs black.

COLYMBUS SUBCRISTATUS, *Jacq. Vog. p.* 37. *pl.* 18.

COLYMBUS PAROTIS, *Sparrm. Muf. Carls. pl.* 9.

PODICEPS RUFICOLLIS. *Lath. Gen. Syn.* 5. 288-7.—*Supp.* 260. 7.

RED-NECKED GREBE. *Lath.*—*Arct. Zool. p.* 499. *C.*

Le Grebe à joues grifes, ou le Jougris, *Buf. Oif.* viii. *p.* 241. *Pl. Enl.* 931.

Suppofed

PLATE VI.

Suppofed to inhabit fome parts of *Denmark* and *Norway*; has been difcovered, though very rarely, near the *Cafpian Sea*; and was once received by *Mr. Pennant*, from *Copenhagen*.

It is probably a fcarce bird in every part; in this country it has been only difcovered by a few individuals, and that very lately; we believe the moft perfect yet taken, to be that fpecimen of which Mr. Latham has given a figure in the fupplement to his General Synopfis; our figure is alfo copied from the fame bird.

It is on the authority of this author, that we include it as a Britifh bird; in his defcription he fays, " I received a perfect fpecimen of the *Male* of this bird from Major *Hammond*, who informed me, that the end of April, the year 1786, two of them alighted in a farm-yard, near his houfe, in *Eaft Kent*, and were taken alive."

" I have alfo met with two other fpecimens; the firft fent to me *January* 28, 1786, by *Mr. Martin*, of *Teignmouth*, a gentleman to whom I owe many other obligations: his fpecimen had not come to perfection, as the colours on the head and neck were much blended, and the ferruginous on the neck only juft breaking forth. *Mr. Boys*, of *Sandwich*, alfo obliged me with a third, the beginning of laft *October*, (1787): his bird, he informed me, weighed nineteen ounces and a half; the length twenty-one inches and a half; breadth eight. The bill yellow at the bafe, dufky olive towards the tip: lore dufky: irides pale brown: head quite fmooth. The defcription differed not much; but the ferruginous colour of the neck was much blended with dufky; the white on the under parts greatly mottled with the fame: Legs dufky; within, greenifh yellow. The middle toe united to the

inner

PLATE VI.

inner as far as the firſt joint; and to the outer, to the middle of the ſecond *."

" The two laſt mentioned are, no doubt, birds not in full plumage. That deſcribed by *Dr. Sparrman* is clearly under the ſame predicament; perhaps a ſtill younger bird than either of the others, as the cinereous parts on the throat appear white, with three or four lines of black, and acroſs the lower part of the neck is a band of white. The bird figured in *Jacquin* ſeems an adult."

That mentioned by *Buffon* was ſeventeen inches in length; had the breaſt mottled with ferruginous; and a white ſpot on the quills.

---

* This ſpecies was unknown to Linnæus, but according to his definition, evidently belongs to the genus COLYMBUS: *Latham* obſerves, that Linnæus has erroneouſly included the *Grebes*, *Divers*, and *Guillemots* into that genus without even a diviſion, though they very materially differ from one another; eſpecially in the legs: thoſe of the *Grebes* are not webbed; the *Guillemots*, though web-footed, have only three toes, all placed forwards; and the *Divers* have three toes before, and one behind.

He therefore thinks that they ſhould be ſeparated; and as the form of the feet of this ſpecies appears to prevail throughout the genus, recommends its being included with the other parts of its eſſential character; he has named his new genus PODICEPS.

Pennant has alſo ſubmitted to a diviſion of the Linnæan *genus*. " The *Grebes* and *Divers* are placed in the ſame genus, *i. e.* of *Colymbi*, by *Mr. Ray* and *Linnæus*; but the difference of the feet forbade our judicious friend, *M. Briſſon* †, from continuing them together; whoſe example we have followed." *Brit. Zool.* 2.496.

† See *Briſ. Orn.* vol. vi. p. 33. 70. 104.

PLATE

7

PLATE VII.

ORIOLUS GALBULA.

GOLDEN ORIOLE.

PICÆ.

Bill compreffed, convex.

GENERIC CHARACTER.

Bill ftrait, conic, fharp pointed; edges cultrated, inclining inwards; mandibles of equal length. Noftrils fmall; at the bafe of the bill, and partly covered. Tongue divided at the end. Toes three forward, one backward; the middle joined near the bafe to the outmoft one.

SPECIFIC CHARACTER,

AND

SYNONYMS.

Bill brownifh-red. Irides red. General colour of the plumage fine golden yellow; between the bill and eye a ftreak of black. Wings black, with a patch of yellow on the middle. Tail yellow except the two middle feathers; all the reft black, from the bafe to the middle black, and thence to their tips yellow. Legs black inclining to a lead colour: claws black.

ORIOLUS GALBULA. *Linn. Syft.* i. *p.* 160. N° 1.
    Faun. Suec. N° 95.
    Georgi Reife. p. 165.
    Sepp. Vog. pl. in. p. 19.
    Lath. Gen. Syn. ii. 449. 43.—*Suppl.* 89.
    Pennant Brit. Zool. ii. 626. 4.
CORACIAS ORIOLUS. *Scop. Ann.* i. *p.* 41. N° 45.—*Faun. arab. p.* 7.
ORIOLUS. *Gefner. av.* 713.—*Aldr. av.* I. 418.

D   GOLDEN

PLATE VII.

Golden Oriole. *Latham.—Pennant.—Brit. Muf.—Lev. Muf.*

Golden Thrush. *Edw. pl.* 185.

Yellow Bird from Bengal. *Albin* iii. *pl.* 19.

Witwall. *Will. orn. p.* 198.

Le Loriot. *Brif. orn.* ii. *p.* 320. N° 58.

   Buf. Oif. iii. *p.* 254. *pl.* 17.—*Pl. enl.* 26. the male.

Widewal, Pyrold, *Frifch. pl.* 31. the male and female.—*Kram. el. p.* 360.

Galbula, feu Picus nidum fufpendens, *Raii. Syn. p.* 68. N° 5.

Size of a Blackbird. The body of the female is of a dull greenifh colour; the wings are dufky inclining to green alfo; and the tail is nearly of the fame obfcure colour, except the two middle feathers which are of a pale yellow.

But the male is evidently one of the moft beautiful birds that has ever been difcovered in this ifland: the whole of the body which is a dull green in the female, is a lovely golden yellow, inclining to an orange colour in the male; the wings are black and form a moft ftriking contraft; and the black ftripe from the beak to the eye is no inconfiderable addition to its beauty.

It is rarely met with in England: Pennant obferves in the *Britifh Zoology* *, that he only knew one inftance of its being fhot in *Great Britain*, and that in *South Wales :* Latham fays " it is now and then met with in England † ;" and adds, in his Supplement, " Since the publication of my *Synopfis* ‡, this bird has been twice fhot in England. One of the fpecimens is now in my collection."

It

PLATE VII.

It * is common in several parts of Europe, but supposed to be most frequent in *France*, where it spends the summer, and propagates its species. It is scarcely ever seen so far north as *Sweden*; and consequently is rare in *England*; is mentioned as a bird of *Ruffia*, though perhaps it only inhabits the warmer parts; comes twice in a year into *Switzerland*, and is found also in *Carniola*; observed in *Malta* in *September* on its passage southward, and returns in spring to the north through the same track; comes into *Conftantinople* in *spring*, and leaves it in *September*, but stays in *Alexandria* till the beginning of *November*, when it takes its leave; from this we must suppose that it winters in *Africa* and *Afia*, especially as this very bird has been brought from *China* and *Bengal* †, as well as the *Cape of Good Hope*.

A variety of this species, with the head and throat of a full black colour, is common in *India*, where it is called the *Mango-bird*, as it appears first at the ripening of that fruit, and is at that season in great plenty ‡.

" The nest is of a curious conftruction, but perhaps not quite so as some of the *Orioles*, though built after the same fashion. It is of the shape of a purse, fastened to the extreme divarications of the outmost twigs of tall trees, and composed of fibres of hemp or straw, mixed with fine dry stalks of grass lined within with moss and liverwort. The *female* lays four or five eggs, of a dirty white, marked with small dark-brown spots, which are thickest about the largest end: she sits three weeks, and is observed to be very tender of her young,

---

* Latham. *Gen. Syn.*

† This bird must have been very little known in England at the time Albin published his History of Birds (1740) for he says in the description annexed to his figure " a drawing from the life of this curious bird was brought from *Bengal* to Mr. *Dandridge*, who was pleased to let me have a drawing from it."

‡ *Lady Impey.*

fearing

PLATE VII.

fearing nothing for their defence; not unfrequently will fuffer herfelf to be taken with the eggs and neft, and continue to fit upon them till fhe dies."

" The food which this bird is moft fond of is *grapes* and *figs*, in the feafon, alfo *cherries*, &c. but at other times is contented with infects, and what elfe it can get."

" It has a loud cry that may be heard far off; but I do not hear it remarked by any one for the leaft fong, though *Gefner* fays it whiftles before rain." *Latham, Gen. Syn.*

" Its note is loud and refembles its name." *Brit. Zool.*

Willughby faw thofe birds expofed for fale in the poulterers fhops in Naples, where the flefh is efteemed as delicate food.

It is fufpected that the *yellow* and *buff Jay* * of Ray are the male and female of the Golden Oriole.

---

* *Raii Syn. av. p.* 194. N° 7, 8.

PLATE

8

PLATE VIII.

ALCA ARCTICA.

PUFFIN AUK.

ANSERES.

Bill obtufe; covered with a thin membrane, broad, gibbous below the bafe. Swelled at the apex. Tongue flefhy. Legs naked. Feet webbed, or finned.

GENERIC CHARACTER.

Bill ftrong, thick, convex, compreffed on the fides. Noftrils linear, placed parallel to the edge of the bill. Tongue almoft as long as the bill. Toes three in number, all placed forward.

SPECIFIC CHARACTER.

Bill compreffed, triangular, fharp-pointed, red, bafe grey, furrows four, oblique. Noftrils long and narrow. Eyelids callous; edges crimfon; on the upper eyelid is a protuberance of a triangular form which projects over the eye. Irides grey. Above black. Cheeks, chin, belly, white. Collar black. Legs orange and near the tail.

ALCA ARCTICA. *Linn. Syft.* 1. *p.* 211. 4.

   Faun. Suec. N° 141.

   Brun. N° 103.

   Muller. N° 140.

   Frifch, t. 192.

   Latham. Gen. Syn. 5. 314. 3.

PUPHINUS ANGLICUS. *Gefner av.* 725.

PICA MARINA. *Aldr. av.* III. 92.

ANAS ARCTICA. *Clufii Exot.* 104.

            PUFFIN,

PLATE VIII.

Puffin, Coulternee. *Raii. Syn. p.* 120. *A.* 5.—*Will. Orn.*
P. 325. *pl.* 65.—*Hift. Groenl.* ii. *pl.* 1.—*Albin.* ii.
pl. 78, 79.—*Edw. pl.* 358. *fig.* 1.—*Brit. Zool.* 11.
N° 232.—*Arct. Zool.* N° 427.—*Tour in Wales,*
pl. 20. *Brit. Muf.*—*Lev. Muf.*

Fratercula, le Macareux. *Briff. av.* VI. 81. Tab. 6. fig. 2.
Buf. Oif. IX. p. 358. pl. 26.—*Pl. Enl.* 275.

Ipatka, *Hift. Kamts.* p. 153.

See Papagey, or See Taucher. *Frifch.* II. 192.

Length from the point of the bill to the end of the feet twelve inches; breadth twenty-one. Weight twelve ounces. The bill is an inch and a quarter long, and of a very fingular fhape, much compreffed on the fides, and near an inch and an half deep at the bafe; from whence both mandibles tend to a point, which is a little curved; acrofs the upper are four oblique furrows; on the under three: half next the point is red; that next the bafe blue grey; and at the bafe is a *cere* full of minute holes: the noftrils are a long and narrow flit on each fide, near the edge of the upper mandible, and parallel to it: the top of the head, the neck, and upper parts of the body black; beneath white: legs orange.

The bill, which gives fuch an appearance of novelty to this bird, varies confiderably according to its age; in the firft year it is fmall, weak, deftitute of any furrow, and dufky; in the fecond year it is larger, ftronger, of a paler colour, and difcovers a faint veftige of a furrow near the bafe; but thofe of the third or more advanced years, have a bill of great ftrength and vivid colours as before defcribed. Thofe birds are fuppofed to be imperfect until the third year; or at leaft not

3   to

PLATE VIII.

to breed before that period: not a single one has ever been discovered at *Priestholm* which had not the bill of an uniform size *.

The *male* very nearly resembles the *female*: in the former the white cheeks are sometimes obscured with a mixture of dark feathers, and in others a patch of the same colour has been observed on each side of the under jaw.

This species is very common in several parts on the coasts of *England*; they are seen in flocks innumerable at *Priestholm Isle* off the coast of *Anglesea*; in great numbers about the *Needles*, in the *Isles of Wight*, *Man*, *Bardsey*, *Caldey*, *Farn*, *Godreve*, and other small, and desert islands near the shore. A few about the rocks of *Dover*.

They are frequent in *Ireland*; on the island of *Sherries*, three leagues N.N.W. of *Holyhead*; and in the S. *Stack*, near *Holyhead* they breed in abundance. Inhabit *Iceland* and *Greenland*, and breed in the extreme part of the islands, especially on the west part of *Disco*, and the island *Orpiksauk* †.

In the different parts it frequents, it has received a variety of appellations, but generally expressive of the singular shape of the beak; as *Coulter-neb* in the *Farn isles*; *Guldenhead*, *Bottle-nose*, and *Helegug* in *Wales*; at *Scarborough*, *Mullet*; at *Cornwall*, *Pope* ‡, and in the *Ferroe isles*, *Lunda*.

To what part those birds emigrate on the approach of winter is very imperfectly known; it is probable when they retire from those northern regions, their flight is directed to some more temperate climate; perhaps they live at sea, and form those multitudes of birds that navigators

---

* See *Tour in Wales*, p. 252; and figures of the different growth of the bill in pl. 20.
† Latham V. p. 316.   ‡ *Will. orn.*

have

PLATE VIII.

have obferved in many parts of the ocean; they are always found there at certain feafons, but retire at the breeding time to the northern latitudes, and during that time are found as near the *Pole* as navigators have ever penetrated *.

In *America*, they are faid to frequent *Carolina*; and have been met with in *Sandwich Sound*, by our late voyagers: the natives ornament the fore parts and collars of their *feal-fkin* jackets with the beaks of them; and in *Aoonalafhka*, they make gowns of their fkins, along with thofe of other birds.

On the coaft of *Kamtfchatka* and the *Kurilfchi* iflands they are very common, even on the *Penfchinfki Bay*, almoft as far as *Ochotka*: the nations of the two firft wear the bills about their necks faftened to ftraps; thefe are put on by their *Shaman* or *Prieft*, for the people are perfuaded that by putting them on with a proper ceremony, they will procure good fortune to all their undertakings †.

" About the fifth or tenth of April, they arrive in vaft quantities at *Prieftholm ifle*; but quit the place again, and return twice or thrice before they fettle to burrow, which they do the firft week in *May,* when many of them diflodge the *rabbits* from their holes, by which they fave themfelves the trouble of forming one of their own: in the laft cafe, they are fo intent on what they are about, as to fuffer themfelves to be taken by the hand. It has been obferved that this tafk falls chiefly to the fhare of the *males*, and that thefe laft affift alfo in incubation: this has been proved on diffection. The *female* lays one white egg ‡. The young are hatched the beginning of *July :* and about the

eleventh

---

* *Pen. Brit. Zool.*   † *Hift. Kamtfch.*

‡ Albin obferves " they build no neft, but lay their eggs on the bare ground"— " They lay but one egg apiece (which is efpecially remarkable)" " The eggs are very

large

PLATE VIII.

eleventh of *Auguſt* they all go off, to a ſingle bird *, and ſo completely as to deſert the young ones that are late hatched; leaving them a prey to the *Peregrine Falcon*, who watches at the mouth of the holes for them, as they through hunger, are compelled to come out. Notwithſtanding the neglect of their young at this time, no bird is more attentive to them in general, as they will ſuffer themſelves to be taken by the hand, and uſe every means of defence in their power to ſave them; and, if laid hold of by the wings, will give themſelves moſt cruel bites on any part of the body they can reach, as if actuated by deſpair; and when releaſed, inſtead of flying away, will often hurry away into the burrow to their young." *Lat. Gen. Syn.* 5. 316.— *Arct. Zool.*

Their fleſh is exceſſively rank, as they feed on fiſh, particularly *ſprats*, or on ſmall *crabs, ſea-weeds*, &c. yet that of the young birds is often pickled and preſerved with ſpices, and is much admired by ſome for its peculiar flavor †. *Dr. Caius* writes, that in his days the church allowed them in Lent, inſtead of fiſh; and alſo that they were taken by means of *ferrets*, as now they are by *rabbits*: at preſent they

---

large for the bigneſs of the bird, even bigger than *hens* or *ducks*, of a *reddiſh* or *ſandy colour*, much ſharper at one end than *hen's* eggs, and blunter at the other." *vol. 2. p.* 78, 79.

But it appears very probable that Albin was miſtaken as to the colour of the eggs, if we may judge by the concurrence of the beſt informed naturaliſts of the preſent time; " I muſt add," ſays Pennant, " that they lay only one egg, which differs much in form; ſome have one end very acute; others have both extremely obtuſe; all are *white*. *Brit. Zool.*

* " The Reverend Mr. *Hugh Davies*, of *Beaumaris*, informed me, that on the 23d of *Auguſt* (1776) ſo entire was the migration, that neither Puffin, Razor-Bill, Guillemot or Tern was to be ſeen there." *Brit. Zool.* 2. 515.

† " They are potted at St. *Kilda* and elſewhere, and ſent to *London* as rarities. The bones are taken out, and the fleſh wrapped in the ſkin; are eaten with vinegar, and taſte like baked *herrings*. *Lat. Gen. Syſ.*

E

are

are either dug out, or drawn from their burrows by a hooked ftick: they bite extremly hard, and keep fuch faft hold on whatever they faften, that it is with difficulty they can be difengaged; when they are taken, their noife is very difagreeable, being like the efforts of a dumb perfon to fpeak.

It flies with great ftrength and fwiftnefs when it gets on the wing, but meets with many falls before that can be effected: the legs are placed fo far behind, that it cannot ftand except quite erect; and at that time it refts not only on the feet, but on the whole length of the legs alfo.

PLATE

9

PLATE IX.

UPUPA EPOPS.

COMMON HOOPOE.

P I C Æ.

Bill compreſſed, convex.

GENERIC CHARACTER.

Bill long, ſlender and bending. Noſtrils ſmall, placed near the baſe. Tongue ſhort, ſagittal. Toes three before, and one behind; the middle one connected at the baſe to the outmoſt.

SPECIFIC CHARACTER,

AND

SYNONYMS.

Bill black, ſlender. Tongue triangular; placed low in the mouth. Creſt compoſed of a double row of feathers; of a pale reddiſh brown. Breaſt and belly white. Back ſcapulars and wings, barred with black and white. Tail of ten feathers; black marked with white, in the form of a creſcent, with the horns pointing towards the end of the feathers. Legs ſhort and black.

Upupa Epops. *Lin. Syſt. Nat.* 1. *p.* 183 Nº 1.
  Scop. Ann. 1. *p.* 53. Nº 62.
  Muller. p. 13. 103.
  Brun. Nº 43.
  Georgi. Reiſe. p. 165.
  Sepp. Vog. pl. in. p. 129.
  Faun Arag. p. 74.
  Kolb. Cap. ii. *p.* 157.
Upupa. *Raii. Syn. p.* 48. A. 6.
  Geſner. av. 776.
  Kramer. elen. p. 337.

F

UPUPA;

PLATE IX.

UPUPA; arquata ftercoraria; gallus lutofus. *Klein Stem av.* 24. tab. 25.

HOOP or COMMON HOOPOE. *Will. orn. p.* 145.

> *Albin.* 2. *pl.* 42. 43.
>
> *Edw.* 7. *pl.* 345.
>
> *Br. Zool.* N° 90.
>
> *Arct. Zool.* ii. *p.* 283. *A.*
>
> *Lath. Gen. Syn.* ii. *p.* 687.—*Supp.* 122.
>
> *Br. Muf.—Lev. Muf.*

DUNG BIRD. *Charlton ex.* 98. *tab.* 99.

La Hupe, ou Puput. *Brif. orn.* ii. *p.* 455. N° 1. *pl.* 43. *f.* 1.

> *Buff. ois* VI. *p.* 439.—*Pl. enl.* 52.

La Huppe. *Belon. Av.* 293.

Bubbola. *Olin. uccel. p.* 36.

Wied-hopf. *Frifch. t.* 43.

Harfogel, Pop. *Faun. Suec. fp.* 105.

Her-fugl. *Brunnich.* 43.

Smerda kaura. *Scopoli.* N° 62.

Ter-chaous, or Meffenger Bird. *Pococke's Trav.* 1. *p.* 209.

Linnæus in the *Fauna Succica* * obferves that this elegant bird receives its name from the found of its note; but by other naturalifts it has been fufpected that its name was originally derived from the French *huppè* or crefted; as the creft is of a very curious ftructure, and alone is fufficient to diftinguifh it from every other European bird. It is the only fpecies of the *Hoopoe* genus that is peculiar to the continent of Europe.

---

* 2d edit. 37.

It

PLATE IX.

It inhabits *Asia* and *Africa*, and is said to be met with in the large forests of *Sweden* [*], and in *Austria* [†]; and has been found as far north as the *Orknies* and *Lapland*, as well as in many of the intermediate places between them; at the *Cape of Good Hope* on one hand, and quite to *Ceylon* [‡] and *Java* on the other. In *Europe* it is considered as a bird of passage, and is said not to winter even in *Greece* [§]. In *England* [||] it is far from common, being seldom seen, and at uncertain times: it has been observed in *Kent*, *Surrey*, *Northumberland*, and *Moyston* in *Flintshire*, as well as in several other counties. A gentleman of veracity in *Essex* informs us that one was discovered last year in a hole in his garden wall, but being frightened away did not return again to that place. Among other proofs of its migrating into, and even breeding in *England*, Mr. *Latham* has mentioned several. "The year 1783 seems to have been more abundant in these birds than any I have yet heard of; one being shot near *Oxford*, on the coast of *Suffolk*, in *May*, and another seen near the same place the 24th of *June* following: these no doubt had bred thereabouts. The place where these were seen was a remarkable barren spot. In the month of *September* of the same year two were shot at *Holderness*, and many were seen in various parts of *Yorkshire*, and as far north as [¶] *Scotland*. One was shot the 10th of *September* at *Cam* in *Gloucestershire*, another on *Epping Forest*, and a third in *Surrey*. A few years since a pair had begun to make a nest in *Hampshire*; but being too much disturbed, forsook it, and went elsewhere [**]. The last year (1786) a young bird was sent to me, the 10th of *May*, full-fledged, shot near *South-*

---

* *Fauna Suecica*, p. 37.   † *Scopoli*.   ‡ *Edwards*.

§ The *Hoopoe* and *Roller* are said to come into *Constantinople* in *August*, from the north, to return in spring. *Faun. Arab.* p. 7. —— "The *Hoopoe* and *Bee-eater* come in the spring, and remain all the summer and autumn." *Russel*. Alep. p. 70.

|| *Latham*, *Gen. Syn.* 688. 1.   ¶ Mr. *Turnstall*.   ** *Ditto.*

fleet, in *Kent**; but the old birds had not been obferved."—It was well known, as a vifitor in *England,* at the time *Albin* publifhed his Hiftory of Birds; his obfervations deferve notice. "The hen of this bird was fhot in the garden of Mr. *Starkey Mayos,* at *Woodford* on *Epping Foreft,* where they had obferved it fome time, and ufed all the means to take it they could; but it was fo fhy, that it avoided all their traps which were laid for it; which the gentleman obferving, ordered it to be fhot: it was fent to me to be preferved for him.

"The cock of this kind I drew from a picture done in *Germany,* by a great mafter there, now in the poffeffion of Mr. *Nifbet,* a gentleman, who had it drawn from the bird when alive.

"There is fome difference in the colours of the hen, and this bird which was a cock, I was credibly informed by *Robert Briftow,* Efq; who faw both the drawings of the cock and hen, and told me his fon fhot the cock, which was like the drawing at his feat at *Micheler,* near *Winchefter* in *Hampfhire."* *Albin,* Vol. II. 42, 43.

Latham obferves, "it is a folitary bird, and feldom more than two are feen together; though it is faid that in *Egypt* it affembles in fmall troops. It is very common in *Cairo,* where it builds in the ftreets, on the terraces of houfes, &c. It is alfo common in the deferts of *Ruffia* and *Tartary,* though fcarce beyond the river *Ob;* however fome are found beyond the *Lake Baikal.* Dr. *Pallas* confirms the account of the filthy manners of this bird, as he met with an inftance of a pair breeding in the privy of an uninhabited houfe in the fuburbs of *Tzaritzn* †.

"I am informed by colonel *Davies,* that they every year are feen in *Gibraltar* in *March,* in fmall flocks of ten or twelve; hence are

---

* *By Mr. Godden of that place.*   † *Arct. Zool.*

6

called

PLATE XI.

called there *March Cocks*. They are fuppofed to come from *Africa*, and to be on their paffage north to fome other place, as they only ftay a few hours to reft themfelves: and it is not uncommon to fee five or fix flocks in a week, during the time of their paffage. He did not obferve them to have any note; but that they had a dipping kind of flight not unlike a *Woodpecker*. I have obferved this Bird to be among paintings both from *China* and *India*; it is therefore no doubt common to both thofe parts." *Lath. Gen. Syn.*

In Sweden the appearance of this Bird is regarded as a prefage of war; and in England its vifits were formerly confidered as ominous by the vulgar.

In Turkey it is called *Tir Chaous*, or the Meffenger Bird from the refemblance its creft has to the plumes worn by the *Chaous*, or *Turkifh* couriers.

Latham fays, the *female* is like the *male*, and lays from two to feven eggs; but for the moft part four or five. Thefe are fomewhat lefs than thofe of a Partridge, but longer and afh-coloured. This Bird is faid to have two or three broods in a year, and to lay the eggs in the holes of trees, like the *Woodpecker*, but in general to make no neft: notwithftanding which, *Buffon* obferved, that two out of fix nefts, which were brought to him for infpection, had a foft lining of mofs, wool, leaves, feathers, and the like; and he is of opinion, that when this is the cafe, the bird has made ufe of the old neft of fome other bird. It will alfo lay, and hatch the young in holes of walls, and even on the ground. The food of this bird is infects; and it is the exuviæ of the large beetles, and fuch like, with which the neft is crouded, that caufe the neft to ftink fo horribly; infomuch that former writers afferted the neft to be made of excrement.

In

PLATE IX.

In *Sepp's* plate the neft is placed in the hollow of a tree; it is compofed of foft bents, and fmooth within. The eggs, four in number, of a blueifh white, marked with pale brown fpots.

Buffon mentions one of this fpecies which lived with a lady for three months, fubfifting only on bread and cheefe; and, contrary to the common opinion, drank frequently, and that by gulps. Another was kept for eighteen months on raw meat, and would not eat any thing elfe.

Olina fays, that this bird lives three years.

In fome countries it is efteemed as good eating. It feldom perches on trees, unlefs they are very low. It does not erect its creft, except when agitated by furprize: in a natural ftate the creft falling behind the neck *; but whenever it alights on the ground, it is faid to fpread its creft beautifully.

Some authors mention a variety of this fpecies. *Kolben* † mentions one at the *Cape of Good Hope*, which is fmaller; the bill fhorter in proportion; and the legs longer: the creft is not fo long, and has no trace of white in it throughout: and in general the plumage is lefs variegated. Another fpecimen from the fame place, had the upper part of the beak of a deep brown, and the belly varied with brown and white; but as this was lefs in every refpect, it was probably a young bird.

Gerini mentions one which he faw at *Florence*, and again on the *Alps*, which had the creft bordered with fky-blue. *Orn. Ital. Hift. des ois.* VI. p. 462.

---

* Pics ke,

† *Kolben. Hift. du Cap. I. p. 152.*

PLATE

10

PLATE X.

SYLVIA DARTFORDIENSIS.

DARTFORD WARBLER.

PASSERES.

Bill conic, pointed. Noftrils oval, broad, naked.

GENERIC CHARACTER.

Bill weak, flender *. Noftrils fmall, a little depreffed. Tongue cloven. The exterior toe joined at the under part to the bafe of the middle one.

SPECIFIC CHARACTER,

Bill black, with a white bafe; the upper mandible a little curved at the tip. Irides red; eyelids deep crimfon. The upper parts of the head, neck, and body, dufky reddifh brown. Breaft and belly deep ferruginous; middle of the belly white. Quills dufky edged with white. Baftard wing white. Exterior web of the outer tail feather white; the reft dufky. Legs yellow.

SYLVIA DARTFORDIENSIS. *Lath. Gen. Syn.* iv. *p.* 435. N°. 27.
DARTFORD WARBLER. *Suppl. p.* 181.
    Pennant. Brit. Zool. i. N° 161. *pl.* 56.
   —— *Arct. Zool.—Lev. Muf.—Berken. oyt. Nat.*
    Hift. Vol. i. 52. 14.
Le Pitchou de Provence. *Buf. ois.* v. *p.* 158. *Pl. enl.* 655. 1.

---

* The Linnæan genus *Motacilla* has been feparated by Pennant, and his method adopted by Latham; by this feparation the Wagtails conftitute one genus, and the Warblers another: the latter are diftinguifhed from the former in feveral refpects; they perch on trees, proceed by leaps, not running, and feldom emit any noife in flight.

This

PLATE X.

This bird meafures five inches from the tip of the bill to the end of the tail: it is of a lively appearance, though not very beautiful in the colours of its plumage; and deferves our immediate attention as one of the leaft known fpecies we have in this country.

It is a native of *France* as well as of *England*. In *Provence* it is commonly found among cabbages: it feeds on the infects that harbour among thofe vegetables, and not unfrequently conceals itfelf under the fhelter of the leaves during the night.

A friend of Mr. Latham's fhot a pair of thofe birds on *Bexley Heath*, near *Dartford* in *Kent*, *April* the 10th, 1773, as they were fitting on a furze bufh: they fed on flies; fpringing from the bufh every time one approached near, and returning to the fame place repeatedly; thereby imitating, as he obferves, the manners of our *Cinereous Flycatcher.*

This fpecies refides with us in the winter. Several fpecimens, which are now preferved in the Leverian Mufeum, were fhot on a common near *Wandfworth* in *Surrey*, 1782.

Mr. Latham appears to entertain fome doubt, whether this fpecies ever breeds in France *. He fays an intelligent obferver of *Englifh* Birds † has informed him, that he never met with this fpecies in the neighbourhood of *London*, except in *winter*; and that it difappears before the end of *April*. Should this be the general fact, he can by no means reconcile the circumftance of its breeding in *France*, as all migratory birds retire northward to breed, not to a warmer climate; and fhould rather fuppofe, that if it does not quit *England* in fummer, it will hereafter be found in the northern parts of it.

---

* Hift. des Ois. v. p. 158.　　† Mr. Green.

PLATE

11

PLATE XI.

AMPELIS GARRULUS.

WAXEN CHATTERER.

PASSERES.

Bill conic, pointed. Noftrils oval, broad, naked.

GENERIC CHARACTER.

Bill ftrait, convex, bending towards the point; near the end of the upper mandible a fmall notch. Noftrils hid in the briftles. Tongue cartilaginous, bifid.

SPECIFIC CHARACTER.

Length eight inches. Bill black, irides reddifh: the feathers on the crown of the head elongated into a creft: the head and upper parts reddifh afh colour. Rump fine cinereous. From the noftrils over each eye, paffes a ftreak of black. Forehead chefnut. Chin black. Breaft pale purplifh chefnut; belly paler, inclining to white near the vent. Leffer wing coverts brown; the greater, fartheft from the body, black with white tips, forming a bar: quills black, the third and fourth tipped on the outer edges with white, the five following with yellow: fecondaries afh-colour, tipped on the outer edge with white; feven of the fecondary feathers have the ends of their fhafts continued into a flat horny appendage, of the colour of fine red fealing-wax. Tail black, tipped with yellow. Legs black.

AMPELIS GARRULUS, *Linn. Syft.* 1. *p.* 299. 1.
Faun. Suec. N° 82.
Kram. el. p. 363. 1.
Frifch. pl. 32.

G

LANIUS

PLATE XI.

LANIUS GARRULUS. *Scop. Ann.* 1. *p.* 20.

GARRULUS BOHEMICUS. *Albin.* 2. *pl.* 26.

Gefn. av. 703.

BOHEMIAN CHATTERER. *Will. orn. p.* 132. *pl.* 20.—*Albin.*

WAXEN CHATTERER. *Latham. Syn.* iii. 91. 1.

Pennant. Brit. zool. 1. 314.

Br. Zool. Nº 112. *pl.* 48. *Arct. Zool.*—*Br.*
Muf.—*Lev. Muf.*

SILK TAIL. *Raii. Syn. av.* 85. A.

Phil. Tranf. Vol. xv. *p.* 1165. *pl.* 1. *f.* 9.

Ray's letters 198. 200.

Le Jafeur de Boheme.

Bombycilla Bohemica. *Bris. orn.* 11. *p.* 333. 63.

Buff. ois III. *p.* 429. *pl.* 26.—*Pl. enl.* 261.

Siden-fuantz, Snotuppa. *Faun. Suec. fp.* 82.

Sieden vel Sieben Suands. *Brunnich.* 25.

Zuferl, Geidenfchweiffl. *Kramer.* 363.

Seiden-fchwantz. *Frifch.* 1. 32.

This bird is fuppofed to breed in *Bohemia* and other parts of *Germany*, but its fummer refidence is perhaps more northward; it is feen in plenty both at *St. Peterfburgh* and *Mofcow*, in the winter; but comes from the north and departs again to the *Arctic* circle in fpring; never known to breed in *Ruffia*, is fcarce in *Siberia*, and has not been obferved beyond the river *Lena* *.

All the birds of this genus are natives of *America*; this fpecies excepted; they wander from their native place all over *Europe*, and at uncertain times vifit the *Southern* parts of *Britain*. They are obferved in the *Northern* parts; about *Edinburgh* in *February* they come an-

---

* Pennant.

3

nually

PLATE XI.

nually and.feed on the berries of the mountain afh: they alfo appear as far fouth as *Northumberland* and *Yorkfhire* frequently, and like the fieldfare make the berries of the white-thorn their food *. They have alfo been met with feveral times near *London* †. They difappear in fpring ‡. In *France* and *Italy* they are not unfrequent.

The nefts of thofe birds are faid to be conftructed in the holes of rocks §, but as we can fcarcely determine even their native country, we need not expect any fatisfactory information relative to its eggs and neft, until fome future traveller fhall be fo fortunate as to difcover them.

The general food is *berries* of all kinds, efpecially *grapes*; in countries where they are plenty they are efteemed good food.

It is faid that the *females* want the red appendages at the end of the fecond quills ‖, as well as the yellow marks on the back **.

A variety of this bird is alfo found in *America* from *Carolina* to *Mexico*, it is the AMPELIS GARRULUS of *Linnæus*, Le jafeur de la Caroline, of *Briffon* and *Buffon*; Caquautototl, *Raii*; and Chatterer of *Carolina*, of *Edwards*, *Catefby*, &c.

This bird is lefs than the *European* kind, is much like it, except that the belly is of a pale yellow inftead of reddifh; both fexes have

---

G 2   the

PLATE XI.

the wings of a plain colour without the marks of yellow: the *female* has no appendages at the ends of the fecond quills, and the plumage is lefs lively than in the male.

This variety is called the *Recolleƈt* at *Quebec*; our late voyagers met with this bird at *Aoonalſhka* *.

---

* Ellis's voyage II. *p.* 15.

PLATE

12

PLATE XII.

TETRAO LAGOPUS.

PTARMIGAN GROUS.

OR

WHITE GAME.

GALLINÆ.

Bill convex: the upper mandible arched. Toes connected by a membrane at the bottom. Tail feathers more than twelve.

GENERIC CHARACTER.

Bill convex, ftrong, and fhort, a naked fcarlet fkin above each eye *. Noftrils fmall, hid in the feathers. Tongue pointed at the end. Legs ftrong, feathered to the toes, and fometimes to the nails. Toes of fome fpecies pectinated on the fides.

* With four toes.

SPECIFIC CHARACTER,

Length fifteen inches. Bill black. Plumage pale brown or afh-colour, croffed or mottled with fmall dufky fpots, and minute bars; the head and neck with broad bars of black, ruft colour, and white. Wings white: Shafts of the greater quills black. Belly white. Winter drefs pure white, except a black line between the bill and eye, and fhafts of the firft feven quills black, in the *male.* Tail of fixteen feathers, the two middle ones afh-coloured in fummer, white in winter, two next flightly marked with white near the ends; the reft entirely black. The upper tail coverts almoft cover the tail.

---

* Three or four fpecies excepted.

TETRAO

PLATE XII.

TETRAO LAGOPUS. *Lin. Syft.* i. *p.* 274. 4.

        Suec. 203.—*Scop. Ann.* i. N° 170.

        Raii. Syn. p. 55. 5.—*Baun. p.* 59.

        Phil. Tranf. vol. lxii. *p.* 390.—*Frifch.*

        pl. 110. 111.—*Kram. el. p.* 356.

        Faun. Groenl. N° 80.—*Georgi. Reife. p.* 172.

LAGOPUS. *Gefner. av.* 576.

        Plinii. lib. x. c. 48.

Perdrix alba feu Lagopus, Perdice alpeftre.

        Aldro. av. 11. 66.

WHITE GAME *. *Will. orn. p.* 176. *pl.* 32.

PTARMIGAN. *Br. Zool.* 1. N° 95.—*Gent. Mag.* 1772.

        pl. in p. 74.—*Sib. Scot.* 16.

        Pen. Zool.—Arct.—Br. Muf.—Lev. Muf.—Lath.

        Gen. Syn. IV. 741. 10.

La Gelinote blanche. *Brif. orn.* 1. *p.* 216. 12.—*Pl. enl.* 129.

    (Winter drefs).—*Pl. enl.* 494. *(Summer drefs).*

La Perdrix blanche. *Belon. av.* 259.

Le Lagopède. *Buff. ois.* ii. *p.* 264. *pl.* 9.

Snoripa. *Faun. Suec. fp.* 203.

Schneehuhn. *Frifch.* 1. 110.

Schneehun. *Kram.* 359.

This fpecies meafures fourteen or fifteen inches from the tip of the bill to the extremity of the tail: extent twenty three, weight nineteen ounces. Its fummer drefs varies exceedingly from that which it affumes in winter; in the former the general colour is pale brown, or afh-colour, not inelegantly marked, or mottled with dufky bars, fpots,

---

* Erroneoufly called the White Partridge

ruft

PLATE XII.

ruft colour, &c. in both fexes, but in the latter, the female is entirely of a very beautiful white; the male of the fame colour, but is diftinguifhed by a dark dafh, or line which paffes from the bill to the eye, and by the fhafts of the firft feven quill feathers being black; the twelve extreme feathers of the tail are of the fame black colour.

Our figure is copied from a male bird which has not wholly affumed its winter appearance, but is in the laft ftage of changing its fummer drefs, as appears from the flight intermixture of dark feathers on its breaft and back.

It inhabits moft of the northern parts of *Europe*, even as far as *Groenland*, in *Ruffia* and *Siberia* it is very frequent; it is feen in plenty on the *Alpine* mountains of *Savoy*, on the *Alps*, and mount *Cenis*.

In *Great Britain* it is met with on the fummits of the higheft hills in the *Highlands* of *Scotland. Hebrides*, and *Orknies*, and a few yet inhabit the lofty hills near *Kefwick* in *Cumberland*, as well as *Wales* *. They live amidft the rocks, and perch on the grey ftones, the general colour of the ftrata in thofe fituations.

Willughby has defcribed the *Ptarmigan* under the name of the White Game. M. *Briffon* † joins it with the White Partridge of *Edwards*, but *Pennant* has given as his decifive opinion that they are two diftinct fpecies. " I have received both fpecies at the fame time from *Norway*, and am convinced that they are not the fame." *Penn.*

The female lays eight or ten eggs, fpotted with red-brown, the fize of thofe of a Pigeon, on the earth, in a ftony fituation, about the middle of June ‡.

---

* *Latham——Pennant.*   † *Tom. 1. p. 216.*   ‡ Latham.

Authors

PLATE XII.

Authors agree that they are ftupid filly birds, and are fo tame as to be drawn into any fnare; or fuffer themfelves to be taken by the hand; if the hen is killed the *male* will not forfake her. The *Groenlanders* take them with noofes tied to a long line, which being carried by two men is drawn over their heads.

* Their food confifts of the buds of trees, young fhoots of *pine*, *heath*, *fruits*, and *berries* which grow on the mountains: on the continent they feed on the *Dwarf Birch* and *Black-berried Heath*, or fometimes on the various kinds of *Liver-wort*.

" They tafte fo like a Grous as to be fcarcely diftinguifhed; like the Grous they keep in fmall packs; but never like thofe birds take fhelter in the heath; but beneath loofe ftones."

In winter they lie in heaps, in lodges which they form under the fnow.

---

* Pennant.

PLATE

13

PLATE XIII.

PICUS MARTIUS.

GREAT BLACK WOODPECKER.

PICÆ.

Bill compreſſed, convex.

GENERIC CHARACTER.

Bill ſtrait, ſtrong, angular, and connected at the end. Noſtrils covered with briſtles reflected down. Tongue very long, ſlender, cylindric, bony, hard and jagged at the end. Toes two forward, two backward. Tail of ten hard, ſtiff, ſharp-pointed feathers.

SPECIFIC CHARACTER.

Bill aſh colour, blending to black, whitiſh on the ſides. Irides yellow. Whole bird black except the crown on the head, which is vermilion. Legs lead colour; covered with feathers on the fore part half their length.

PICUS MARTIUS.   *Linn. Syſt.* 1. *p.* 173. N° 1.

                  Scop. Ann. 1. *p.* 46. N° 51.

                  Brun. N° 38.

Picus niger maximus.   *Raii. Syn. p.* 42. 1.

GREAT BLACK WOODPECKER.   *Will. Orn.* 135. *pl.* 21.

                  Albin. 2. *pl.* 27.

                  Amer. Zool.

                  Lath. Gen. Syn. 11. *p.* 552. 1.

                  Arct. Zool. 11. *p.* 276. A.

H   Lc

PLATE XIII.

Le Pic Noir. *Brif. Orn.* IV. *p.* 21. N° 6.

Buf. Oif. VII. *p.* 41. *pl.* 2.—Male, *Pl. enl.*
596.

Orn. de Salern. pl. 10. *f.* 2.

Schwartz Specht. *Frifch. t.* 34.

This fpecies is near feventeen inches in length; the plumage is entirely black, except the crown of the head, which is of a vermilion colour, rather inclining to crimfon; the bill, and claws, are of confiderable ftrength, particularly the latter, which are curved in a more formidable manner than thofe even of many rapacious birds of equal magnitude.

The *female* differs from the *male* in the general colour of the plumage; that of the *female*, having a ftrong caft of brown on the back, and the vermilion coloured feathers, with which the whole crown of the *male* is invefted, being only fparingly diffufed on the crown of the female, though they terminate in a rich tuft on the hind part of the head.

Both *male* and *female* are very liable to variations in the red on the crown; fome are adorned with a profufion of thofe feathers, while others have fcarcely any; and fpecimens have been met with entirely black, without even a trace of the vermilion colour on their heads.

As an *Englifh* Woodpecker it is the largeft we have; it even confiderably exceeds the fize of the GREEN WOODPECKER, *Picus Viridis*. It is very rare in this country, and generally believed to have been only obferved in the fouthern parts, and in *Devonfhire* *.

---

* Mr. *Latham* writes, " Mr. *Tunftall* tells me, that he has been informed by a fkilful Ornithologift, of its being fometimes feen in *Devonfhire*." *Gen. Syn.*

5

It

PLATE XIII.

It is found in almoſt every part of *Europe*, but is plenty only in *Germany*; it is rarely ſeen in *France*; never in *Italy*; and only during the ſummer in *Sweden*, *Switzerland*, and *Denmark*. Extends to *Ruſſia*, where it is common in the woods from *St. Peterſburgh*, to *Ochotſk* on the *eaſtern* Ocean, and to *Lapmark* on the *weſt* *.

" This ſpecies is ſo very deſtructive to Bees, that the *Baſchirians* in the neighbourhood of the river *Ufa*, as well as the inhabitants of other parts, (who form holes in the trees twenty-five or thirty feet from the ground, wherein the *Bees* may depoſit their ſtore), take every precaution to hinder the acceſs of this bird; and in particular are cautious to guard the mouth of the hive with ſharp thorns; notwithſtanding which, the *Woodpecker* finds means to prove a very deſtructive enemy: and it is obſerved to be in moſt plenty where the Bees are in the greateſt numbers †." *Latham.*

Its food does not conſiſt entirely of *Bees*; *Albin* writes of the bird he has figured, " The guts are ſeventeen inches long, great and lax; the ſtomach alſo lax and membraneous, full of *Hexapods* and *Ants*. It wants the *appendices* or blind guts as the reſt of this tribe."

Its neſt is capacious and deep, and is ſaid to be uſually built in old *Aſh* or *Poplar trees*; *Friſch* obſerves, that they often ſo excavate a tree, that it is ſoon after blown down with the wind; and that under the hole of this bird may often be found a buſhel of duſt and bits of wood.

The *female* lays two or three white eggs; which colour, according to *Willughby* ‡, is peculiar to the whole genus, or at leaſt to all thoſe which have come under his inſpection.

---

* *Arct. Zool.*   † *Dec. Ruſſ.* IV. p. 9. 17.   ‡ *Zool. Danic.*

H 2   PLATE

14

PLATE XIV.

MOTACILLA TROCHILUS.

WILLOW WREN.

PASSERES.

Bill conic, pointed. Noftrils oval, broad, naked.

GENERIC CHARACTER.

Bill flender, weak. Noftrils fmall, a little depreffed: Tongue cloven. The exterior toe joined at the under part to the bafe of the middle one.

SPECIFIC CHARACTER.

Upper parts pale olive green; under parts pale yellow. A ftreak of yellow over the eyes. Wings and tail brown, edged with yellowifh green. Legs yellowifh.

MOTACILLA TROCHILUS. *Linn. Syft.* I. *p.* 338. N° 49.

       Faun. Suec. N° 264.—*Scop. Ann.* I. N° 238.

       Kram. el. p. 378. N° 22.—*Brun.* N° 286.—

       Muller 281.—*Frifch. t.* 24. *f.* 2.

MOTACILLA HISPANICA. *Haffelq. Voy.* 287. 52.

TROCHILUS. *Gefner av.* 726.

ASILUS. *Aldrov. av.* II. 293.

SMALL YELLOW BIRD. *Raii Syn. p.* 80. A. 10.

LITTLE YELLOWISH BIRD. *Will. Orn. p.* 228.

GREEN WREN.

Regulus non criftatus. *Albin.* II. 59.

YELLOW

PLATE XIV.

YELLOW WREN. *Latham Gen. Syn.* IV. 512.

Penn. Brit. Zool. N° 151.

Arɛ̃. Zool.—Br. Muſ.—Lev. Muſ.

Le Pouillot, ou Chantre. *Briſ. Orn.* iii. *p.* 479. N° 45.

Buff. Ois. V. *p.* 344.—*Pl. enl.* 651. *f.* 1.

Chofti, ou Chanteur. *Belon av.* 344.

Schnee Rienig (Snow King). *Friſch*, I. 24.

Schmittl. *Kramer.* 378.

The Yellow Wren ranks among the leaſt of the Britiſh Birds; it meaſures only four inches and three quarters from the tip of the bill to the extremity of the tail. The colours of its plumage are not attractive, neither do we introduce it as a rare bird, being one of our moſt frequent ſpecies; but it is a very delicately formed creature, exceedingly active, and by concealing itſelf immediately among the thickeſt of the foliage when any noiſe approaches, it may not be ſo generally known as ſome leſs timid birds.

It chiefly frequents large woods, which abound with willows; and builds its neſt at the roots of trees, or in the hollows of dry banks; it is conſtructed in the form of an egg, with a hole at the top for its entrance, the outſide is compoſed of moſs and hay, or ſtraw; and the inſide is lined with ſoft feathers, wool, or hair. It lays ſeven white eggs *, or, according to *Latham* and *Albin*, only five; they are freckled all over with reddiſh ſpots. Its note is low and plaintive, ſcarcely more than *twit, twit* †, which it utters when it is running up and down the branches of trees in ſearch of inſects on which it feeds. It is ſaid that the male has a ſong during incubation, far from unpleaſing, and is ſoft, though weak. It is migratory, but viſits us early.

---

* *Pennant Br. Zool.* i. 151.   † *Latham.*

Albin

PLATE XIV.

Albin fays it fings like a grafhopper, and frequents woods and folitary places, fitting on the tops of oaks.

Pennant obferves, that the breaft, belly, and thighs, vary in colour in different birds; in fome thofe parts are of a bright yellow, in others they fade almoft into white. The legs alfo appear to admit of variation, thofe of our fpecimens are yellowifh in both fexes, *Albin* defcribes thofe of his male fpecimen to be pale amber colour, and thofe of the hen to be black.

Latham, in his *Gen. Syn.* has given a defcription of four other birds, which he confiders only as varieties of the MOTACILLA TROCHILUS. Among thofe are included the GREATER NON-CRESTED REGULUS of *Willughby*, and the MOTACILLA ACREDULA of *Linnæus*. This latter bird appears in the *Britifh Zoology* of *Pennant* as a new fpecies, (the Scotch Wren;) it has been alfo confidered as a diftinct kind in the writings of fome, and the fynonymas of others, as *Briffon*, *Buffon*, *Ray*, *Sloane*, *Catefby*, and *Edwards*, but as it differed from our fpecies only in the colour of the upper parts, inclining more to brown than to green, and the lower parts more to yellow, Mr. *Latham* concluded it was only a variety. It is a native of *Jamaica*, *Carolina*, and *America*; but one was communicated to Mr. *Latham* by *E. S. Frafer*, Efq. who informed him that it was fhot in the *Highlands of Scotland*.

PLATE

PLATE XV.

MOTACILLA FLAVA.

YELLOW WAGTAIL.

PASSERES.

Bill conic, pointed. Noſtrils oval, broad, naked.

GENERIC CHARACTER.

Bill weak and ſlender; ſlightly notched at the tip. Tongue lace‑
rated at the end. Legs ſlender. Tail frequently in motion, ſeldom
perch; have a twittering noiſe in flight *.

SPECIFIC CHARACTER.

Bill black: head and upper parts of the body olive green, rump
paleſt: under parts from the throat bright yellow; on the throat a
few black ſpots; above the eye a ſtreak of yellow, through the eyes
another of duſky colour: beneath the eye alſo a ſtreak of duſky.
Leſſer wing coverts as the back; the others duſky, edged with pale
yellow: quills duſky. Tail black except two of the outer feathers,
which are partly white. Legs dark brown; hind claw very long.

MOTACILLA FLAVA. *Linn. Syſt.* I. *p.* 331. N° 12.
    Faun. Suec. 253.
    Scop. ann. 1. N° 226.
    Brun. N° 273. 274.

---

* Vide Latham's diviſion of the Motacilla genus.

      I   *Muller.*

PLATE XV.

Muller. N° 273.

Kram. el. p. 374. 2.

Frifch. pl. 23.

Georgi Reiffe. p. 174.

Sepp. Vog. pl. in. p. 103.

Faun. Arag. p. 88.

Gefner. av. 168.

YELLOW WATER WAGTAIL. *Raii. Syn.* 75. A. 2.

Will. Orn. p. 238. *pl.* 68.

Edw. pl. 158. (the female).—258 (the male).

Br. Zool. 1. N° 143.

Arɛt. Zool.—Br. Muf.—Lev. Muf.

Latham Gen. Syft. IV. 400. 6.—*Suppl.* 179.

1 a Bergeronette de Printemps. *Brif. Orn.* iii. *p.* 468. N° 40.

Buff. Oif. V. 265. *pl.* 14. *f.* 1.—*Pl. enl.* 674.

N° 2.

Sufurada. *Belon. obf.* 11.

Codatremola. *Zinan.* 51.

Gelb-brüftige. Bachfteltze. *Frifch.* I. 23.

Gulfpink. *Brunnich.* 273.

Gelbe Bachftelze. *Kram.* 374.

The Yellow Wagtail is not equal in fize to the Common, or White Wagtail, it meafures only fix inches and a quarter in length. It is a bird of diftinguifhed beauty, particularly the male, whofe plumage is for the moft part of a very lovely yellow, by no means inferior to that of the male Golden Oriole; the yellow colour on the

I   breaft

PLATE XV.

breaſt of the female is paler, the ſtreak over the eye whiter, and it wants the black markings on the throat.

It is uſually obſerved in moiſt meadows, and corn-fields in this country in the ſummer-time; but migrates, or ſhifts its ſituation in the winter: *Pennant* ſays it continues in *Hampſhire* the whole year.

It makes its neſt in the *corn-fields* on the ground, the outſide is compoſed of bents and fibres of the roots, the inſide is lined with hair. They are commonly found with five eggs in them, of a whitiſh colour, varied with red brown ſpots.

Is ſeen in France at all times of the year, except the winter is un-commonly ſevere. Is ſaid to inhabit *Sweden, Ruſſia, Siberia,* and *Kamtſchatka.*

Latham deſcribes the legs, black: thoſe of our ſpecimens are brown.

I 2 PLATE

16

PLATE XVI.

PARUS CAUDATUS.

LONG-TAILED TITMOUSE.

PASSERES.

Bill conic, pointed. Noftrils oval, broad, naked.

GENERIC CHARACTER.

Bill ftrait, a little compreffed, ftrong, fharp-pointed, briftles reflected over the noftrils. Tongue terminated by three or four briftles. Toes divided at their origin; back toe very large and ftrong.

SPECIFIC CHARACTER.

Bill fhort, thick, and black. Top of the head white, furrounded by a broad ftreak of black, like a crown, it paffes down the hind part of the neck, and back to the rump. Side of the head white. Sides of the back, the rump, belly, fides, and vent, dull rofe colour. Wing black. Tail very long, feather of unequal lengths, fome black, others black with white tips.

PARUS CAUDATUS. *Linn. Syft.* 1. *p.* 342. N° 11.
      Scop. Ann. 1. *p.* 164. N° 247.
      Kram. el. p. 379. N° 6.
      Sepp. Vog. pl. in. p. 49.
      Frifch. t. 14.
      Raii. Syn. p. 74.

LONG-

PLATE XVI.

Long-tailed Titmouse. *Will. Orn. p.* 242. *pl.* 43.

Albin 11. *pl.* 57. *f.* 1.

Raii. Syn. p. 74.

Br. Zool. 1. N° 166.—*Arct. Zool.—Br. Muf.*

—*Lev. Muf.*

Latham. Gen. Syn. IV. 550. 18.—*Suppl.* 190.

La Mefange à longue queue. *Brif. Orn.* III. 570. N° 13.

Buff. Oif. V. *p.* 437. *pl.* 19.—*pl. enl.* 502. *f.* 3.

Monticola. *Aldro. av.* II. 319.

Pendolino, Paronzino. *Zinan.* 77.

Alhtita. *Faun. Suec. fp.* 83.

Gaugartza. *Scop.* N° 247.

Belzmeife Pfannenftiel. *Kram.* 379,

Langfchwaentzige Meife. *Frifch.* 1. 14.

This bird is very common in *England*; is faid to inhabit *Sweden*, and thence to extend even to *Italy*; the fulnefs of its plumage enables it to bear the inclemencies of the northern regions in winter, but it admits of fome furprize that fuch as are found in warmer countries are not clothed with a plumage more adapted to the climate.

The length is five inches and a quarter, the breadth feven inches, the tail is remarkabiy long in proportion to the fize of the body; in form it is like that of a magpie, confifts of twelve feathers of unequal lengths; thofe in the middle are the longeft, thofe on each fide grow gradually fhorter. The legs are generally black, but of fome fpecimens are brown.

The

PLATE XIV.

The form of the neft is almoft peculiar to this fpecies only, it is of an oval fhape, with a fmall hole or entrance in the fide; the materials of the external part are mofs, liverwort, and wool, curioufly interwoven, the infide is lined with a thick bed of the fofteft feathers. The neft is not fufpended from a bough as is ufual with fome of the tribe, but is built between the forked branches of low wood, about three feet from the ground: they generally contain from ten to feventeen or even twenty eggs of greyifh colour, fpeckled with pale red-brown.

Thofe birds are moft frequent in gardens and orchards, to which they do much injury by devouring the tender fhoots; they are very active, and fly to and fro with great facility, or run up and down the branches in every direction. The parents and their offspring remain together the whole winter, but feparate in the fpring *.

The male has more of the rofe colour than the female; in the former however it is fubject to much variation.

---

* " The young follow the parents the whole winter; and from the flimnefs of their bodies, and great length of tail, appear, while flying, like fo many darts cutting the air," *Pennant.*

PLATE

17

PLATE XVII.

LOXIA ENUCLEATOR.

PINE GROSBEAK.

PASSERES.

Bill conic, pointed. Noftrils oval, broad, naked.

GENERIC CHARACTER.

Bill ftrong, convex above and below, very thick at the bafe. Noftrils fmall and round. Tongue as if cut off at the end. Toes placed three before and one behind.

SPECIFIC CHARACTER.

Bill ftout at the bafe, the upper mandible hooked at the tip; Noftrils covered with recumbent feathers. Head, neck, breaft, and rump, rofe-coloured crimfon. Back and leffer wing coverts black, edged with reddifh; greater wing coverts black, tipt with white; quills black; fecondaries have the outer borders white, primaries have grey. Belly and vent afh-coloured. Tail rather forked. Legs brown.

LOXIA ENUCLEATOR. *Linn. Syft.* I. *p.* 299. N° 3.
     Faun. Suec. 223.
     Brun. N° 239.
     Muller, N° 246.
GREATEST BULFINCH, *Edw. pl.* 123, 124. M. & F.
PINE GROSBEAK, *Arɛt. Zool.* 2. N° 209. *Br. Muf.—Lev. Muf.*
     Latham's Gen. Syn. iii. *p.* 111. N° 5.
     Pennant's Brit. Zool. I. N° 114. *pl.* 49. *fig.* 2.

    K   Gros-

PLATE XVII.

Gros-bec de Canada, *Brif. Orn.* iii. *p.* 250. N° 15. *pl.* 12. *f.* 3.—
    Pl. enl. 135. 1.
Le Dur-bec, *Buf. Ois.* iii. *p.* 457.
Tallbit. Natt-waka. *Faun. Succ.*
Coccothrauftes Canadenfis. *Brif.*

———————————————————

The male Pine Grofbeak is certainly one of the moft beautiful of
the feathered tribe that inhabit either of the fifter countries of *Great
Britain.* It meafures nine inches from the tip of the bill to the end
of the tail, its weight two ounces; the general colour of its plumage is rofe-coloured crimfon, and black, elegantly marked with white
on the edges of the feathers: the bill, which is remarkably ftout, and
curved at the tip, is well adapted for the purpofe of dividing the
cones of the pines to obtain the feeds.

The female has not the beautiful appearance of the male; the
principal colour of her plumage is dirty green, inclining to brown,
the crown of the head varied only with a few reddifh or yellowifh
teints, and fome feathers of the fame colour flightly difperfed over
the back, breaft, and belly.

In *England* this fpecies is found only in the moft northern parts, or is
probably entirely confined to *Scotland;* like the Crofsbill it inhabits the
pine forefts in the *Highlands; Pennant* fufpects that they breed there,
as he has obferved them flying above the great pine forefts of *Invercauld, Aberdeenfhire,* in the month of Auguft.

I

It

PLATE XVII.

It is found in the pine forefts alfo of *Sweden*, the northern parts of *Ruffia* *, of *Siberia*, and *Lapland*; they are alfo common in the northern parts of *America* : from April to September they are frequent at *Hudfon's Bay*; the fouthern fettlements are inhabited by them throughout the year. It has been met with at *Aoonalafhka* †, and in *Norton Sound*.

Latham obferves, that at *Hudfon's Bay* it frequents the groves of *pines* and *junipers* in May; and makes a neft in the trees with fticks lined with feathers, at a fmall height from the ground. The eggs are four in number, and white; the young are hatched the middle of June: he adds, " though this bird, when adult, is beautiful in colour, the young brood for fome time remain of a plain dull blue." The natives of the *Bay* call it *Wufcunithow* ‡.

---

* " Common about St. *Peterfburgh* in autumn, and is caught in great plenty at that time for the ufe of the table, returning north in fpring." *Pennant.*

† Ellis's Narr. vol. ii. p. 15. ‡ Mr. Hutchins.

K 2 PLATE

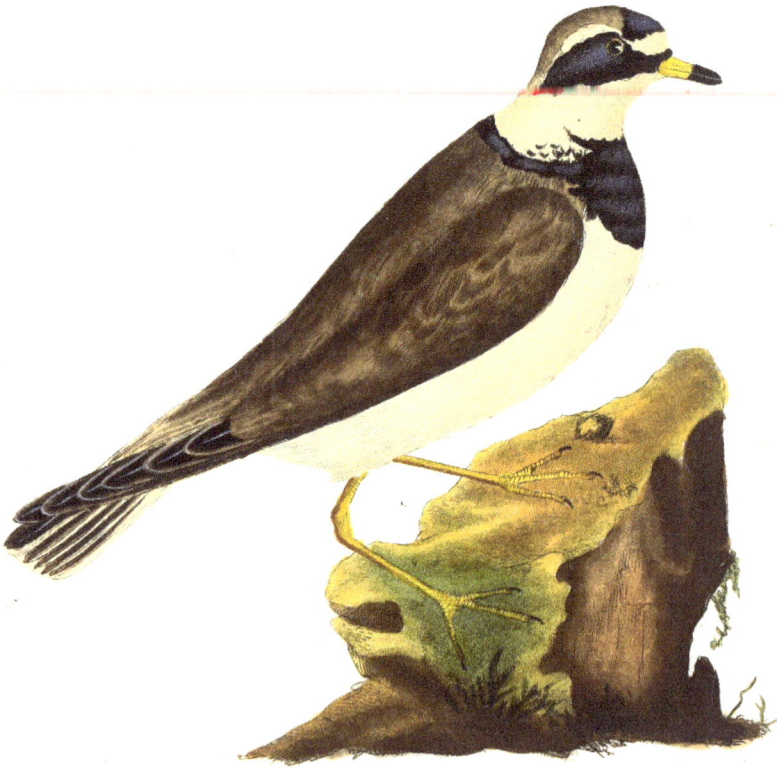

18

PLATE XVIII.

CHARADRIUS HIATICULA,

SEA LARK, or RINGED PLOVER.

GRALLÆ.

Bill roundifh. Tongue entire, flefhy. Thighs naked. **Toes** divided.

GENERIC CHARACTER.

Bill ftrait, roundifh, obtufe. Noftrils linear. Toes three in number, all placed forwards.

SPECIFIC CHARACTER.

Length feven inches. Bill orange, black at the tip. From the bafe of the upper mandible to the eyes, a black line: another from one eye to the other. Crown of the head brown. Chin and throat white, paffing round the neck in a broad collar: beneath this, on the lower part of the neck, is a fecond line of black, encircling the neck behind, but becoming narrower as it paffes backward. Breaft and under parts white. Back and wing coverts pale brown. Two middle feathers of the tail greyifh brown, growing almoft black towards the ends; the three next on each fide the fame, with white tips; the laft but one is white, with a brown band: the outer one white. Legs orange; claws black.

CHARADRIUS HIATICULA. *Lin. Syft.* I. *p.* 253. 1.
    Faun. Suec. 187.
    Scop. Ann. 1. N° 147.

            Brun.

PLATE XVIII.

Brun. N° 184.

Georgi Reise. p. 172.

Faun. Groenl. N° 78.

Sea Lark.   Raii Syn. p. 112. A. 6. 190, 13.

Sloan. Jam. p. 319. 13. pl. 269. fig. 2.

Albin. 1. pl. 80.

Will. Orn. p. 310. pl. 57.

Br. Zool. 11. 383.

Ringed Plover.   Pennant's Br. Zool. N° 211.

Arct. Zool. N° 401.

Br. Muf.—Lev. Muf.

Lath. Gen. Syn. V. 201. 8.

Le petit Pluvier-à Collier, Brif. Orn. V. p. 63. 8. pl. 5. fig. 2.—

Pl. enl. 921.

Le Pluvier à Collier, Buff. Oif. viii. p. 90. pl. 6.

Griefs hennl. Kram. 354.

Strandpipare, Grylle, Trulls, Lappis Pago. Faun. Suec. fp. 187.

Bornholmis Præfte-krave, Sand-Vrifter. Brun. 184. Frifch. 11. 214.

Thofe Birds migrate to our fhores in the fpring, but are never very numerous; they remain with us during the fummer, and depart in autumn. They run lightly, and with much fwiftnefs, and when difturbed take fhort flights; at the fame time they make a loud twittering noife.

The female makes no neft, but depofits four eggs on the ground, under fome convenient fhelter; the eggs are about one inch and an half in length, of a dull whitifh colour, fpotted and blotched with black.

The

PLATE XVIII.

The same species is found in several parts of the Continent; in *Greenland*, and in *America*. Latham obferves that it vifits *Hudfon's Bay* the middle of June, and departs in September. He adds, " it is a folitary bird; and obferved, on any one's approaching near the eggs, to ufe many ftratagems to decoy the perfon from it, by drawing off its attention. Called at *Hudfon's Bay*, *Kifqua*, *the napi Shifh*."— *Gen. Syn.*

The fame writer alfo mentions a variety which inhabits *Cayenne*; the length of this variety is fix inches and an half. Bill black: forehead, and before as far as the breaft, white, paffing round the lower part of the neck as a collar: the reft of the plumage pale dufky afhcolour: the end half of the tail dufky black, the tip fringed with rufous: legs pale.

PLATE

19

PLATE XIX.

TRINGA PUGNAX.

R U F F.

G R A I L Æ.

Bill roundifh. Tongue entire, flefhy. Thighs naked. Toes divided.

GENERIC CHARACTER.

Bill roundifh, ftrait, about the length of the head. Noftrils narrow, Toes four.

SPECIFIC CHARACTER.

Feathers of the neck remarkably long. General colour brown, commonly marked with fpots, or concentric circles of black. Legs dull yellow. Female has no ruff.

TRINGA PUGNAX.   *Linn. Syft.* 1. *p.* 147. 1.
                       Faun. Suec. 175.
                       Scop. Ann. 1. Nº 140.
                       Brun. 168. 169.
                       Kram. p. 352.
                       Frifch. t. 232. 235.
                       Georgi Reifc. p. 172.

AVIS PUGNAX.   *Aldr. av. III.* 167.

PLATE XIX.

RUFF and REEVE. *Albin.* I. *pl.* 72. 73.
Penn. Brit. Zool. N° 192. *pl.* 69.
Arct. Zool. p. 479. A.
`Lath. Gen. Syn.* V. 159. I.
Br. Muf.—Lev. Muf.

RUFFE. *Raii Syn. p.* 107. A. 3.
Will. Orn. p. 302. *pl.* 56.

Le Combattant, ou

Paon de Mer, *Brif. Orn.* V. *p.* 240. 18. *pl.* 22. *fig.* 1. 2.
Buff. Oif. vii. *p.* 521. *pl.* 29. 30.
Pl. enl. 305. 306.

Krofler. *Kram.* 352.

Brufhane. *Faun. Suec. fp.* 175.

Bruufhane. *Brunnick*, 168.

Streitfchnepfe, Ramphæhnlein. *Frifch.* 11. 232. 235.

The length of the male is twelve inches, of the female ten inches ; the bill of the former is yellow in fome, in others black, or dark brown ; the face is covered with yellow pear-fhaped pimples ; the back part of the head and neck are furnifhed with long feathers, which expand in a very fingular manner on each fide of the neck, and impend loofely over the breaft, like the ruff anciently worn in this country. A portion or tuft of thofe feathers project alfo juft beyond each eye, and have the appearance of long ears.

As the moft remarkable peculiarity of thofe Birds are that no two fpecimens are ever found to agree in the colours of the plumage, it is impoffible to give any defcription, except of its form, that may affift the unfkilful ornithologift to determine the fpecies, when it

I

has

PLATE XIX.

has attained the ruff; it cannot, however, be miftaken, as no Bird of this country refembles it in the fmalleft degree. The ground colour is generally brown, but it varies in different Birds to every hue between the lighteft teint that can deferve that name, and the deepeft chocolate colour; fometimes we find the ruff of a fine tender buff colour, without the flighteft appearance of fpots, except on the breaft and back, which may be of a deep black, intermingled with a few white feathers, and gloffed with fhining purple; others we find that have the ruffs of a deep brown, barred with black; fome with white ruffs fpotted with brown, or brown fpotted with white; and indeed with every variation that it is poffible to defcribe.

The *females*, or *Reeves*, Pennant afferts, never change their colours, which he fays are pale brown; the back fpotted with black, flightly edged with white; the tail brown; the middle feathers fpotted with black; the breaft and belly white; the legs of a pale dull yellow: but I have two fpecimens that do not well agree with his defcription, or correfpond with each other; and in the Leverian mufeum a variety of that fex is preferved that is wholly white, except the wings, on which the ufual markings are vifible in a very pale colour.

The female has no ruff, and the male does not attain that appendage until the fecond feafon; at the time of incubation the plumage of the latter is in the full perfection, and the pimples break out on his face: but after that time they fhrink beneath the fkin, the long feathers of the ruff fall off, and he again affumes the plain appearance of the female.

Thefe Birds inhabit the North of *Europe* in fummer, as far as Iceland, as well as the northern marfhes of *Ruffia* and *Siberia*. In this country they are found in *Lincolnfhire*, the Ifle of *Ely*, and in the eaft

L 2

riding

riding of *Yorkſhire* * ; they arrive at thoſe places early in the ſpring, and diſappear about *Michaelmas.*

The Reeve lays four eggs in a tuft of graſs the beginning of May; they are white, marked with large ruſty ſpots.

Soon after their arrival, the males begin to *hill*; that is, to collect on ſome dry bank near a ſplaſh of water, in expectation of the females. Each male keeps poſſeſſion of a ſmall piece of ground, round which it runs ſo often as to form a bare circular path; the inſtant a female alights among them, the males are in motion; a general battle enſues, and the fowlers, who have been waiting for the advantage of ſuch au event, catch them in their nets in great numbers †.

In the fens each male remains within his circle, and defends himſelf againſt every invader with much reſolution; the leaſt infringement on his poſſeſſion by another male is reſented with the greateſt violence; and if any farther attack is made, a battle is the conſequence: in fighting they have the ſame action as a cock, ſpread their ruffs, and place their bills to the ground.

" It is uſual to fat theſe birds for table by means of *bread* and *milk* mixed with *hemp-ſeed*, and ſometimes boiled *wheat*; to theſe by many *ſugar* is added; which laſt in a fortnight's time will cauſe them to be one lump of fat, when they will fetch from two ſhillings to half a crown each."—*Lath. Gen. Syn.*

---

* Brit. Zool.

† They viſit a place called *Martin-mere* in *Lancaſhire*, the latter end of March, or beginning of April, but do not continue there above three weeks.—*Brit. Zool.*

P L A T E

20

PLATE XX.

MERGUS ALBELLUS.

SMEW.

ANSERES.

Bill obtufe, covered with a thin membrane, broad, gibbous below the bafe, fwelled at the apex. Tongue flefhy. Legs naked, feet webbed, or finned.

GENERIC CHARACTER.

Bill flender, a little depreffed, furnifhed with a crooked nail; edges of the mandibles very fharply ferrated. Noftrils near the middle of the mandible, fmall and fubovated. Feet furnifhed with four toes; three forwards, and one behind; the outer toe before longer than the middle one.

SPECIFIC CHARACTER.

Bill lead colour. General colour of the plumage white. Head crefted at the back part; on each fide of the head an oval black fpot, beginning at the bill, and encircling the eye. On the lower part of the neck, on each fide, are two curved black ftreaks, pointing for-ward. Inner fcapulars, back, coverts on the fide of the wing and the greater quill feathers, black. Tail cinereous. Legs grey.

MERGUS ALBELLUS. *Lin. Syft.* 1. *p.* 209. 5.
     Faun. Suec. N° 137.
     Brun. N° 97.
     Kram. El. p. 344. 3.
     Frifch. t. 172.

MERGUS

PLATE XX.

MERGUS ALBULUS. *Scop. Ann.* I. N° 91.

MERGUS RHENI. *Raii Syn. p.* 135. 5.

Will. Orn. p. 337.

MERGUS RHENANUS. *Gefner. av.* 131.

SMEW. *Albin.* I. *pl.* 89.

Penn. Brit. Zool. ii. N° 262.

Lath. Gen. Syn. 6. 421. 5.

Arct. Zool. N° 468.

WHITE NUN. *Will. Orn.* 337. *pl.* 64.

Raii Syn. p. 135. A. 3.

WEESEL COOT. *Albin.* I. *pl.* 88.

RED-HEADED SMEW. *Br. Zool.* II. 263.

Br. Muf.—Lev. Muf.

Le petit Harle huppé, ou la Piette, *Brif. Orn.* vi. *p.* 243. 3. *pl.* 24.
fig. I.

Buf. Ois. viii. *p.* 275. *pl.* 24.

Pl. enl. 449.

L'Harle étoilé, *Brif. Orn.* vi. *p.* 252. 6.

Brun. N° 98.

Kreutz-Ente, (Crofs-Duck) *Frifch.* II. 172.

———————————————

The Smew is about fixteen inches in length, and twenty-four inches in breadth; its weight thirty-four ounces; our figure is copied from a fpecimen of the male. The colours of the female do not exactly correfpond with thofe of the male; the head of the former is ferruginous, and flightly crefted; cheeks, chin, and throat, white; between the bill and the eye the fame oval fpot as in the male; breaft clouded with grey; belly white; legs pale afh. It is generally called the *Lough Diver.*

It

PLATE XX.

It vifits this country only in the winter; on the Continent it is found as far fouth as Carniola; is alfo found in Iceland, and is fuppofed to breed and remain there during the winter; or that it paffes to fome other arctic region. It has been obferved with the Merganfers, Ducks, and other Water Birds in their migratory courfe up the *Wolga* in February *.

It alfo inhabits *America*, having been fent from New-York †.

Latham, in his fupplement, fays that he once difcovered a few fhrimps in the belly of one of thofe birds, and fuppofes them to be its chief food.

---

* Dec. Ruff. ii. p. 145.
† Arct. Zool.

PLATE

21

PLATE XXI.

ANAS QUERQUEDULA.

GARGANEY.

ANSERES.

Bill obtufe, covered with a thin membrane, broad, gibbous below the bafe, fwelled at the apex. Tongue flefhy. Legs naked. Feet webbed or finned.

GENERIC CHARACTER.

Bill convex above, flat beneath, hooked at the apex with mem- branous teeth.

SPECIFIC CHARACTER.

Bill lead colour. Head dufky with oblong ftreaks. From the cor- ner of each eye a white line paffes to the back of the neck. Cheeks and upper part of the neck, brown-purple, marked with minute ob- long white lines, pointing downwards. Breaft light brown, with femi- circular bars of black. Belly white. Wing coverts grey; firft quills afh coloured, exterior webs of the middle quills green. Legs lead colour.

M ANAS

PLATE XXI.

ANAS QUERQUEDULA. macula alarum viridi, linea alba fupra oculos.

Fn. Sv.—Linn. Syft. 1. p. 203.

Scop. Ann. 1. N° 75.

Brun. N° 81.

Muller, N° 125.

Kram. El. p. 343. 18.

Frifch. pl. 176.

QUERQUEDULA Varia. Gefner. av. 107.

QUERQUEDULA Prima. Will. Orn. 291. t. 74.

Raj. av. 148. 8.

GARGANEY. Ditto.

Br. Zool. N° 289. pl. 101.

Arct. Zool. p. 576. O.

Lath. Gen. Syn. 5. 550. 87.

Br. Muf.—Lev. Muf.

La Sarcelle· Brif. Orn. VI. 427. tab. 39. 1. 2.

Buff. Oif. 9. p. 260.—Pl. enl. 946. (male)

Belon. av. 175.

Scavolo, Cervolo, Garganello. Aldr. av. 3. 89. 90.

Krickantl. Kramer. 343.

Kriech-Ente. Frifch. 2. 176.

Norvegis Krek-And. Quibufd. Saur-And. Brunnich, 81.

This fpecies is found in *England* in the Winter ; at that time alfo it is feen in *France*. In April it departs, and migrates to the North as the Summer advances, to breed.

Iu

PLATE XXI.

In Europe it is found as far as *Sweden*; it is very common throughout *Ruſſia* and *Siberia*, and as far as *Kamtſchatka*.

Our figure is of the male bird; the female has an obſcure white mark over the eye, the reſt of the plumage is of a browniſh aſh colour.

PLATE

22

PLATE XXII.

MUSCICAPA ATRICAPILLA.

PIED FLYCATCHER.

PASSERES.

Bill conic, pointed. Noftrils oval, broad, naked.

GENERIC CHARACTER.

Bill flatted at the bafe; almoft triangular; notched at the end of the upper mandible, and befet with briftles. Toes divided as far as their origin.

SPECIFIC CHARACTER.

Bill black. Upper parts of the body, wings, and tail, black. Forehead and under part white. Several white feathers in the Wing. Upper tail coverts black and white mixed. Legs black.

MUSCICAPA ATRICAPILLA. *Lin. Syft.* i. *p.* 236. 9.
                                        Frifch. pl. 24.
                                        Kram. El. p. 377. 16.
Atricapilla five ficedula.   *Aldr. av. II.* 331.
COLDFINCH.   *Raii Syn. p.* 77. A. 5.
                                        Will. Orn. p. 236.
                                        Edw. pl. 30.
                                        Br. Zool.—Lond. 1766.

PIED

PLATE XXII.

PIED FLYCATCHER.　　　　　*Penn. Brit. Zool.* I. N° 135. *Lond.* 1776.
　　　　　　　　　　　　　Arct. Zool.
　　　　　　　　　　　　　Lath. Gen. Syn. III. 324. 2.
Le Traquet d'Angleterre.　*Brif. Orn.* iii. *p.* 436. 27,
Rubetra Anglicana.　　　　*Buff. Ois. V. p.* 222.
Meerfchwartz puffle.　　　*Kram. Auft.* 377.

The Coldfinch, or according to Latham and Pennant, the Pied Fly-catcher, is found in *Yorkfhire, Lancafhire,* and *Derbyfhire*; in thofe parts of the kingdom it is not very frequent, in every other it is extremely rare.

It is unneceffary for us to defcribe the many varieties that are known of this fpecies; in England the colours of its plumage varies confiderably at different feafons of the year; but fuch as are natives of foreign countries, are again fo diffimilar to ours, that different authors have alternately defcribed them as varieties, or new fpecies.

The bird that *Latham* defcribes, was white on the outer web of the exterior tail feather; the two exterior tail feathers of *Willoughby's* bird was marked with white; and on the contrary, we have a fpecimen which does not exhibit the leaft trace of white on either. The upper tail coverts are black and white mixed, in fome fpecimens; in others they are wholly black; and *Le Gobe-mouche noir* of *Briffon*, which is only another variety, differs in having a mixture of grey on the

PLATE XXII.

the upper parts, the thighs brown and white, and *three* of the exterior tail feathers white on the margins.

A more pleasing variety than either, is found in *Lorraine* and *Brie*; it corresponds in size with those found in England; but the white of the breast which terminates under the cheeks in the latter, passes quite round the neck like a collar in the former *.

The plumage of the female is brown in those parts where the male is black; it has no white on the forehead; the white spot on the wing is more obscure; and the under parts of the body is of a dusky white. The male possesses only the full black during the summer; as that season declines, its plumage alters, and it gradually assumes so perfectly the appearance of the female, that he cannot be distinguished from her.

The nest is usually built in the hole of a tree, not very near the ground; it is composed of fibres, mixed with mofs, and contains six eggs. It feeds on Insects.

* This variety is called by Buffon *Le Gobemouche noir à Collier*. Hift. des Oif. 4. p. 520. pl. 25. f. 1.

PLATE

23

PLATE XXIII.

STERNA HIRUNDO.

COMMON TERN.

ANSERES.

Bill obtufe, covered with a thin membrane, broad gibbous below the bafe, fwelled at the apex. Tongue flefhy, legs naked. Feet webbed or finned.

GENERIC CHARACTER.

Bill ftrait, flender, pointed. Noftrils linear. Tongue flender and fharp. Wings very long. Tail forked, back toe fmall.

SPECIFIC CHARACTER.

Bill and feet red. Crown and tip of the bill black. Neck, and underfide white. Back and wings fine grey.

STERNA HIRUNDO. *Linn. Syft. Nat.* 1. *p.* 227. 2.
Faun. Suec. N° 158.
Haffelq. p. 272. N° 40.
Scop. Ann. 1. N° III.
Brun. N° 151. 152.
Mull. p. 21.
Faun. Groenl. N° 69.
Kram. El. p. 345. (*Larus*).
Frifch. 2. 219.

N

THE

PLATE XXIII.

THE SEA-SWALLOW. *Raii. Syn. p.* 131. *A.* 1. 191. 7.

Will. Orn. p. 352. *pl.* 68.

Albin. 11. *pl.* 88.

COMMON TERN. *Lath. Gen. Syn.* 6. 361. 14.

GREAT TERN. *Br. Zool.* N° 254. *pl.* 90.

Lev. Muf. Br. Muf.

THE KERMEW. *Marten's Spitzberg.* 92.

Le Grande Hirondelle-de-Mer. *Brif. Orn.* VI. *p.* 203. 1. *pl.* 19. *fig.* 1.

Buff. Oif. 8. *p.* 331. *pl.* 27.—*Pl. Enl.* 987.

Tarna. *Faun. Suec.*

Sterna (Stirn, Spyrer, Schnirring). *Gefn. av.* 586.

Grauer fifcher. *Kram.* 345.

Iflandis Kria. *Norvegis* Tenne, Tende, Tende-lobe, Sand-Tolle, Sand-Tærrne. *Danis* Tærne. *Bornholmis* Kirre, Krop-Kirre. *Brunnich.* 151. Makauka. *Scop.* N° 3.

Schwartz plattige Schwalben Moewe. *Frifch.* 11. 219.

The length of this fpecies is fourteen inches; its breadth thirty; and its weight four ounces and a quarter. It is very common on the fea-coafts, banks of lakes and rivers in this country during fummer; it quits the breeding places at the approach of winter, and returns in fpring.

It is found in various parts of *Europe* and *Afia*; in the fummer as far as *Greenland* and *Spitzbergen*. It is alfo found in *America*; arrives at *New England* in May, and difappears in Autumn. At *Hudfon's Bay* it is known by the name of *Black-head**.

---

* *Lath. Gen. Syn.*

D r

PLATE XXIII.

Dr. Forfter mentions a variety at *Hudfon's Bay*, having the Legs black; Tail fhorter and lefs forked; and the outer feathers wholly white *: The Bird Albin has figured in his plate 88, vol. 2. appears alfo to be a variety; the legs are black, and the bill is of the fame colour, except the tip, which is red.

Thefe Birds breed among tufts of rufhes, grafs, or mofs near the water fide; they lay three or four eggs, about an inch and three quarters in length, of a dull olive colour, marked with irregular black fpots, and fprinkled with fpecks of an obfcure brown in June; the young birds are hatched in July, and quit the neft foon after.

They feed on fmall fifh and water infects; are very clamorous and daring; and during the time of incubation, will dart on any perfon who may pafs by their neft, though they fhould neither provoke nor difturb them.

It appears to have all the actions over the water which the Swallow has on land, fkimming and defcribing vaft circuits over the furface of the waves when feeking its prey, diving with intrepidity the inftant it difcovers it, and inftantly appearing again on the wing with the fifh in its mouth. Notwithftanding the affinity of its actions with thofe of the Swallow, *Pennant*, in the Britifh Zoology, has altered the name to *Tern*, " a name," he obferves in a note, " thefe birds are known by in the *North of England*; and which we fubftitute inftead of the old compound one of *Sea-Swallow*; which was given them on account of their forked tails."

---

* *Phil. Tranf. vol.* lxii. *p.* 421.

24

PLATE XXIV,

STURNUS CINCLUS,

WATER-OUZEL

PASSERES.

Bill conic, pointed. Noftrils oval, broad, naked.

GENERIC CHARACTER.

Bill ftrait, fubulate, and fomewhat angular.

SPECIFIC CHARACTER.

Bill compreffed on the fide, black. A white fpot above and another beneath the eye. Upper part of the head and neck deep brown. Back, Wings, Tail, black; feathers with brownifh edges. Chin, fore part of the breaft pure white, Belly rufous brown; next the tail black. Legs black.

STURNUS CINCLUS, niger, pectore albo. *Linn. Syft. Nat.* 2. 168.

4. editio Decima.

MOTACILLA pectore albo, corpore nigro. *Fn. Suec.* 216.

MOTACILLA CINCLUS. *Scop. Ann.* 1. Nº 223.

Kram. el. p. 374. 3.

MERULA

PLATE XXIV.

MERULA AQUATICA. *Gefn. av.* 608.

WATER-OUZEL, or

WATER-CRAKE. *Will. Orn.* 149.

          Raii. Syn. p. 66. *A.* 7.

          Albin. 2. *pl.* 39.

          Br. Zool. 1. *N⁰* 111.

          Arĉt. Zool.

          Lath. Gen. Syn. 3. 48. 50.

          Br. Muf.—Lev. Muf.

WATER-CRAW. *Turner.*

Le Merle d'Eau. *Brif. Orn.* V. *p.* 252. 19.

          Buff. oif. 8. *p.* 134. *pl.* 11.—*Pl. enl.* 940.

Watnftare. *Faun. Suec. fp.* 214.

Merlo Aquatico. *Zinan.* 109.

Providni Kofs. *Scop. N⁰* 223.

Norvegis Foffe Fald, Foffe Kald, Quærn Kald, Stroem-Stær, Bække

          Eugl. *Brun.* 203.

Waffer-amfel, Bach-amfel. *Kra.* 374.

Lerlichirollo. *Aldr. av.* 3. 186.

The Water-Ouzel is a very fhy and folitary bird, and though well known as a Britifh fpecies, is generally confined to fuch parts as abound with fmall rivulets, or with waters that courfe between the craggy fragments of mountains; it is therefore that we find it plenty only in *Wales, Cumberland, Yorkfhire,* and *Weftmoreland.*

2

It

PLATE XXIV.

It feeds on fmall fifh and infects: its neft is built among the ftones on the ground near the water fide; beneath the fmall fhelving rocks that over-hang the ftreams it frequents; or in holes contrived in fteep and perpendicular banks; it is compofed of hay and fibres of roots, is lined with dead oak leaves, has a covering of green mofs, and contains five eggs of a white colour with a blufh of red. In young birds the belly is wholly white.

Moft authors have noticed the very fingular manner in which it fearches for its prey, it not only dives under the water, but will fly and run after them at the bottom in the fame manner as on land * : Kramer fays, that one of them had been caught under water by means of a line and hook, which had been baited to catch fifh †.

Thefe birds are fmaller than the Ring-Ouzel, their length is feven, and breadth eleven inches, weight two ounces and an half; they are found in Europe as high as Feroe and Finmark ‡; as far as Kamtf-chatka in the Ruffian dominions; in Chriftianfoe and Norway.

---

* Hift. def. Oif.—Dacouv. Ruf. vol. i. p. 307. 314.

† Albin fays, " it feeds on fifh, yet refufeth not infects; fitting on the banks of rivers it now and then flirts up its tail; although it be not web-footed, yet it will fometimes dive or dart quite under water. It is a folitary bird, accompanying only with its mate in coupling and breeding time."

‡ Arct. Zool.

PLATE

INDEX TO VOL. I.

ARRANGEMENT

ACCORDING TO THE

SYSTEM OF LINNÆUS.

* Not deſcribed by Linnæus.

O   ORDER

INDEX.

ORDER IV.

GRALLÆ.

ORDER V.

GALLINÆ.

ORDER VI.

PASSERES.

VOL.

I N D E X.

V O L. I.

A R R A N G E M E N T

ACCORDING TO

LATHAM's SYNOPSIS of BIRDS.

DIVISION I. LAND BIRDS.

ORDER II. PIES.

GENUS IV.

ORDER III. PASSERINE.

GENUS XXXI.

O 2   GENUS

I N D E X.

O R D E R IV. GALLINACEOUS.

D I V I-

INDEX.

VOL.

INDEX.

VOL. I.

ARRANGEMENT

ACCORDING TO

PENNANT's BRITISH ZOOLOGY.

* Not mentioned by Pennant as British Birds.

GENUS

INDEX.

INDEX.

VOL. I.

ALPHABETICAL ARRANGEMENT.

THE

NATURAL HISTORY

OF

BRITISH BIRDS;

OR, A

SELECTION OF THE MOST RARE, BEAUTIFUL, AND INTERESTING

BIRDS

WHICH INHABIT THIS COUNTRY:

THE DESCRIPTIONS FROM THE

SYSTEMA NATURÆ

OF

LINNÆUS;

WITH

GENERAL OBSERVATIONS,

EITHER ORIGINAL, OR COLLECTED FROM THE LATEST
AND MOST ESTEEMED

ENGLISH ORNITHOLOGISTS;

AND EMBELLISHED WITH

FIGURES,

DRAWN, ENGRAVED, AND COLOURED FROM THE ORIGINAL SPECIMENS.

VOL. II.

By E. DONOVAN.

LONDON:

PRINTED FOR THE AUTHOR; AND FOR F. AND C. RIVINGTON,
No. 62, ST. PAUL'S CHURCH-YARD. 1795.

25

PLATE XXV.

TRINGA VANELLUS.

Lapwing, or Tewit.

GRALLÆ.

Bill roundifh. Tongue entire, flefhy. Thighs naked. Toes divided.

GENERIC CHARACTER.

Bill roundifh, ftrait, about the length of the head. Noftrils narrow,
Toes four.

SPECIFIC CHARACTER.

Bill, Crown, Creft, Throat, black; a black line under each Eye;
at the back part of the Head a Creft of about twenty narrow feathers
of unequal length; fides of the Neck white; fore part as far as the
Breaft black; the Back and Wings green, moft beautifully gloffed
with fine purple, brown and blue. Quills black. Breaft and Belly
white. Upper Tail Coverts and Vent pale rufous. Tail white from
the bafe half up, extreme half black. Legs red.

Tringa Vanellus. Pedibus rubris, crifta dependente, pectore nigro.
>   *Fn. Suec.* 148.—*Linn. Syft. Nat. Editio Decima.*
>   I. 148.
>   *Scop. Ann.* I. N° 141.
>   *Brun.* N° 170.

A 2

Mull.

PLATE XXV.

Mull. N° 192.

Kram. El. p. 353.

Frifch. II. 213.

Olin. Uc. pl. in p. 21.

Georgi Reife. p. 172.

LAPWING, BASTARD

PLOVER. PEWIT. *Raii. Syn. p.* 110. *A.* 1.

Will. Orn. 307. *pl.* 57.

Albin. 1. *pl.* 74.

Arƈt. Zool. p. 480. *D.*

Br. Zool. 190.

Lev. Muf.

Lath. Gen. Syn. V. 161.

Le Vanneau. *Brif. Orn. V. p.* 94. 1. *pl.* 8. *fig.* 1.

Buff. Oif. 8. *p.* 48. *pl.* 4.—*Pl. enl.* 242.

Le Vanneau, Dixhuit, Papechieu. *Belon. Av.* 209.

Zweiel. *Gefner. Av.* 765.

Pavoncella. *Olina.* 21.

Pavonzino. *Aldr. Av.* 111. 202.

Kiwik. *Kram.* 353. *Frifch.* 11. 213.

Wipa, Kowipa, Blæcka. *Faun. Suec. Sp.* 176.

Danis Vibe, Kivit. *Brunnich.* 170.

The length of this fpecies is about thirteen inches and a half; the breadth more than two feet; the weight eight ounces; the female is exaƈtly like the male both in form and colour, but is rather fmaller.

It

PLATE XXV.

It lays four eggs, of a dirty olive caſt, ſpeckled with black *, in a ſlight neſt compoſed of bents, or on a bed of dried graſs, ſcraped together on the ground; the hen ſits about three weeks; the young as ſoon as hatched run together like chickens.

The old birds ſhew a remarkable ſolicitude for their young, flying with great anxiety about them if diſturbed; and uſing every ſtratagem to decoy the diſturber from the neſt; feigning to flutter as if wounded on the ground at a little diſtance, or running along as if lame: ſhould thoſe artifices prove uſeleſs they become deſperate, and will ſtrike at the perſon or animal whom they cannot entice away.

Towards winter both young and old aſſemble in flocks of four or five hundered on the heaths or marſhy places, at which time they are caught in nets for the uſe of the table in the ſame manner that *Ruffs* are.

They are common in moſt parts of *Europe*, as far as *Iceland*; change place according to the ſeaſon; are met with in *Perſia* and *Egypt* in winter, and *Latham* ſays he has ſeen a ſpecimen from *China*.

They are eſteemed a delicacy as their nouriſhment is only ſlugs and worms; theſe they draw out of the ground morning and evening. They are ſometimes kept in gardens, and by good treatment become both uſeful and familiar.

---

* Pennant obſerves that " the eggs are held in great eſteem for their delicacy; and " are ſold by the *London* Poulterers for three ſhillings the dozen."

PLATE

26

PLATE XXVI.

PARUS CRISTATUS.

CRESTED TITMOUSE.

PASSERES.

Bill conic, pointed. Noftrils oval, broad, naked.

GENERIC CHARACTER.

Bill fhort, ftrong, entire, briftles at the bafe. Tongue blunt, with briftles at the end.

SPECIFIC CHARACTER.

Forehead and fides of the Head white; on the head a creft of black pointed feathers with white edges. Chin and Throat black; with a collar of the fame colour bounding the Cheeks. Back, Wings, and Tail, rufous grey. Under parts of the Body white, fides with a rufous tinge. Legs lead colour.

PARUS *Criftatus*, Capite Criftato.   *Linn. Syft.* 1. p. 340. N° 2.

        Scop. Ann. 1. p. 162. 243.

        Raii. Syn. p. 74. N° 6.

        Muller. p. 34. N° 282.

        Georgi Reife. p. 175.

        Frifch. t. 14.

        Kram. el. p. 379. N° 2.

CRESTED TITMOUSE.   *Raii. Syn.* p. 74. N° 6.

        Albin. 2. pl. 57.

        Will. Orn. p. 242. t. 43.

        Arct. Zool. Br. Muf.

9   Le

PLATE XXVI.

Le Mefange Puppèe. *Brif. Orn.* 3. *p.* 558. N° 8.

Buff. Oif. V. *p.* 447.—*Pl. Enl.* 502. *f.* 2.

The Crefted Titmoufe is fo very rarely found in this country, that neither *Latham* or *Pennant* has defcribed it as a Britifh fpecies; nor has the latter included it in the Appendix of the Britifh Zoology, among the foreign birds which vifit us at unftated periods.

It is met with in many parts of *France*, particularly in *Normandy*, and the intermediate country between that and *Sweden*; writers concur that it is of a very folitary difpofition, never mixing with other birds, nor in numbers even with its own fpecies; it is chiefly found among the ever-green trees in the deep and gloomy receffes of extenfive forefts, and is therefore little known even in thofe parts where the fpecies is moft frequent.

Walcot mentions that they are fometimes feen in *Scotland*, and once vifited that country in a large flock; the fpecimen from which our figure is copied was fhot in *Scotland* alfo, in company with feveral others in the year 1792.

PLATE

27

PLATE XXVII.

EMBERIZA NIVALIS?

TAWNY BUNTING.

PASSERES.

Bill conic, pointed. Noftrils oval, broad, naked.

GENERIC CHARACTER.

Bill conic, angular on each fide; a hard knob within the upper mandible.

SPECIFIC CHARACTER.

Bill yellow tipt with black. Head and round the Neck tawny. Back brown, marked with black. Legs black. Rump tawny. Tail twelve feathers, rather forked, exteriors white.

EMBERIZA NIVALIS *Faun. Suec.* 227. B.

EMBERIZA FRIGIDA. *Lath. Cat. Englifh Birds, Suppl.*

TAWNY BUNTING. *Pen. Br. Zool.* 121.

Lath. Gen. Syn. III. 164.

GREAT PIED MOUNTAIN FINCH or BRAMBLING. *Will. Orn.* 225.

L'Ortolan de Neige, Hortulanus Nivalis. *Brif. av. III.* 285.

Schnee-ammer (Snow-hammer). *Frifch.* 1. 6.

———

Thefe birds are fometimes met with in the northern parts of England, but are not common; three males and one female were fhot in the garden of Mr. Slade, Vauxhall-road, about a fortnight fince. I care-

B

fully

PLATE XXVII.

fully examined them, and found they varied very much in their colours; it is not indeed furprifing that the accurate Linnæus fhould confider the Tawny and Mountain Buntings with their varieties, as the Snow Bunting in its different approaches to its fummer appearance.

Pennant is of a different opinion, and has defcribed it under the Englifh name Tawny Bunting, as a diftinct fpecies. Latham has alfo defcribed it as a different fpecies in his general Synopfis; and in the lift of the Birds of Great Britain, in the fupplemental volume, he adds " EMBERIZA FRIGIDA," and refers to the defcription of the Tawny Bunting in the Britifh Zoology, Nº 121.

The name " Emberiza glacialis" has been alfo given to the fame, or a mere variety of this bird.

Our fpecimens are about fix inches and three quarters in length, twelve inches and three quarters in breadth; weight an ounce.

PLATE

28

PLATE XXVIII.

COLYMBUS TROILE.

FOOLISH GUILLEMOT.

ANSERES.

Bill obtufe, covered with a thin membrane ; broad, gibbous below the bafe, fwelled at the apex. Tongue flefhy. Legs naked. Feet webbed, or finned.

GENERIC CHARACTER.

Bill ftraight, flender, pointed. Noftrils linear, at the bafe of the bill. Legs near the tail. Feet webbed.

SPECIFIC CHARACTER.

Bill black. Infide of the mouth yellow. Tips of the fmall quills, breaft, and belly white. The reft deep moufe colour.

COLYMBUS TROILE. *Linn. Syft. I. p.* 220.

2.—Fn. Sv. N° 149.

Brun. N° 108.

Mull. N° 152.

Frifch. t. 185.

GUILLEMOT, or SEA HEN.

Lonruvia Hoieri. *Raii. Syn. p.* 120. *A.* 4.

Will. Orn. p. 324. *pl.* 65.

Albin. 1. *pl.* 84.

Edw. pl. 359. *Fig. I.*

B 2

FOOLISH

PLATE XXVIII.

FOOLISH GUILLEMOT. *Br. Zool.* N° 234.

Arct. Zool. N° 436.

Br. Muf.— Lev. Muf.

THE LAVY. *Martin's Voyage,* St. Kilda, 32.

Le Guillemot. *Brif. Orn.* VI. *p.* 70. 1. *pl.* 6. *fig.* 1.

Buff. Oif. 9. *p.* 350. *pl.* 25.—*Pl. Enl.* 903.

Lommia. *N. Com. Petr. IV.* 414.

Sea-Taube, or Groenlandifcher Taucher. *Frifch.* 11. 185.

The Guillemot is found in immenfe numbers on feveral of the Englifh coafts in fummer; they continue in the Orknies * the whole year; chiefly breed in the uninhabited *Ifle of Prieftholm,* near the *Ifle of Anglefea*; the *Farn Ifles* near the coaft of Northumberland, and among the high cliffs in the neighbourhood of *Scarborough* †, Yorkfhire. Like the Auk, (which are alfo found with them in vaft numbers). They lay only one egg, more than three inches in length, of a blueifh white or pale fea-green colour, moft elegantly ftreaked with black lines crofling each other in every direction.

They are very filly; for though they fee their companions killed by their fide, they only make a fhort circuit, and alight in the fame place to be fhot at in turn.

Our bird is feventeen inches in length, weight twenty-five ounces, breadth twenty-feven inches and a half. *Brunnich* mentions a variety,

---

* *Penn. Br. Zool.*

† *Willoughby.*

with

PLATE XXVIII.

with a broader and fhorter bill, and yellow margins. *Muller* fpeaks of a variety, with a white ring round the eyes, and a line of the fame colour behind.

They are found in moft of the northern parts of Europe to *Spitzbergen* ‡, the coaft of Lapmark, along the *White* and *Icy* Sea to Kamtfchatka; are found at *Newfoundland* and in fome parts of *North America*.

It is called *Guillem* by the Welch, *Guillemot* or *Sea Hen*, at Northumberland and Durham; in the fouthern parts, *Willocks*.

---

‡ Both Pennant and Latham has feparated the Guillemots from the Divers, and Grebes; to the firft Latham has given the generic title *Uria* after Briffon; the fecond he continues under the Linnæan genus (*Colymbus*); and the third he calls *Podiceps*.

PLATE

29

PLATE XXIX.

COLYMBUS AURITUS.

EARED GREBE.

ANSERES.

Bill obtufe, covered with a thin membrane, broad, gibbous below the bafe, fwelled at the apex. Tongue flefhy. Legs naked. Feet finned, or webbed.

GENERIC CHARACTER.

Bill ftraight, flender, pointed. Noftrils linear, at the bafe of the bill. Legs near the tail. Feet webbed.

SPECIFIC CHARACTER.

Bill curved a little upwards at the point. Lore and irides crimfon. Head black, with an orange-coloured tuft of feathers behind each eye. Breaft filvery white. Ridge and tips of the wings white. Legs olive.

COLYMBUS AURITUS. *Linn. Syft. I. p.* 222. 8.

Fn. Sv. 152.

Scop. Ann. I. N° 100.

Muller. p. 20.

EARED GREBE. *Br. Zool.* N° 224. *pl.* 79.

Arct. Zool. p. 499. *B.*

Lath. Gen. Syn. 5. 284. 4.

EARED DOBCHICK. *Edw. pl.* 96. *fig.* 2.

La Grebe à oreilles. *Brif. Orn.* 6. *p.* 54. 6.

C

Le

PLATE XXIX.

Le petit Grebe huppé. *Buff. Oif.* 8. *p.* 235.
Novegis Sav-Orre, Soe-Orre.
Bornholmis Soe-Hoene.
Iflandis Flauefkitt. *Brun.* 136.

The length of this Species is twelve inches; they inhabit the fens near Spalding, where they breed; they are found in the northern parts of Europe, and in the temperate parts of Siberia and Iceland. Said by *Bougainville* to be met with in *Falkland Iflands*, where it is called the *Diver with Spectacles.—Boug. Voy. p.* 61.

The neft, like moft others of the fame Genus, is compofed of twigs, roots and ftalks of water-plants, and is ufually found floating among the reeds and flags, nearly filled with water. The female lays four or five fmall white eggs, which are hatched in the water.

PLATE

PLATE XXX.

FALCO APIVORUS.

HONEY BUZZARD.

ACCIPITRES.

Birds of prey. Bill and claws ſtrong, hooked. An angle in each margin of the upper mandible. Body muſcular. Females larger and more beautiful than the males.

GENERIC CHARACTER.

Bill arched from the baſe, which is covered with a wax-like membrane, or cere.

SPECIFIC CHARACTER.

Bill and Cere black. Legs yellow; Claws black. Head aſh-coloured. Back and Wings dark brown. Breaſt and Belly white, ſpotted. Tail barred.

FALCO APIVORUS, Cere nigra, pedibus feminudis flavis, capite cinereo, caudæ faſcia cinerea apice albo. — *Fn. Sv.* 66. — *Linn. Syſt. Nat. I.* 91. 23. *edit.* 12.

BUTEO APIVORUS, *Raii Syn. p.* 16. N° 2.

HONEY BUZZARD, *Will. Orn. p.* 72. *t.* 3.

Albin I. t. 2.

Pen. Br. Zool. I. N° 26.

Latham's Gen. Syn. I. p. 52. N° 33.

Arſt. Zool. 2. *p.* 224. *I.*

C 2

La

PLATE XXX.

La Bondrèe. *Brif. Orn. I. p.* 410. N° 33.

Buff. Oif. I. p. 208.

Pl. enl. 420.

Le Goiran, ou Bondrèe. *Belon av.* 101.

Frofch-geyerl. *Kram.* 331.

Slag-Hok. *Faun Suec. fp.* 65.

Mufe-Hoeg, Mufe-Baage. *Brun. p.* 5.

Though the Honey Buzzard inhabits various parts of the continent of Europe it is no where common except in the open parts of *Ruffia* and *Siberia*; is feen as far north as *Sondonor* in *Norway*. In England it is fcarcely ever met with.

The length of our Specimen is twenty-three inches; weight when taken thirty ounces. In its colours it precifely correfponds with the defcription of the Honey Buzzard in *Latham*'s General Synopfis of Birds, but differs very materially from that either of Linnæus *, Briffon †, Pennant ‡, or Albin ‖.

Albin fays, " This bird builds its neft of fmall twigs, laying on them wool, and upon the wool its eggs. Some of them have been found to

---

* *Linnæus* defcribes the tail with only one cinerous band, the tip white

† *Briffon* fays " the fide tail-feathers are banded with white on the inner webs, and are fpotted with brown."

‡ *Albin* defcribes the tail of his fpecimen " plain without bars."

‖ *Pennant*, in the Br. Zool. defcribes the Honey Buzzard, " Chin, breaft, and belly white; the two laft marked with dufky fpots, pointing downwards; and three dufky bars on the tail." He mentions a variety " entirely of a deep brown; had much the fame marks on the wings and tail as the male; and the head tinged with afh-colour."

make

make ufe of an old neft of a kite to breed in, feeding their young with the *nymphæ of wafps,* the combs of wafps being found in the aforefaid nefts, in which were two young ones, covered with a white down, fpotted with black; their feet of a pale yellow, their bills between the noftrils and the head white, their craws large: in the crops were found lizards, frogs, &c. In one of them were found two lizards entire, with their heads towards the bird's mouth, as if they fought to creep out."—" This bird runs very fwiftly like a hen."—*Alb.* 1. *t.* 2.

The eggs of the Honey Buzzard are varioufly defcribed by different authors; the fpecimens formerly preferved in the Portland Mufeum were of a very deep red brown, with ferruginous blotches of chefnut; *Mr. Latham* fays he was informed by *Mr. Boys,* " that they are of a blueifh white, marked with irregular rufous fpots; the fhape of the egg almoft globular; ufually three in the neft." *Mr. Pennant* fays he was favoured with a defcription of the eggs by *Mr. Plumly;* " they were blotched over with two reds, fomething darker than thofe of the Keftril."

PLATE

31

PLATE XXXI.

LANIUS COLLURIO.

RED-BACKED SHRIKE,

BUTCHER-BIRD, or FLUSHER.

ACCIPITRES.

Birds of prey. Bill and claws ftrong, hooked. An angle in each margin of the upper mandible. Body mufcular. Females larger and more beautiful than the males.

GENERIC CHARACTER.

Bill hooked towards the end, with a notch near the tip of the upper mandible; bafe not furnifhed with a cere. Tongue jagged at the end.

SPECIFIC CHARACTER.

Bill black. A black ftroke through the eyes. Head light grey. Upper parts of the back, and wing coverts, ferrugineous. Breaft, belly, and fides, bloffom-coloured. Legs black. Tail black; all the feathers, except the two middle ones, more or lefs white at the bafe.

LANIUS COLLURIO. Cauda fubcuneiformi, dorfo grifeo, rectricibus quatuor intermediis unicoloribus, roftro plumbeo.—*Lin. Syft. Nat.* 1. 94. 3. *edit.* 12.

> *Faun. Arag. p.* 71.
> *Scop. Ann. I. p.* 24. N° 19.
> *Kram. p.* 363.
> *Muller. p.* 11.
> *Sepp. Vog. pl. in p.* 127.

D LANIUS

PLATE XXXI.

Lanius Tertius. *Aldr. av. I.* 199.

Lanius minor ruffis feu 3ttus Aldrovandi, Raii Syn.
 p. 18. A. a.

Merulæ congener alia, Raii Syn. p. 67. N° 13 ?

Lesser Butcher-Bird. *Will. Orn. p.* 88.
      Albin. vol. II. pl. 14.

   Flusher, *in Yorkſhire.*

Red-backed Shrike, *Br. Zool. I.* N° 72.
     Lath. Gen. Syn. I. 167. 15.—*Suppl.*
   52. 15.
     Arɛt. Zool. N° 131.

Le petite Pie grieſche griſe. *Belon av.* 128.

L'Ecorcheur. *Briſ.* 2. *p.* 151. N° 4.
    Buff. Oiſ. I. p. 304. *pl.* 21.—*pl. enl.* 31.
   fig. 2.

Danis Tornſkade. *Norv.* Hantvark. *Br.* 23.

Dorngreul, Dornheher. *Kram.* 363.

Bufferola, Ferlotta roſſa. *Zinan.* 91.

Mali Sokrakoper. *Scopoli,* N° 19.

The length of this Species is ſeven inches and a half, breadth
eleven inches; the female is of a dull ferruginous, mixed with grey:
the breaſt, belly, and ſides, dirty white, croſſed with ſemicircular duſky
lines : the tail deep brown, except the outer feather on each ſide,
whoſe exterior webs are white; the female is rather larger than the
male.

 9   It

It vifits this country in the fpring and departs in autumn; it is common in *France* and *Italy*, as well as in the temperate parts of Ruffia.

It builds its neft in a hedge or low bufh, and lays fix white eggs, encircled at the largeft end with a rufous brown circle; it not only feeds on infects, but will devour the young of other birds, taking hold of them by the neck and ftrangling them, then tearing out the eyes, brain, &c. and when fatisfied fticks the remainder on a thorn for another meal; when confined in a cage it will do the fame againft the wires with beetles, grafhoppers, or pieces of fheep's kidney.

It is faid to imitate well the notes of other birds, though it has none of its own.

D 2   P L A T E

32

PLATE XXXII.

TRINGA CINCLUS.

Ox-Eye, Purre,

or

Stint.

GRALLÆ.

Bill roundifh. Tongue entire, flefhy. Thighs naked. Toes divided.

GENERIC CHARACTER.

Bill roundifh ftrait, about the length of the head. Noftrils narrow. Toes four.

SPECIFIC CHARACTER.

Bill flender, black. Head, neck, back and tail, afh-coloured, or brown with dark fpots. Breaft, belly, and lower parts of the quill feathers white. Legs greenifh brown.

TRINGA CINCLUS. *Linn. Syft. I. p.* **251.** 18.—*Georgi Reife, p.* 172.

Cinclus five Motacilla.

Maritima, Lyfsklicker. *Gefn. av.* 616.

PURRE. *Br. Zool.* N° 206. *pl.* 17.

Arct. Zool. p. 390.

Lath. Gen. Syn. 5. 182. 30.

SANDERLING. *Albin.* 3. *pl.* 88.

LEAST

PLATE XXXII.

LEAST SNIPE, *Raii* .*n. p.* 190. 11.

   Sloan. am. p. 320. 14. *pl.* 265. 4.

STINT, or OX-EYE. *Raii Syn. p.* 110. *A.* 13.

   Will. Orn. p. 305.

WAGTAIL, *Kolb. Cap. I. p.* 152. ?—*Brown Jam. p.* 477.

L'Allouette de Mer. *Brif. av.* 5. 211. *tab.* 19. *fig.* 1:

   Belon av. 213.

   Buff. Oif. 7. *p.* 548.—*Pl. enl.* 851,

Giarolo. *Aldr. av.* 3. 188.

Length feven inches and a half; extent fourteen inches; weight an ounce and a half.

This Species is very common in moft parts of *Europe*, and is faid to be found at the *Cape of Good Hope*; in *Jamaica* and other Weft-India Iflands. They frequent our coafts in the winter in vaft flocks, alternately fwimming and flying in large circles with the greateft regularity: they leave our fhores in fpring, and retire to fome unknown place to breed. *Mr. Latham* fufpects that they breed on the coaft of Kent, having received fome birds which fcarcely differed from the defcription, from *Mr. Boys* of Sandwich; they were fhot at *Romney*, in the month of Auguft.

Le Cincle of *Buffon* and L'Alloutte de Mer à Collier of *Brifson* has much affinity to this Bird, and is fuppofed to be only a difference of fex or age, as they are often taken in company.

PLATE

33

PLATE XXXIII.

CORACIAS GARRULA.

GARRULOUS ROLLER.,

P I C Æ.

Bill compreffed, convex.

GENERIC CHARACTER.

Bill ftrait, bending towards the tip, edges cultrated. Noftrils narrow and naked.

SPECIFIC CHARACTER,

AND

S Y N O N Y M S.

Head, neck, breaft, and belly light bluifh green. Back and fcapulars reddifh brown; tail forked; black, blue and green. Legs dirty yellow.

CORACIAS GARRULA. cærulea, dorfo rubro, remigibus nigris. *Lin. Syft. Nat.*

CORVUS dorfo fanguíneo remigibus nigris, rectricibus viridibus. *Fn. Sv. 73.*

CORNIX cærulea. *Gefn. av. 335.*

GARRULUS argentoratenfis. *Raj. av. 41.*

GARRULOUS ROLLER. *Lath. Gen. Syn. I. p. 406, Nº 1.*

Suppl. 815. 1.

Arct. Zool. ii. p. 253. G.

E

ROLLER

PLATE XXXIII.

ROLLER *Wil. Orn.* 131. *pl.* 20.

Raii. Syn. p. 41. N° 3. *p.* 42:

Pen. Br. Zool. appen. p. 624. *pl.* 2.

Edw. Pl. 109.

Le Rollier. *Brif. Orn.* ii. *p.* 64. *pl.* 5. *f.* 2. *Pl. enl.* 486.

Le Rollier d'Europe. *Buff. Oif.* 3. *p.* 135. *pl.* 10.

The Shagarag. *Shaw's Travels.* 252.

Spranfk Kraka, Blakraka, Allekraka. *Faun. Suec. fp.* 94.

Ellekrage. *Brun.* 35.

Blave racke, Birck-heher, *Frifch. t.* 57.

On the authority of Mr. Pennant we have ventured to introduce this fpecies. " Of thefe birds," fays Mr. Pennant, in the appendix to the Britifh Zoology, " we have heard of only two being feen at large in our Ifland; one was fhot near *Helfton-bridge, Cornwal,* and an account of it tranfmitted to us by the Reverend Doctor *William Borlafe.*"

Thefe birds are frequent in moft parts of *Europe;* in *Germany, Sicily* and *Malta* they are fo common as to be fold in the markets *. *Edwards* mentions one fhot on *Gibraltar Rock;* it is alfo met with from the fouthern parts of *Ruffia* to the neighbourhood of the *Irtifh* †.

It makes its neft in woods, moft frequently in *Birch trees* ‡; never lays more than five eggs, which are of a clear green, fprinkled with innumerable dark fpecks §. It does not come to its colour till

---

* *Willughby.*   † *Arct. Zool.*

‡ *Frifch.*   § *Latham.*

9   the

the fecond year; flies in troops in autumn, and is often feen in tilled ground, with rooks and other birds, fearching for worms, fmall feeds, and roots ‖; it feeds alfo on frogs and beetles ¶.

By one author it is faid fometimes to make the neft in holes in the ground, in one of which two eggs were found **; by another, it is obferved never to be feen on the ground ††.

" It is remarkable for making a chattering noife; from which it is called by fome *Garrulus.*" *Pennant.*

‖ *Frifch.*   ¶ *Faun. Suec.*

** *Hift. des Oif.* iii. *p.* 139.   †† *Dec. Ruff. I. p.* 108.

E 2   PLATE

34

PLATE XXXIV.

LE GEAY BLANC.

WHITE JAY.

Brif. av. 2. *p.* 51. *A.*

In the defcription of CORVUS GLANDARIUS, COMMON JAY, we mentioned the fpecimen from which our prefent figure is taken; it was found in a neft with four other Jays of the common fort, and can only be confidered as a variety.

We have introduced it into this work, as a folitary example how far the plumage of birds will fometimes vary from local circumftances.

PLATE

35

PLATE XXXV.

STRIX BRACHYOTOS.

SHORT-EARED OWL.

ACCIPITRES.

Birds of prey. Bill and claws ſtrong, hooked. An angle in each margin of the upper mandible. Body muſcular. Females larger and more muſcular than the males.

GENERIC CHARACTER.

Bill ſhort, hooked, without cere. Head large. A broad diſk ſurrounding each eye. Legs feathered to the toes. Tongue bifid. Nocturnal.

SPECIFIC CHARACTER.

Horns or ears a ſingle feather. Above dark brown intermixed with pale yellow colour. Beneath pale yellow longitudinally ſtreaked with dark brown; feathered to the toes. Tail yellow brown barred with dark colour, tip white. Wings when cloſed reach beyond the tail.

STRIX OTUS.
SHORT-EARED OWL. *Pennant's Br. Zool.* N° 66. *t.* 31.

E   STRIX

PLATE XXXV.

STRIX BRACHYOTOS.   *Dr. Forſter's Phil. Tranſ.* Vol. lxii. *p* 384.
                   N° 2.

                   Lath. Gen. Syn. I. 124.   *Suppl.* 43.

SHORT-EARED OWL.   *Amer. Zool.*

Length fourteen inches, breadth when the wings are extended three feet, weight fourteen ounces.

Mr. Pennant appears to be the firſt author who has deſcribed this ſpecies *; he ſays it is a bird of paſſage, has been obſerved to viſit *Lincolnſhire* the beginning of *October*, and to retire early in the ſpring; he ſuppoſes its ſummer retreat is Norway. It conceals itſelf in the long graſs in the day-time; when diſturbed it will ſeldom fly far, but will light and ſit looking at one, at which time the horns may be ſeen very diſtinctly. Mr. Pennant further adds, " it is found frequently on the hill of *Hoy* in the *Orknies*, where it flies about and preys by day like a hawk. I have alſo received this ſpecies from *Lancaſhire*, which is a hilly and wooded country, and my friends have alſo ſent it from *New England* and *Newfoundland*." *Penn. Br. Zool.*

---

* Mr. Latham has made this ſevere, though not entirely unmerited animadverſion on the remarks of M. de Buffon, in *Hiſt. des Oiſ.* Vol. I. p. 353, note (a) " M. de Buffon ſeems to think that this bird is the *Scops*, than which no two ſpecies differ more. We have not the *Scops* in *England*, neither do I think the above-deſcribed bird to be a native of *France*. It would therefore have appeared candid in the abovementioned author, to have ſuſpended his opinion of the matter till he had been better informed, as he ſeems to bear ſomewhat hard upon Mr. *Pennant*, who, I am clear, is the firſt who has deſcribed it."—*Gen. Syn.*

Dr. Forſter

6

PLATE XXXV.

Dr. Forſter gave it the ſpecific name Brachyotos, in the *Philoſo-phical Tranſactions*; he ſays it is called *Mouſe Hawk* at *Hudſon's Bay*. It viſits that part in *May*, and makes a neſt of dry graſs on the ground: The eggs are white; it departs ſouth in *September*; is called by the natives *Thothoſecauſew* †. It is very common in the northern and woody parts of *Siberia* †.

Is known in *England* by the name of *Woodcock Owl*, as it is ſup-poſed to perform its migrations with the Woodcock. Feeds on mice.

---

† *Latham Gen. Syn.*

PLATE

36

PLATE XXXVI.

PICUS MINOR.

Lesser Spotted Woodpecker.

Picæ.

Bill compreffed, convex.

GENERIC CHARACTER.

Bill angular, ftrait. Noftrils covered with recumbent briftles. Tongue very long and round, with a fharp, hard, barbed point. Two fore and two hind claws,

SPECIFIC CHARACTER.

Crown crimfon. Above black barred with white. Beneath pale brown. In the female the crown is white.

Picus Minor. Albo nigroque varius vertice rubro, ano albido.
    Linn. Syft. Nat.
   Picus albo nigroque varius, rectricibus tribus latera-
    libus feminigris. *Fn. Suec.* 83. *Haffelqu. iter*
    242.
Picus varius Minor. *Alb. av.* I. *p.* 20. *t.* 20.
Picus varius tertius. *Raj. av.* 43.
Lesser Spotted Woodpecker. *Will. Orn.* 138. *pl.* 31.
    Alb. av.
    Lath. Gen. Syn. 2. 566. 14.
    Suppl. 107.

          Penn.

PLATE XXXVI.

Penn. Br. Zool. N° 89. *pl.* 37.

Amer. Zool.

Le petit pic varié. *Brif. Orn.* iv. *p.* 41. N° 15.

Le petit Epeiche. *Buff. Oif.* 7. *p.* 62.—*Pl. enl.* 598.

Kleiner bunt. Specht. *Frifch. t.* 37.

Baumbackterl. *Kramer.* 336. N° 5.

This is the fmalleft European fpecies of the Woodpecker genus we have any knowledge of at this time ; its length is fix inches, breadth eleven inches, and weight one ounce.

It vifits orchards, and feeds on the larva of infects, which it fometimes pecks out of the trunks of trees or decayed wood. It builds in an hole of a tree. Our figure reprefents the male ; the female has the crown of the head white.—This fpecies is not commonly met with in *England.*

Buffon fays it inhabits moft of the provinces of *France* *, and *Linnæus* obferves it inhabits the higher parts of *Afia* ; it is faid alfo to be feen as far *north* as *Denmark, Ruffia,* and *Siberia.*

Pennant remarks it has all the characters and actions of the greater kind, but is not fo often met with.

----

* Salerne denies its being found in France. *Orn. p.* 107.

PLATE

37

PLATE XXXVII.

PICUS VIRIDIS.

GREEN WOODPECKER.

PICÆ.

Bill compreffed, convex.

GENERIC CHARACTER.

Bill ftrait, ftrong, angular. Noftrils covered with recumbent briftles. Tongue very long, flender, armed with a fharp bony point. Two fore and two hind claws.

SPECIFIC CHARACTER,

AND

SYNONYMS.

Crown crimfon. Back green. Rump yellow, beneath pale green. Legs and feet greenifh, inclining to lead colour.

PICUS VIRIDIS. P. viridis, vertice coccineo. *Fn. Suec.* 80.
Linn. Syft. Nat.
Gefn. av. 710. *Scop. Ann. I. p.* 47. N° 52.
Brun. N° 39. *Sepp. Vog. pl. in. p.* 43.
Raii Syn. p. 42. A.
Piço Verde. *Aldr. av. I.* 416.

G

GREEN

PLATE XXXVII.

GREEN WOODPECKER. *Albin.* I. *pl.* 18.

Br. Zool. I. N° 84.

Arɛt. Zool. II. *p.* 277. B.

Lath. Gen. Syn. II. *p.* 577. N° 25.

Woodſpite, Rain fowl.

High-hoe, Hew-hole. *Will. Orn. p.* 135. *t.* 21.

Le Pic verd. *Briſ. Orn.* 4. *p.* 9, N° 1.

Buff. Oiſ. 8. *p.* 7. *pl.* 1.—*Pl. enl.* 371. 879.

Le Pic mart., Pic verd,

Pic jaulne. *Belon. av.* 299.

Grun-ſpecht, *Friſch. t.* 35. *Kramer.* 334.

Wedknar, Gronſpik.

Grongjoling. *Faun. Suec. ſp.* 99.

This ſpecies is thirteen inches in length, weight ſix ounces and an half. The female has no red mark on the lower jaw; Friſch and Klein obſerve they have no red on the crown of the head; but Latham, in his Synopſis of Birds, ſays, he has had them when they could ſcarcely fly, the red was then mixed with brown, but became full red after the firſt moult.

It is common in many parts of *Europe,* and is found as high north as *Lapmark* ; in *England* it is met with in moſt woody places.

They build in the hollow trunks of trees, fifteen or twenty feet from the ground: with their bills, which are very ſtrong, hard, and formed like a wedge, they can bore through the living part of the wood, till they come to that which is rotten; the hole thus formed is

as

PLATE XXXVII.

as perfectly in the form of a circle as if made with the affiftance of a pair of compaffes, and is hollowed out to a proper depth before the eggs are depofited. They lay generally five, fometimes fix * eggs; the young birds climb up and down the trees before they can fly.

According to *Pennant* the eggs are of a beautiful femitranfparent white; greenifh, with black fpots, *Latham*; and greyifh or yellowifh white, marked with irregular pale yellow brown lines in the figure of the egg in *Sepp*'s plate †.

It feeds on Infects, which it fometimes extracts from beneath the bark of trees, or from the folid wood by means of its ftrong, though flender barbed tongue; is faid to make great havock among bees.

In the Leverian Mufeum there is a variety of this fpecies, entirely of a ftraw colour, except the crown, which is faintly marked with red. It was fhot at *Belvoir chafe*.

---

* *Willoughby. Pennant.*   † *Sepp. Vog. pl. in p. 43.*

G 2   PLATE

38

PLATE XXXVIII.

MERGUS SERRATOR,

RED-BREASTED MERGANSER.

ANSERES,

Bill obtufe, covered with a thin membrane, broad, gibbous below the bafe, fwelled at the apex. Tongue flefhy. Legs naked; feet webbed, or finned.

GENERIC CHARACTER.

Bill long, roundifh, ferrated, hooked at the apex *.

SPECIFIC CHARACTER,

AND

SYNONYMS.

Irides red. Head and upper part of the neck black, crefted; lower part white. Breaft brown, mottled. Belly white. Back black. Wings, exterior fcapulars black; interior white.

---

* Noftrils near the middle of the mandible, fmall, and fubovated. Feet furnifhed with four toes, three forwards, and one behind; the outer toe before longer than the middle one. *Lath. Gen. Syn.*

MERGUS

PLATE XXXVIII.

Mᴇʀɢᴜs sᴇʀʀᴀᴛᴏʀ. Crifta dependente, capite nigro maculis ferru-
gineis. *Faun. Suec.—Linn. Syſt. Nat.*
Georgi Reiſe. p. 169.*—Muller,* N° 134*.*
Mergus albellus. *Scop. Ann.* I. N° 89.
Anas Longiroftra. *Geſn. av.* 133. *Aldr. av.* 3. 113.
Mergus criftatus capite caftaneo, &c. *Kram. El. p.* 343. 2. (female.)
———— cirratus fufcus. *Raii Syn. p.* 135. A. 4. *Will. Orn. p.* 336.
(Mergus cirratus minor.) *pl.* 64. (female.)
Rᴇᴅ-ʙʀᴇᴀsᴛᴇᴅ Gᴏᴏsᴀɴᴅᴇʀ. *Edw. pl.* 95.
Albin. 2. *pl.* 101.
Rᴇᴅ-ʙʀᴇᴀsᴛᴇᴅ Mᴇʀɢᴀɴsᴇʀ. *Penn. Br. Zool.* 2. 261.
Lath. Gen. Syn. 6. 423. 3.
Lᴇssᴇʀ Tᴏᴏᴛʜᴇᴅ Dɪᴠᴇʀ. *Morton's Northampton,* 429.
L'Harle hupé. *Briſ. av.* 6. 237.
Buff. Oiſ. 8. *p.* 273. *pl. enl.* 207.
Braun kopfiger Tilger.
Taucher. *Kram.* 343.
Pracka. *Faun. Suec. ſp.* 136.

———————————————————

Length twenty-one inches, breadth thirty-three inches, weight two
pounds. We have reprefented the male, the female has only the rudiment of a creft : the head and upper part of the neck, dull ferruginous : chin white : fore part of the neck and the breaft ferruginous,
mottled with black and white : upper part of the neck, back, rump,
and fcapulars, cinerous : the lower part of the breaft and belly white.

Both male and female are very liable to variation in the colour of
their plumage ; in fome the white fpace on the neck is much more
diffufed than in others ; the fame has been obferved of the portion of

I

white

PLATE XXXVIII.

white on the wings; and the females differ in the brightnefs of their colour frequently.

Mr. *Latham* mentions a fpecimen which was fhot near *Sandwich* in *Kent*; but it is chiefly found in the northern parts of *Great-Britain*; it is obferved to breed on *Loch Mari*, in the county of *Rofs*, and in the *Ifle of Ilay* *.

The neft is made of withered grafs, and is lined with the down of the bird's breaft; it lays from eight to thirteen eggs, like thofe of a wild duck, but fmaller and whiter; the young are of a dirty brown like goflings †.

It is found in moft of the northern parts of *Europe*, in *Ruffia*, about the great rivers of *Siberia*, and the lake *Baikal*; alfo frequent in *Greenland*, where it breeds on the fhores; in *Newfoundland* and *Hudfon's Bay*.

---

* *Pennant's Zool.* † *Latham Gen. Syn.*

PLATE

39

PLATE XXXIX.

LOXIA CURVIROSTRA.

COMMON CROSSBILL.

PASSERES.

Bill conic pointed. Noftrils oval*, broad, naked.

GENERIC CHARACTER.

Bill ftrong, convex above and below, very thick at the bafe. (Noftrils fmall and round †). Tongue as if cut off at the end.

SPECIFIC CHARACTER,

AND

SYNONYMS.

Both mandibles curve oppofite ways and crofs each other. Male red. Female green.

LOXIA CURVIROSTRA. Roftro forficato. *Faun. Suec.* 177.—*Linn. Syft. Nat.* 2. 171. 96. I. *edit.* 10. *Kram. El.* 365. Nº 2. *Brun. p.* 66. N° 238.

---

* *Linnæus.* † *Latham's Synopfis.*——*Pennant's Br. Zool.*

H *Muller,*

PLATE XXXIX.

Muller, N° 244.

Georgi Reife, p. 174.

Frifch. t. 11.

LOXIA.   Gefn. av. 591.

SHELL APPLE or CROSS BILL.   Raii Syn. p. 86. A.

Will. Orn. p. 248. t. 45.

Albin. 1. pl. 61.

Penn. Br. Zool. I. N° 115. pl. 49.

Arct. Zool.

Lath. Gen. Syn. 3. 106. I.

Edw. pl. 303.

Le Bec-croifé,   Brif. Orn. 3. p. 329. N° 1. pl. 17. f. 3.

Buff. Oif. 3. p. 449. pl. 27. f. 2.——

·Pl. enl. 218.

Korffnaff, Kinlgelrifvare.   Faun. Suec. fp. 224.

Krumbfchnabl, Kreutzvogel.   Kram. 365.

Kreutz-Schnabel.   Frifch. I. 11.

The length of this bird is fix inches and **three** quarters; it is diftinguifhed from other fpecies of the fame genus by the very fingular ftructure of its bill, both mandibles of which curve acrofs each other. The male is generally of a fine orange red inclining to rofe-colour, mixed more or lefs with brown, the female of a dull green; but both fexes are very liable to variations: the male is fometimes of a yellowifh orange; of a deep red; or even inclining to a dark purple hue, intermixed with yellow, red, brown, green, &c. the female varies alfo, but feldom acquires more than a dull intermixture of other colours on the olive-green according to the different feafons. The males are like the females when young, and gradually change to a fine red.

Mr.

PLATE XXXIX.

Mr. Pennant fays there are two varieties of this fpecies, our prefent fpecimen, and another which is very rare; of the latter he fays he received a male and female from Shropfhire; they were fuperior in fize to the former, the bill remarkably thick and fhort, more encurvated than that of the common kind, and the ends more blunt *.

The Crofs-bill is common in *Sweden*, *Germany*, and *Switzerland*; is found alfo in *Ruffia* and *Siberia*, in *North America*, *Greenland*, &c. It is not fuppofed to breed in *England*, but to vifit us generally in fmall flocks, though it has been feen in vaft multitudes in fome feafons. As the feeds of the Fir, or Pine, is their natural food, they always retire to forefts where thofe trees grow in moft abundance: they feed alfo on Hemp-feed; and are faid to do great damage in orchards, by tearing the apples to pieces to eat the pips or feeds.

It is obferved, in *North America*, to build its neft in the higheft part of the Fir-trees, faftening it to the branches by the refinous matter which exudes from the trees †.

---

* *Pennant's Br. Zool.* † *Latham.*

H 2 PLATE

40

PLATE XL.

MOTACILLA BOARULA.

GREY WAGTAIL.

PASSERES.

Bill conic, pointed. Noſtrils oval, broad, naked.

GENERIC CHARACTER.

Bill weak, and ſlender; ſlightly notched at the tip. Tongue lacerated at the end. Legs ſlender *.

SPECIFIC CHARACTER.

Crown, neck, back, aſh-colour. Throat black in the male. A pale ſtreak over the eye. Rump yellow. Breaſt and belly pale yellow. Wings brown; feathers edged with yellow. Tail-feather black, edged with yellow or brown; exteriors white.

MOTACILLA BOARULA.   *Linn. Mant.* 1771. *p.* 527.
        Scop. Ann. I. N° 225.
        Faun. Arag. p. 89.

---

* *Latham, Gen. Syn.*

I   Motacilla

PLATE XL.

Motacilla flava altera.   *Raii Syn.* 75. 3.

YELLOW WAGTAIL.   *Albin.* 11. *pl.* 58. (female.)

GREY WAGTAIL.   *Will. Orn. p.* 238.

                       Edw. pl. 259. (male.)

                       Br. Zool. 1. N° 144.——*Arct. Zool.*

                       Lath. Gen. Syn. 4. 178. 4.

La Bergerette.   *Belon. av.* 351.

La Bergoronette jaune, Motacilla flava.   *Briff. av. p.* 3. 471. *t.* 23.

                                   fig. 3. (male.)

Three kinds of Wagtails are found in this country, the Common, or White, the Yellow, and the Grey; the two former we have already figured; the latter is a very elegant bird, and appears to be the rareft of the three fpecies, it breeds in the north of England; fuppofed not nearer than *Cumberland* *, and departs fouthward in October.

In the male only the chin and throat are black. Length feven inches and an half.

All the birds of this genus frequent watery places; are very lively, and have a brifk motion in their tails. They feed on Infects. The neft of the Grey Wagtail is made on the ground; it is compofed of dried fibres and mofs, lined with wool or feathers within; it contains from fix to eight eggs, of a dirty white, marked with yellow fpots.

---

* *Latham, Gen. Syn.*

PLATE

41

PLATE XLI.

CUCULUS CANORUS.

COMMON CUCKOW.

P I C Æ.

Bill compreffed, convex.

GENERIC CHARACTER.

Bill roundifh and curved a little. Noftrils bounded by a fmall margin. Tongue fhort, pointed. Toes two forward, two backward. Ten feathers in the Tail.

SPECIFIC CHARACTER,

Above afh-colour. Beneath white, waved with tranfverfe black lines. Tail cuneated, black, with white fpots.

CUCULUS CANORUS. cauda aequali nigricante albo punctata. —
　　　　　　　　Linn. Syft. Nat. I. 110. 52. I. *edit.* 10.
　　　　　　　　Scop. Ann. I. *p.* 44. N° 48.
　　　　　　　　Brun. N° 36.
　　　　　　　　Georgi Reife, p. 165.
　　　　　　　　Sepp. vog. pl. in p. 117.
　　　　　　　　Faun Arag. p. 73.
CUCKOW.　　　　*Raii. Syn. p.* 23.
　　　　　　　　Will. Orn. p. 97. *pl.* 10. 77.
　　　　　　　　Albin. I. *pl.* 8.
　　　　　　　　Br. Zool. 1. N° 82. *pl.* 36.

　　　　　　I　　　　　　　　　　　· COMMON

PLATE XLI.

Common Cuckow.	*Lath. Gen. Syn.* 2. *p.* 509.
	Suppl. 98. I.
Le Coucow.	*Brif. Orn.* 4. *p.* 105. N° 1.
	Buff. Oif. 6. *p.* 305.—*pl. enl.* 811.
Le Coqu.	*Belon. av.* 132.
Ruckuk.	*Frifch. pl.* 40. 41.
Ructuct.	*Kram.* 337.
Gjok.	*Faun. Suec. fp.* 96.

The earlieft appearance of the Cuckow in this country is fuppofed to be in *February* *, it is rarely in *March*, but more commonly in *April:* it has been emphatically called the harbinger of Summer, or the meffenger of Spring; and its note, when heard early in the year, fhould never fail to invite the rural œconomift to his ufeful occupation. With *Stillingfleet* and *Pennant* we acknowledge the fallibility of human Calendars, for the purpofes of hufbandry; and with them muft conclude, that " fome attention fhould be given to thofe feathered guides, who come heaven-taught, and point out the true commencement of the feafon; their food being the Infects of thofe feafons they continue with us †."

The Cuckow is fo well known in this, and every other country of *Europe*, that we are not furprifed to find its Natural Hiftory has engaged the particular attention of every writer on Ornithology, in whofe works it could be introduced with propriety: it may hence be difficult, if not impoffible, to treat of its peculiar habits with an elegance of language fuperior to the defcription of *Buffon*, to felect more judicioufly the beautiful fictions of Antient Bards, than has been done by *Pennant*

---

* Mentioned in Br. Zool. Pen.   † *Br. Zool.*

7   and

PLATE XLI.

and other preceding writers; who have thus embellifhed its hiftory; or to add to general information any material circumftance that has evaded the vigilance and accuracy of *Latham*.

The note of this bird is a call to love, and is peculiar to the male; who, perched on the branch of a tree, or the fummit of an eminence, thus invites the female from the coppice in which fhe fits in filence: in a calm evening his note may be heard among the trees far off; and when difappointed of its mate, the neighbouring woods reecho his hollow note at the diftance of a quarter of a mile.

All Authors have allowed that the Cuckow does not hatch its own eggs, but depofits them in the neft of fome other bird, generally in that of a Hedge-Sparrow, Water-Wagtail, or a Yellow-Hammer; fome Writers fay the Cuckow lays only one egg, others two * in the neft; the fofter-parent attends them with the fame care as her own, and when the brood is hatched, fhe fhews no diflike to the fpurious offspring; fhe treats them with equal tendernefs, and toils with the fame affiduity to fupply them with food; the young Cuckow, when fledged, follows its little inftructor for a fhort time; but as its appetite encreafes, and the fmall Infects it collects, in imitation of its fuppofed parent, foon become infufficient for its fubfiftence, they feparate. Its ingratitude is proverbial among the *French* †, from a ridiculous fuppofition that it changed into a Hawk, and devoured its nurfe.

About the end of *June* the call of the male ceafes, though it does not take its final departure till the end of *September* or beginning of

---

* The egg figured by *Sepp* is like the Jackdaw's, both in fize and fhape, of a greenifh white, fpeckled with brown.—*Latham* fays it is certainly not that of the Cuckow, which he defcribes not much bigger than that of the Hedge-Sparrow, greatly elongated in fhape, the ground colour not unlike it, and mottled all over with ferruginous purple.

† " Ingrat comme un Coucou."

October.

PLATE XLI.

October. *Latham* obferves he has heard it call at midnight more than once or twice in the courfe of the Summer, and adds it was bright moon-light every time.

They feed on Infects, flefh, &c. in the ftomach of feveral that have been diffected the Caterpillars of the Fox * and Buff-tip † Moth have been found; in others vegetable matter, egg-fhells, Beetles, &c.

They are fuppofed to migrate to *Africa* and *Aleppo*, and to vifit feveral countries in their paffage; and are known in the northern parts of the world, even to *Kamtfchatka.*

Le Coucou roux, of *Briffon,* is a variety of a young bird, having the upper parts varied with rufous, where the other is white. Birds of the firft year are very liable to variation, fcarce two being found alike; the bars are much more numerous in fome than in others, and the ground colour more or lefs varied with ferruginous, according to the age.

On diffection, the ftomach has been difcovered to be very capacious and long; protruding far beyond the *fternum,* that part being fo very fhort, as not to be fufficient to take off the preffure in incubation, whereby digeftion may be impeded. This has been affigned as the reafon why it does not hatch its own eggs.

Length of the adult bird is fourteen inches, breadth twenty-five inches, weight two ounces and an half.

---

* *Phal. Rubi.* † *Phal. Bucephala.*

PLATE

42

PLATE XLII.

CHARADRIUS MORINELLUS.

DOTTEREL.

GRALLÆ.

Bill roundifh. Tongue entire flefhy. Thighs naked. Toes divided.

GENERIC CHARACTER.

Bill ftrait, roundifh, obtufe. Noftrils linear. Toes three, all placed forwards.

SPECIFIC CHARACTER.

Bill, Head, Belly black. Legs black brown. A broad white band above the eye; another acrofs the breaft. Breaft and fides dull orange. Back and Wings olive brown.

CHARADRIUS MORINELLUS.—Pe&ore ferrugineo, fafcia fupercili-
orum pe&orifque lineari alba, pedi-
bus nigris. *Faun. Suec.* 158, 160.
Linn. Syft. Nat. 2. 150. 79. 6.
edit. 10.
Brun. 185.
Morinellus avis anglica. *Gefner av.* 615.
DOTTEREL. *Raii Syn. p. III. A. 4.*

Will.

PLATE XLII.

Will. Orn. p. 309. *pl.* 55. 57.

Albin. 11. *pl.* 62.

Br. Zool. N° 210. *pl.* 73.

Arct. Zool. p. 487. *A.*

Pluvialis minor, five Morinellus, le petit Pluvier, ou le Guignard.—

Brif. av. V. 54. *tab.* 4. *fig.* 2.

Buff. Oif. 8. *p.* 87.

Pl. enl. 832.

Lappis Lahul.   *Faun. Suec.*

The Male of this fpecies is about nine inches in length; its weight four ounces; the Female is rather larger, the colours are in general more obfcure, the white ftripe over the eye is narrower, the black on the belly is intermixed with white, and the white line acrofs the breaft is wanting.

They are found in plenty in fome parts of *England,* in others are unknown. Are moft common in *Cambridgefhire, Lincolnfhire,* and *Derbyfhire,* about the latter end of *April,* in *May* and *June;* during which time they are very fat, and are much efteemed for their delicate flavour. In *April* and *September* they are taken on the *Wiltfhire* and *Berkfhire* downs * : they are alfo feen on the fea fide at *Meales,* in *Lancafhire,* in *April;* where they continue about three weeks; from thence they remove northward to *Leyton Haws,* where they ftay about a fortnight †. It is fuppofed that they breed in the mountains of *Cumberland* and *Wefimoreland,* as they appear there in *May,* and are obferved there after the breeding feafon. They breed alfo on feveral of the Highland hills *. Are proverbially ftupid birds, and eafily taken in a net, or fhot.

---

* *Latham.*   † *Pennant.*

Le

PLATE XLII.

Le Guignard d'Angleterre of *Brifon* * is confidered as a variety of this fpecies; the weight and fize correfpond with the former defcription; but the fore part of the Neck, Breaft, Belly, Sides, and Thighs, are pale yellow and white mixed, the Tail white, except the two middle feathers, the Legs and Feet of a fordid green. *Albin* has figured this bird; he fays he received it from *Lincolnfhire*, by the name of Dotterel †.

Thofe birds are common in the northern parts of *Europe*; *Linnæus* fays they are frequent in the *Lapland Alps*, and that they vifit *Sweden* in *May*. Breed in the northern parts of *Ruffia* and *Siberia*.

---

* *Brif. Orn.* V. p. 58. 6.   † *Albin.* pl. 63. Vol. 2.

PLATE

PLATE XLIII.

LOXIA COCCOTHRAUSTES,

GROSBEAK

OR

HAWFINCH.

PASSERES.

Bill conic, pointed. Noftrils oval, broad, naked.

GENERIC CHARACTER.

Bill ftrong, convex above and below, thick at the bafe. Noftrils, fmall, round. Tongue as if cut off at the end. Toes placed three before and one behind.

SPECIFIC CHARACTER.

AND

SYNONYMS,

Bill horn-colour. Irides grey. Crown of the head rufous chefnut; fides the fame colour, paler. Round the eye, and chin black. Breaft pale rufous bloffom colour. Hind part of the neck afh-colour. Back and coverts of the wings deep brown. Four outermoft fecondaries fhaped like fome antient battle-axes. Tail feathers black; on the inner webs white. Legs pale brown.

LOXIA COCCOTHRAUSTES. linea alarum fimplici alba, rectricibus
latere tenuiore bafeos albis. Linn.
Syft. Nat. 2. 171. 96. 2. edit. 10.

K Scop.

PLATE XLIII.

	Scop. Ann. I. N° 1.
	Cramer. el. p. 364. N° I.
	Frisch. t. 4. *M. and* F.
	Olin. uccel. pl. in pl. 37.
GROSBEAK or HAWFINCH.	*Sep. Vog. pl. in p.* 137.
	Raii. Syn. p. 85. *A. I.*
	Albin. 1. *pl.* 56.
	Edwards. pl. 188.
	Pen. Br. Zool. I. N° 113.
	Lath. Gen. Syn. III. 109. 4.
	Suppl. 148. 4.
	Arct. Zool.
Le Grosbeak ou Pinson royal.	*Belon av.* 373.
	Brif. Orn. III. *p.* 219. N° I.
	Buff. Oif. III. *p.* 444. *pl.* 27. *f.* 1.
	pl. enl. 99. 100.
Dlefchk	*Scop.*
Stenkneck.	*Faun. Suec.*
Kernbeis, Nufbeiffer.	*Kram.*

* This beautiful bird is rarely met with in this country except in winter; it is only an occafional vifitor with us, though in *France* it is not uncommon; and in *Germany, Italy, Sweden,* and the fouthern parts of *Ruffia* it is very plenty. It has been feen in *England* in the fummer months once or twice, and *Latham* feems inclined to believe they may fometimes breed here.

They feed on berries, and on the kernels of cherries, almonds, haws, &c. their bills are very large, and fo ftrong that they are able to crack the hardeft ftones of any fruit with the greateft facility.

9

They

They are faid to build the neft in hollow trees; or between the forked branches, about twelve feet from the ground; it is compofed of fmall dry fibres, intermixed with liver-wort; they lay five or fix eggs * of a roundifh fhape, of a bluifh green, fpotted with olive brown, and interfperfed with a few irregular black markings according to *Latham*; in the figure given by *Sepp* the eggs are of a pale purple colour, fpotted with brown; the neft appears of a loofe texture, and is placed on an oak.

The length of this fpecies is feven inches, breadth thirteen, weight two ounces; the colours of the Female are not fo bright as thofe of the Male, and the fpace between the bill and the eye, which is black in the latter, is grey in the other fex.

The general defcription we have given of its colours muft not be fuppofed to conftitute its diftinguifhing character: they vary exceedingly in different fpecimens; in fome the bill is almoft black, the crown of the head in fome is whitifh; in others wholly black: fometimes the white band acrofs the wing inclines to grey; in others no trace of white can be perceived: it has been feen with the body wholly black, and *Scopoli* mentions one entirely white, the quills excepted.

---

* Willughby.

K 2   **P L A T E**

PLATE XLIV,

PODICEPS* NIGRICANS.

DUSKY GREBE.

ANSERES.

Bill obtufe, covered with a thin membrane, broad, gibbous below the bafe, fwelled at the apex. Tongue flefhy. Legs naked. Feet webbed or finned.

GENERIC CHARACTER.

Bill ftrong, flender, and fharp pointed. Noftrils linear. Space between the eye and bill bare of feathers. Tongue flightly cloven at the end. Body depreffed: feathers thick fet, compact, and very fmooth and gloffy. Wings fhort. No tail. Toes furnifhed on each fide with a broad plain membrane.

SPECIFIC CHARACTER,

AND

SYNONYMS.

Bill black. Lore and irides red. Upper parts of the head, neck, and body dufky brown, beneath filvery white. Legs dirty olive.

---

* We have had occafion in a former defcription to mention the alteration made by *Briffon* in the Columbus genus of Ray and Linnæus; and fince adopted by our Englifh ornithologifts, *Pennant* and *Latham*. In the Linnæn genus are included the *Grebes, Guillemots* and *Divers*, which as they differ materially in the form of their feet, have been feparated by thofe later authors into diftinct tribes.—Podiceps is the new generic title given by *Latham* to the Grebes.

PODICEPS

PLATE XLIV.

PODICEPS NIGRICANS.	*Lath. Gen. Syn. Vol.* 5. 286. 5.
COLYMBUS NIGRICANS?	*Scopoli.* N° 101.
COLYMBUS MINOR, la.	
petite grebe.	*Brif. Orn.* 6. 56.
BLACK and WHITE DOBCHICK.	*Edwards av.* 96. *fig.* 1.
DUSKY GREBE.	*Br. Zool.* 225.

This fpecies inhabits the fens of *Lincolnfhire*, where it is perhaps not uncommon, though feldom found elfewhere. Length eleven inches.

In its manners it nearly agrees with the other birds of the fame tribe already figured in this work.

PLATE

PLATE XLV.

CHARADRIUS PLUVIALIS.

GOLDEN PLOVER.

GRALLÆ.

Bill roundifh. Tongue entire, flefhy. Thighs naked. Toes divided.

GENERIC CHARACTER.

Bill ftrait, obtufe. Noftrils linear. Toes three, all placed forwards.

SPECIFIC CHARACTER.

Upper fide of the plumage dufky ; fpotted with greenifh yellow. Beneath white. Legs black.

CHARADRIUS PLUVIALIS.	Pedibus cinereis, corpore nigro viridique maculato, fubtus albido.— *Linn. Syft. Nat.* 2. 151. 79. 8. *edit.* 10.
Pluvialis aurea, le pluvier doré.	*Brif. av. v.* 43. *tab.* 4. *fig.* 1. *Buff. Oif.* 8. *p.* 81. *pl.* 5.—*Pl. enl.* 904.
GREEN PLOVER.	*Raii Syn. p.* 111. *A.* 2. 190. 9. *Albin.* 1. *pl.* 75. *Will. Orn.* 308. *pl.* 57. *Sloan. Jam. p.* 318. 10. *pl.* 269. 2.

L   GOLDEN

PLATE XLV.

Golden Plover.	*Penn. Br. Zool.* 2. 474. 32. 208.
	Lath. Gen. Syn. 5. 193. 1.—*Suppl.* 252.
Brachhennl.	*Kram.* 354.
Rechter Brachvogel.	*Frisch.* 2. 217.
Pivier.	*Aldr. av.* 3. 206.
Piviero verde.	*Zinan.* 102.
Brok-Fugl.	*Brun.* 187.

The Golden Plover is found in small flocks on our moors and heaths, in the winter season; it is not a common species in this country. It breeds on several unfrequented mountains, particularly on those of the *Isle of Rum*, and the loftier *Hebrides* *: and on the *Grampian*, and all the heathy hills of the islands, and *Highlands* of *Scotland* †.

It is an inhabitant of Sweden, Denmark, Lapland, and other countries towards the frozen ocean; and according to *Russel* ‡ extends to the south as far as Aleppo.

It lays four eggs, sharply pointed at the lesser end, two inches and one-eighth in length, of a pale cinereous olive, blotched with blackish spots §.

In some specimens the belly is black, in others black intermixed with white; this is entirely owing to the season; early in March the black on the breast is first seen, it increases till that part becomes full black; but after the time of incubation that colour disappears. Instead of a hind toe some have only a small claw.

---

* *Pennant Br. Zool.*   † *Flor. Scot.* 1. *p.* 35.   ‡ *Russel, p.* 71.   § *Lath. Gen. Syn.*

PLATE

PLATE XLVI.

LARUS CANUS.

COMMON GULL.

ANSERES.

Bill obtufe, covered with a thin membrane, broad, gibbous below the bafe, fwelled at the apex. Tongue flefhy. Legs naked; feet webbed, or finned.

GENERIC CHARACTER.

Bill ftrong, ftrait, bending down at the point, an angular prominence on the under part of the lower mandible. Noftrils narrow, in the middle of the bill.

SPECIFIC CHARACTER.

Bill yellow. Back grey; the reft white. Legs dull green.

LARUS CANUS.	albus, dorfo cano. *Linn. Syft. Nat.* 2. 136. 69. 2. *edit.* 10.
	Scop. Ann. 1. N° 104.
	Brun. N° 141.
	Georgi Reife, p. 170.
COMMON GULL.	*Penn. Br. Zool.* 2. N° 249. *pl.* 89. f. 2.—*Arct. Zool.* N° 458.
	Lath. Gen. Syn. vol. 6. 378. 8.
Common Sea Mall, or Mew.	*Raii Syn. p.* 127. *A.* 3.
	Will. Orn. p. 345. *pl.* 76.

White

PLATE XLVI.

White web-footed Gull.	*Albin*. 2. *pl*. 84.
La grande Mouette cendrée.	*Brif. Orn*. 6. *p*. 182. 10. *pl*. 16. *fig*. 2.
	Buff. Oif. 8. *p*. 428.—*Pl. enl*. 977.
Gabbiano minore.	*Zinan*. 115.

This fpecies is the moft common of all the gulls. It breeds on the rocks and cliffs on our fhores and rivers which are contiguous to the fea, and is feen in vaft numbers on the Thames in fpring and winter, picking up the fmall fifh, worms, &c. left by the tide.

It is feen as far north as *Iceland, Lapland*, and the *Ruffian Lakes*, and alfo on the coaft of *Newfoundland*. It is an inhabitant of the warmer climates of the fouth, as *Greece*, fome parts of *Italy*, and moft of the fhores of the *Mediterranean Sea*.

The length is feventeen inches: breadth thirty-fix inches, and weight one pound. The eggs are two inches and a half in length; of a deep olive brown, marked with irregular deep red reddifh blotches *.

They differ a little in their markings: Mr. Latham mentions one, the head and half the neck of which were marked with fhort dufky ftreaks.

---

* *Lath. Gen. Syn.*

PLATE

47

PLATE XLVII.

FALCO MILVUS.

KITE.

ACCIPITRES.

Birds of prey. Bill and claws ſtrong. Hooked. An angle in each margin of the upper mandible. Body muſcular. Females larger, and more beautiful than the males.

GENERIC CHARACTER.

Bill much arched. A cere or membrane at the baſe.

SPECIFIC CHARACTER

AND

SYNONYMS.

Cere and irides yellow. Head hoary white with daſhes of black. Body ferruginous. Tail forked. Legs yellow.

FALCO MILVUS: Cera flava, cauda forficata, corpore ferrugineo, capite albidiore.—*Fn. Sv.* 59. *Linn. Syſt. Nat.* 1. 89. 10. *edit.* 10.

MILVUS. *Raii Syn. p.* 17. N° *A.* 6.
Geſn. av. 609.

KITE, or GLEAD. *Will. orn. p.* 74. *t.* 6.

M

KITE.

PLATE XLVII.

KITE.	*Penn. Br. Zool.* 1. 185. 53.
	Lath. Gen. Syn. 1. *p.* 61. N° 43.
Le Milan royal.	*Belon. av.* 129.
	Brif. orn. 1. *p.* 414. N° 35. *t.* 33.
	Buff. oif. 1. *p.* 197. *t.* 7.—*Pl. enl.* 422.
Rother Milon.	*Kram.* 326.
Glada.	*Faun. Suec. fp.* 57.
Nibbio.	*Zinan* 82.

Glede, Puttock, Kyte *Turneri.*

The Kite is very common in England, and is well known in fe-veral parts of the continent of Europe*; but it inhabits the more northern countries only during the hotteft months of fummer. *Bofman* mentions it as a native of Guinea†; Linnæus alfo fays it in-habits Europe, Afia, and Africa. No author has yet defcribed it as a native of America.

It breeds in woods. The neft is formed of different materials; the outfide of fticks, the lining of rags, bits of flannel, rope, paper, &c‡. It lays two or three eggs, which are roundifh, and of a whitifh colour, fpotted with dirty yellow. The egg of the Kite is defcribed by Mr. Latham (in his Supplement to the Synopfis of Birds) from the fpe-cimen formerly preferved in the Portland Mufeum; it was of a bluifh

---

* " The flefh is groffe. *Aldrov.* yet it's eaten by the poore people in Germany." *Robt. Lovell, Hift. Animals and Birds,* 1661.

† Bofman, Voy. de Guinee, p. 278.

‡ *Penn. Br. Zool.*

white,

PLATE XLVII.

white, inclining to red at one end, blending itfelf with the white by fmall markings.

As a bird of prey, the Kite is known to be very deftructive among poultry; it devours alfo fmall birds and animals, and Mr. Latham fays it will fometimes eat fifh, as it has been found feeding on the remains of one by the fide of a pond, after having probably beaten off its firft poffeffor.

The forked tail of the Kite diftinguifhes it from every other bird, and ferves to direct its flight with the greateft precifion. It fometimes appears fufpended, and quite motionlefs at a confiderable height, then glides with aftonifhing velocity through the fky, without the fmalleft apparent action of its wings. When it defcends on fmall birds, it generally carries them off in its talons, to devour them.

The length of this bird twenty-fix inches: breadth five feet. They differ very frequently in their colours. Mr. Pennant mentions a beautiful variety entirely of a tawny colour that was fhot in Lincolnfhire.

PLATE

48

PLATE XLVIII.

MOTACILLA SALICARIA.

SEDGE BIRD.

Bill conic, pointed. Noftrils oval broad, naked.

GENERIC CHARACTER.

Bill ftrait, flender. Tongue jagged.

SPECIFIC CHARACTER

AND

SYNONYMS.

Bill black, head brown, marked with dufky ftreaks, a white line over the eye, with a black line above it; cheeks brown. Back, wings, and tail, brown; wing coverts edged with pale brown. Body beneath yellowifh white. Legs dufky.

METACILLA SALICARIA. cinerea, fubtus alba, fupercillis albis.
—Linn. Syft. Nat. 1. 185. 8. *edit.*

10.

Avis confimilis ftoparolæ, & magnaninæ, *Raii Syn.* 81. 6.
Junco minor. *Sepp. vog. pl. in p.* 99 ?
Lucinia falicaria. *Klein. av.* 47 ?
SEDGE BIRD. *Albin.* 3. *pl.* 60.
Penn. Br. Zool. 1. N° 155.

SEDGE

PLATE XLVIII.

SEDGE WARBLER.	*Lath. Gen. Syn.* 4. *p.* 403. N° 21.
WILLOW LARK.	*Br. Zool.* 2. 241: *Lond.* 1766.
LESSER Reed Sparrow.	*Will. Orn.* 144.
La Fauvette de roseaux.	*Brif. Orn.* 3. *p.* 378. N° 5.
	Buff. Oif. 5. *p.* 142.—*Pl. enl.* 581. 2.

This elegant bird is commonly met with in marſhy places, or near rivers where willows, reeds, and ſedges grow. The neſt is generally made among the reeds. It is compoſed of ſtraw, and dried fibres of plants, lined with hair, and contains five eggs, of a dirty white colour, marbled with brown*.

It feeds on flies, ſpiders, &c. which it finds on the willows, or among the ruſhes, where it conceals itſelf. It imitates the note of the ſwallow, ſky-lark, houſe-ſparrow, and other birds, in a pleaſing but hurrying manner, and ſings all night †.

Some authors have ſuppoſed that it leaves us before winter, but that is uncertain.

Length of this bird is ſix inches and an half.

---

* *Latham. Gen. Syn.*   † *Pennant. Br. Zool.*

INDEX

INDEX to VOL. II.

ARRANGEMENT

ACCORDING TO THE

SYSTEM of LINNÆUS.

ORDER I.

ACCIPITRES.

ORDER II.

PICÆ.

INDEX.

ORDER III.

ANSERES.

ORDER IV.

GRALLÆ.

ORDER VI.

PASSERES.

VOL.

INDEX.

VOL. II.

ARRANGEMENT

ACCORDING TO

LATHAM's SYNOPSIS of BIRDS.

DIVISION I. Land Birds.

ORDER I. Rapacious.

GENUS II.

GENUS III.

ORDER II. Pies.

GENUS IV.

GENUS XII.

N 2

GENUS

INDEX.

DIVISION II. WATER BIRDS.

ORDER

INDEX.

VOL.

INDEX.

VOL. II.

ARRANGEMENT

ACCORDING TO

PENNANT's BRITISH ZOOLOGY,

GENUS

INDEX.

INDEX.

VOL. II.

ALPHABETICAL ARRANGEMENT.

4

THE

NATURAL HISTORY

OF

BRITISH BIRDS;

OR, A

SELECTION of the MOST RARE, BEAUTIFUL, and INTERESTING

BIRDS

WHICH INHABIT THIS COUNTRY:

THE DESCRIPTIONS FROM THE

SYSTEMA NATURÆ

OF

LINNÆUS;

WITH

GENERAL OBSERVATIONS,

EITHER ORIGINAL, OR COLLECTED FROM THE LATEST
AND MOST ESTEEMED

ENGLISH ORNITHOLOGISTS;

AND EMBELLISHED WITH

FIGURES,

DRAWN, ENGRAVED, AND COLOURED FROM THE ORIGINAL SPECIMENS.

VOL. III.

By E. DONOVAN.

LONDON:

PRINTED FOR THE AUTHOR; AND FOR F. AND C. RIVINGTON,
No. 62, ST. PAUL'S CHURCH-YARD. 1796.

49

PLATE XLIX.

MERGUS MERGANSER.

GOOSANDER.

ANSERES.

Bill obtufe, covered with a thin membrane, broad, gibbous below the bafe, fwelled at the apex. Tongue flefhy, Legs naked, Feet webbed, or finned.

GENERIC CHARACTER.

Bill long, roundifh, taper, ferrated and hooked at the apex. A creft on the Head. Migrate.

SPECIFIC CHARACTER.

Goofander male. Bill, Irides and Legs red. Head black: Lower part of the Neck, Breaft and Belly white. Wings and Tail black and white.

MERGUS MERGANSER: crifta dependente, capite nigro-cœrulef-
cente, collari albo. *Linn. Syfl. Nat.* 1.
p. 129. 62. 2. *edit.* 10.
MERGUS ÆTHIOPS. *Scop. Ann.* 1. N° 90.
Mergus Cirrhatus (fæm.) *Gefn. av.* 134. Merganfer (Merrach) 135.
MERGANSER, or GOOSANDER. *Will. orn.* 335.
Raii Syn. p. 134. *A.* 1.
Penn. Br. Zool. Vol. 2. *p.* 556.
46. 260.
Lath. Gen. Syn. Vol. 6. *p.* 418. 1.
Suppl. p. 270. 2.

A 2   L'Harle

PLATE XLIX.

L'Harle. *Brif. Orn.* 6. *p.* 231. 1. *pl.* 22.

   Buff. Oif. 8. *p.* 267. *pl.* 23.—*Pl. enl.* 951.

Meer-rache. *Kram.* 343.

See-rache. *Frifch.* 2. 190, 191.

Wrakfogel, Kjorkfogel, Ard, Skraka. *Faun. Suec. fp.* 135.

Pekfok. *Crantz's Greenl.* 1. 80.

The Goofander is never feen in the fouthern parts of Great-Britain, except in very fevere winters. In fummer it retires northward to breed; continues the whole year in the *Orknies*, and has been fhot in the *Hebrides* in fummer: it frequents rivers and lakes, and feeds on fifh. In winter it appears about *Sandwich*, with the Smew, Red-breafted Merganfer, and other water birds.

It is common in the northern parts of the Continent of *Europe* and *Afia*. In fummer is found in *Iceland, Greenland, Lapland*, and other *Arctic* regions, where it rears it's young; but migrates towards the fouth as the winter feafon approaches. In *America* alfo it abandons the more northern parts in winter.

Length of this Bird is twenty-eight inches; breadth forty: weight nearly four pounds.

The Bird we have figured is generally confidered as the male, and the Dun-Diver as the female of the fame fpecies; and among the more refpectable Naturalifts who have held this opinion, we muft place Linnæus, and fince his time Mr. Pennant *; Mr. Latham,

---

* Willoughby does not feem inclined to adopt this opinion implicitly; he fays, " In the *Dun-Diver*, which we take to be the *female* of the *Goofander*, we found a *large labyrinth*—fo that we will not be very confident that the *Goofander* and *Dun-Diver* differ no more than in fex." *Orn. p.* 336.

to

to whofe abilities and attention the fcience of Ornithology is fo much indebted, has, however, endeavoured to prove, not only that they are two diftinct fpecies, but that the Mergus Caftor of Linnæus is a mere variety * of the Dun-Diver, or fuppofed female of the Mergus Merganfer, our prefent fpecies.

Mr. Latham alfo obferves, among other minute particulars, that the Dun-Diver is ever lefs than the Goofander; and individuals of that bird differ greatly in fize: that in one fpecimen the creft of the fuppofed female was longer and fuller than in that thought to be the male; a circumftance obferved in no other bird that is furnifhed with a creft; for in fuch the females in many cafes have not even the rudiment of one. He fays alfo, on the authority of Dr. Heyfham, that the Dun-Diver is infinitely more common in *Cumberland* than the Goofander, at leaft ten or fifteen of the firft, to one of the laft. And he farther adds in the Supplement, " I have been lately informed by Dr. *Heyfham*, that he fome time fince diffected two *Dun-Divers*, the one weighing about two pounds: this proved a *female*; the eggs numerous, and appearing very diftinct. The other bird being much larger, weighed full three pounds. The creft in this was longer than that of the other, and the belly of a bright buff colour. This proved, on diffection, a *male*."

---

* The fpecimen Mr. Latham defcribes was killed on the coaft of *Suffolk*; it was no bigger than the *Smew*, but agreed in the general colour of its plumage with the *Dun-Diver*, except the neck, which had a greater mixture of afh-colour, and that it had a pale ftreak between the noftrils and eye.

P L A T E

50

PLATE L.

EMBERIZA MILIARIA.

COMMON BUNTING.

PASSERES.

Bill conic, pointed. Noſtrils oval, broad, naked.

GENERIC CHARACTER.

Bill conic, angular on each ſide ; a hard knob within the upper mandible.

SPECIFIC CHARACTER

AND

SYNONYMS.

Head and upper part of the body olive brown, with black ſpots. Beneath yellow-white. Wing and Tail feathers edged with pale rufous. Legs pale brown.

EMBERIZA MILIARIA.	*Linn. Syſt.* 1. *p.* 308. 5.
	Faun. Suec. N° 228.
EMBERIZA ALBA.	*Geſn. av.* 654.
COMMON BUNTING.	*Raii Syn. p.* 93.
	Albin. 2. *pl.* 50.
	Penn. Br. Zool. 1. *p.* 324. 118.
	Lath. Gen. Syn. 3. *p.* 171. 8.
Le Proyer, Prier, ou Pruyer.	*Belon. av.* 266.
	Briſ. orn. 3. *p.* 292. 10.
	Buff. oiſ. 4. *p.* 355. *pl.* 16.--*Pl. enl.* 233.
	Strillozo.

PLATE L.

Strillozo. *Olina*, 44.

Korn Larkor. *Lin. it. fcan.* 292. *tab.* 4.

Cimbris Korn-Lærke. *Norveg.* Knotter. *Brun.* 247.

Brafler. *Kramer* 372.

Graue Ammer. *Frifch.* 1. 6.

This fpecies continues with us the whole year: about the latter end of harveft they collect together in flocks, and feparate again in fpring: they feed on corn, oats, and moft other kinds of grain, which they eafily bruife with the hard protuberance with which the infide of the upper mandible is furnifhed.

The female builds the neft among low bufhes, and lays five or fix eggs; fhe can fcarcely be diftinguifhed from the male, except by the colour of her plumage, being fomewhat paler. Length fix inches and an half.

It is common in this country, though much lefs fo than the Yellow-hammer, which alfo is a fpecies of the Bunting genus: they are found in vaft numbers in *Italy*, in *Germany*, and in the fouthern parts of *Ruffia*, *Sweden* and *Denmark*.—They are frequently fhot, or taken in nets, and fold for Larks, or Bunting Larks.

Albin fays, " It *fings* fitting upon the higheft twigs of trees or fhrubs;" but Latham obferves it has no *fong*, only a fcream, or tremulous kind of fhriek, three or four times repeated.

PLATE

51

PLATE LI.

FALCO TINNUNCULUS.

KESTREL.

ACCIPITRES.

Birds of prey. Bill and claws ſtrong, hooked. An angle in each margin of the upper mandible. Body muſcular. Females larger and more beautiful than the male.

GENERIC CHARACTER.

Bill arched from the baſe, which is covered with a wax-like membrane or cere.

SPECIFIC CHARACTER

AND

SYNONYMS.

Male: Cere and feet yellow. Back and wings rufous brick-colour, with black ſpots. Beneath pale ferruginous, with dark longitudinal ſtreaks. Tail rounded at the end; of a pale grey colour, with a black band near the extremity. *Female:* leſs bright, and ſpotted with duſky colour. Tail pale brown, with many dark ſtreaks.

FALCO TINNUNCULUS: cera pedibuſque flavis, dorſo rufo punctis nigris, pectore maculis longitudinalibus fuſcis, cauda rotundata. *Fn. Suec.—Linn. Syſt. Nat.* 1. 15. 90. *edit.* 10.

KESTRIL. *Lath. Gen. Syn. Vol.* 1. *p.* 94. 79.—*Suppl. p.* 25. 79.
The Keſtrel, Stannel, Stone-gall, or Wind-hover, *Will. orn. p.* 84. *t.* 5.
Br. Zool. N° 60.

B

Kiſtrel,

PLATE LI.

Kiftrel, Kaftrel, or Steingal, *Turnew.*

La Crefferelle. *Belon. av.* 125.

Brif. orn. 1. *p.* 393. N⁰ 27.

Buff. oif. 1. *b.* 280. *t.* 18.—*Pl. enl.* 401. 471.

Roethel-Geyer. *Frifch.* 1. 84. *fæm.* Maufe-Falck. *Frifch.* 1. 88.

Kyrko-Falk. *Faun. Suec.*

Gheppio, Acertello, Gavinello. *Zinan.* 88.

Poftoka, Splintza, Skoltfch. *Scopoli,* N° 5.

Windwachl, Rittlweyer, Wannenweher, Kramer, 331.

The male of this fpecies is much more beautiful than the female; the former (of which a figure is given in the annexed plate,) weighs about fix ounces and a half: its length fourteen inches; and meafures, when the wings are expanded, twenty-feven inches between each tip. The female weighs eleven ounces: the colour of the back and wings is more pale and dufky than in the male; and the middle of each feather is marked with an oblong dark ftreak, pointing downwards: the breaft of a dirty yellowifh white; and the tail pale red brown, croffed with numerous black bars.

It is not uncommon in England, and many other parts of Europe; it breeds in the hollows of trees, cavities of rocks, old buildings, &c. It lays four eggs, of a pale ferruginous colour, marked with irregular fpots of a deeper hue *. It feeds on mice, fmall birds, and infects; and throws up the fur and feathers in the form of a round ball.

This bird was formerly ufed in falconry, to catch fmall birds and young partridges.

---

* Portland Mufeum.

PLATE

52

PLATE LII.

MERGUS MINUTUS.

RED-HEADED MERGANSER.

ANSERES.

Bill obtufe, covered with a thin membrane, broad gibbous below the bafe, fwelled at the apex. Tongue flefhy. Legs naked, feet webbed or finned.

GENERIC CHARACTER.

Bill convex above, flat beneath, hooked at the apex, with membranous teeth.

SPECIFIC CHARACTER

AND

SYNONYMS.

Bill lead colour. Head red brown, with a fmall creft. Cheeks, throat, and belly white. Back and tail mottled, dufky and white. Wings dufky, with a patch of white on the coverts, and two bars of the fame below. Legs dufky.

MERGUS MINUTUS: capite grifeo lævi. *Fn. Suec.—Linn. Syft.*
        Nat. 1. *p.* 129. 5.
MERGUS tinus, *Hafelq. It. p.* 269. N°. 37.
———- glacialis, *Brunnich*, N°. 99.
———- pannonicus. *Scop. Ann.* 1. *p.* 392.
THE WEZEL COOT. *Albin. orn.* 1. *p.* 84. *t.* 88.
LOUGH DIVER. *Raii Syn. p.* 135.—*Will. orn. p.* 338.

MINUTE

PLATE LII.

MINUTE MERGANSER. *Lath. Gen. Syn.* 6. 429. 6.
RED-HEADED SMEW. *Pen. Br. Zool.* 263.
L'Harle etoilé. *Brif. av.* 6. 252.—*Buf. oif.* 8. *p.* 278.
Le petit Harle huppé, (femelle.) *Brif. orn.* 6. *p.* 243. 3. *pl.* 24. *fig.* 2.

This bird is found in the fouthern parts of England in winter. It is faid to be found as far fouth as latitude 37, being met with in the ifland of *Fino* ; in the *Archipelago* *, it migrates towards the north in fummer, to breed along with the other *Merganfers* ; and is therefore met with during that feafon in *Iceland*, *Lapland*, and other Arctic regions.

Length fourteen inches and a half: breadth twenty-three inches: weight fifteen ounces.

---

* *Lath. Gen. Syn.*

PLATE

53

PLATE LIII.

FALCO PEREGRINUS.

PEREGRINE FALCON.

Accipitres.

Birds of prey. Bill and claws ſtrong, hooked. An angle in each margin of the upper mandible. Body muſcular. Females larger and more beautiful than the males.

GENERIC CHARACTER.

Bill arched from the baſe, which is covered with a wax-like membrane or cere.

SPECIFIC CHARACTER.

Cere, legs and feet yellow. Beak blue. Above aſh colour bar-red with black and brown. Beneath white tinged with brown, with longitudinal and tranſverſe lines.

Falco Peregrinus. *Raii Syn. p.* 13. *No.* 1.
Peregrine, or Haggard Falcon. *Will. Orn. p.* 76. *t.* 8.
Blue backed Falcon. *Charlton Exer. &c.* 73.
Peregrine Falcon: *Penn. Br. Zool.* N°. 48. *t.* 20.
      Lath. Gen. Syn. 1. 52. *p.* 73.
      ——— *Suppl.* 18.
Le Faucon pelerin. *Briſ. Orn.* 1. *p.* 341. N° 6.
      Buff. Oiſ. 1. *p.* 249.

The Peregrine Falcon is very common in the north of *Scotland*, where it is trained for falconry; it does not, however, appear to be common in *England*. It breeds on ſome high rocks near *Gilſland*

C   in

PLATE LIII.

in *Cumberland*: in the mountains about *Kefwick* * ; and on the rocks
of *Llandidno* in *Caernarvonfhire* †.

It is common on the continent of Europe, in fummer ; inhabits
alfo Kamtfchatka, moft parts of America, &c. It varies very much
at different periods of age ; one mentioned by Pennant, had the
whole under fide of the body of a deep dirty yellow, but marked with
black, as ufual, in other fpecimens. It feeds on fmaller birds, as
partridges, plovers, moor game, &c. The fize of this fpecies is
generally about nineteen inches in length, breadth thirty-eight
inches, weight thirty-fix ounces.

---

* Latham.   † Pennant.

PLATE

PLATE LIV.

ARDEA MINUTA,

LITTLE BITTERN.

GRALLÆ.

Bill roundiſh. Tongue entire, fleſhy. Thighs naked. Toes divided.

GENERIC CHARACTER.

Bill ſtrait, long, acute. Toes connected as far as the firſt joint by a ſtrong membrane.

SPECIFIC CHARACTER

AND

SYNONYMS.

Bill greeniſh yellow. Head, back, quills, tail black gloſſed with green. Neck, breaſt and thighs, buff; thighs feathered to the knees.

ARDEA MINUTA. *Linn. Syſt.* 1. 240. *ed.* 12.
Ardea vertice dorſoque nigris, collo antice et alarum tectricibus
    luteſcentibus, (Stauden Ragerl, Kleine Moofs-kuh) *kram.* 348.
Ardeola. *Sepp. Vog. pl. in p.* 57.
LITTLE BITTERN. *Penn. Br. Zool. Appen.* 8. *p.* 638.
                Lath. Gen. Syn. 5. *p.* 66. 27.
              —— *Suppl.* 235.
Little brown Bittern. *Edw. pl.* 275.
Le Blongios tacheté   *(female) Briſ. Orn.* 5. *p.* 497. 47.
Blongios de ſuiſſe   *(male) Buff. Oiſ.* 7. *p.* 395.
Kleiner Rohrdommel. *Friſch.* 2. 206. 207.

9

The

PLATE LIV.

This elegant Bittern is rarely met with in this country. The firſt inſtance we find on record, on which we can venture to deſcribe it, as a Britiſh ſpecies, is that of a male ſpecimen having been ſhot as it perched on one of the trees, in the quarry, or public walks in *Shrewſbury*, on the banks of the *Severn*. This account, together with a general deſcription and drawing of the bird, was communicated by Mr. *Plymley* of *Lagnor, Shropſhire*, to Mr. Pennant, who inſerted them in the Appendix to his Britiſh Zoology *.

Additional authority is alſo given to this circumſtance by Mr. Latham, in the fifth volume of the Synopſis of Birds; he ſays, " another ſpecimen was killed in 1773, near *Chriſt Church* in *Hampſhire*, now † in the Muſeum of Mr. *Turnſtall*." He obſerves they have been found frequently in Arabia, but are ſcarce in other parts. In *France* are very rare.

This bird does not exceed fifteen inches from the bill to the tip of the tail ; it's form is graceful, it's body not larger than that of a thruſh. According to *Sepp*, it lays four eggs, of a white colour, and ſize of a blackird's ; the neſt is compoſed of bits of ſticks, with ſome flag leaves interſperſed.

---

* Publiſhed in 1776.　　　† 1785.

PLATE

55

PLATE LV.

CHARADRIUS HIMANTOPUS.

LONG-LEGGED PLOVER.

GRALLÆ.

Bill roundifh. Tongue entire, flefhy. Thighs naked. Toes divided.

GENERIC CHARACTER.

Bill ftrait, roundifh, obtufe. Noftrils narrow.

SPECIFIC CHARACTER

AND

SYNONYMS.

Bill black, long. Legs red, very long and flender, and without a back toe. Wings extend beyond the tail. Forehead, breaft, belly, tail, white. Wings and back black, gloffed with green. A few dark fpots on the back of the neck and crown.

CHARADRIUS HIMANTOPUS: fupra niger, fubtus albus, roftro
        nigro capite longiore, pedibus rubris longiffimis.
        Linn. Syft. Nat. I. p. 151. 79. 10. *ed.* 10.
Himantopus, *Raii Syn. p.* 106. 9. *p.* 193. *Pl. I. fig.* 1.
        Will. Orn. 297.
Long-legged Plover, *Penn. Br. Zool. No.* 209.
        Lath. Gen. Syn. 5. 195. 3.
        —— *Suppl.* 252.
Long Legs.   *Raii Syn. p.* 109. 7.
        Sloan. Jam. 2. *p.* 316. 6. *pl.* 267.
        D   Le

PLATE LV.

Le Grand Chevalier d'Italie. *Belon Pontr. d'Oyfeaux.* 53.
L' Echaffe. *Brif. Orn.* 5. *p.* 33. *I. pl.* 3. *fig.* 1.
Buff. Oif. 8. *p.* 114. *pl.* 8.
———— *Pl. enl.* 878.

This bird meafures thirteen inches from the tip of the bill to the end of the tail; to the claws nearly eighteen inches. The extraordinary and very difproportionate length of the legs gives it fuch an uncommon appearance, that we may pronounce it, with ftrict propriety, the moft fingular fpecies ever met with in this country.

It is alfo not lefs rare than fingular. *Sir Robert Sibbald* records a brace that were fhot in Scotland; another was fhot a few years ago on *Stanton-Harcourt* Common near Oxford *: and Mr. *White*, bookfeller, of Fleet-ftreet, has a fpecimen which was fhot out of a flock of fix or feven, in *Frenchman-ponds*, in *Hampfhire*. This bird does not perfectly agree in it's plumage with our fpecimen, but is no doubt only a mere difference in the fex.

According to Latham it is common in *Egypt*; plentiful about the falt lakes, and often on the fhores of the Cafpian Sea; and in the fouthern deferts of *Independent Tartary*. Found alfo at *Madras*, in the *Eaft Indies*; and in the warmer parts of *America*.

---

* Pennant, Br. Zool.

PLATE

56

PLATE LVI.

PODICEPS MINUTUS.

LITTLE GREBE.

ANSERES.

Bill obtufe, covered with a thin membrane, broad gibbous below the bafe, fwelled at the apex. Tongue flefhy, legs naked, feet webbed.

GENERIC CHARACTER.

Bill ftrait, flender, pointed. Noftrils linear. Lore bare of feathers. Tongue flightly cloven at the end. Body depreffed. Wings fhort. No tail. Legs placed far behind. Toes furnifhed on each fide with a broad plain membrane.

SPECIFIC CHARACTER

AND

SYNONYMS.

Above, head, neck, breaft dark brown. Belly greyifh, with a gloffy appearance. Legs dirty greenifh colour.

COLYMBUS AURITUS. *Linn. Syft.* 1. *p.* 223. 8. γ.
Faun. Suec. p. 184.

PODICEPS MINUTUS. *Lath. Gen. Syn. v.* 5. 289. 10. *Suppl. in Lift of Birds of Great Britain.*

Little Grebe. *Latham.*
Penn. Br. Zool. N° 226.

Dipper Didapper, Dobchick, &c. *Raii Syn. p.* 125. *A.* 3.
Will. Orn. p. 340. *pl.* 61.

Le

PLATE LVI.

Le Grebe de la riviere, ou le Caftagneux. *Brif. Orn.* 6. *p.* 59. 9.

Buff. Oif. 8. *p.* 244. *pl.* 20.—*Pl. enl.* 905.

Trapazorola arzauolo, Piombin. *Aldr. av.* 3. 105.

Kleiner Seehahn, or Noerike. *Frifch.* 2. 184.

This Bird frequents marfhy places with other fpecies of the fame tribe. It makes a very large neft of grafs and aquatic plants in the water, without any faftening to the banks, fo that the neft rifes or falls with the water : it lays five or fix eggs of a dirty yellowifh white colour, which are kept conftantly wet by the water that rifes through the neft : the natural warmth of the Bird when fitting excites a fermentation in the vegetables, and ferves to hatch the young brood.

It is a moft expert diver, and is faid by *Salerne* * to be able to ftay a quarter of an hour under water. If purfued it plunges into the water, and feldom appears again within the reach of gun-fhot.

Length of this fpecies is ten inches, weight fix ounces and a half. The male very nearly correfponds in colours with the female, and both vary according to their age. Having no tail, and the legs being placed far behind, give a very aukward and clumfy appearance to this bird. Common in moft parts of *Europe*, and mentioned as a native of *America.*

---

* *Salerne Orn.* p. 377.

PLATE

57

PLATE LVII.

PARUS CÆRULEUS.

BLUE TITMOUSE.

PASSERES.

Bill conic pointed. Noftrils oval, broad, naked.

GENERIC CHARACTER.

Bill fhort, ftrong, entire. Briftles at the bafe. Tongue blunt with briftles at the end.

SPECIFIC CHARACTER

AND

SYNONYMS.

Crown, Wings, Tail, blue. Forehead and Cheeks white. Back greenifh. Beneath yellow. A white bar acrofs the Wings.

PARUS CÆRULEUS: remigibus cærulefcentibus: primoribus margine exteriore albis, fronte alba, vertice cæruleo.
Fn. Sv.—Lin. Syft. Nat. 1. 100. *p.* 190. 4. *edit.* 12.
Scop. Ann. 1. *p.* 163. N° 244.
Kram. el 379. N° 3.
Muller, p. 34. N° 285.
Albin. 1. *pl.* 47.

BLUE TITMOUSE. *Penn. Br. Zool.* 1. N° 163. *pl.* 57. *f.* 2.
Lath. Gen. Syft. vol. 4. 543.

E   La

PLATE LVII.

La Mefange Bleue. *Briſſon. av.* 3. 544.

Buff. oiſ. 5. *p.* 413.

———*Pl. enl.* 3. *fig.* 2.

Blava fnitza, Blau mandlitz. *Scopoli.* N° 244.

Blaumeiſe. *Kramer* 379. *Friſch.* 1. 14.

Blamees. *Faun. Suec. ſp.* 267.

Parozolino, o Fratino. *Zinan.* 76.

———

This is a very beautiful bird, it frequents gardens and orchards where it does much injury to the fruit trees by tearing off the bloſſoms, in ſearch of the eggs and larva of infeƈts The female builds its neſt in holes of walls, or trees, and lays from fourteen or fifteen to twenty eggs ; it is ſaid to deſert its neſt, if any of the eggs are touched or broken : but defends its young when hatched, with much ſpirit. Length of this ſpecies four inches and a half.

PLATE

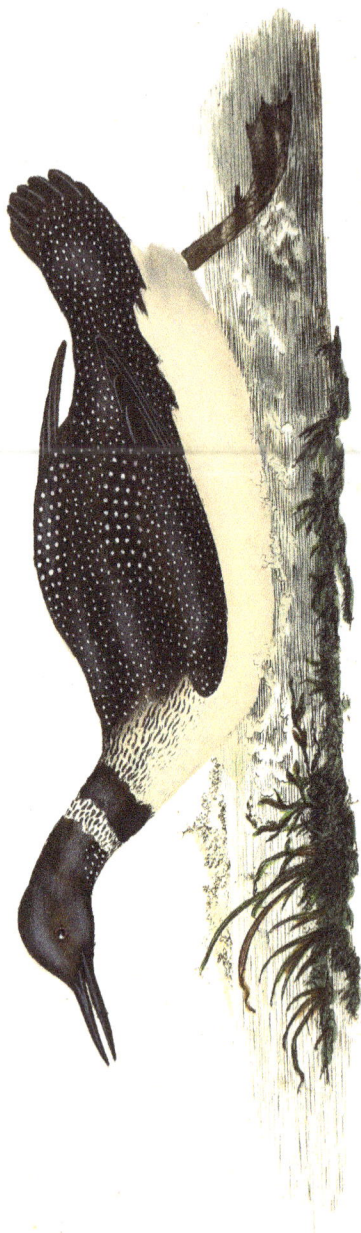

PLATE LVIII.

COLYMBUS GLACIALIS.

NORTHERN DIVER.

ANSERES.

Bill obtufe, covered with a thin membrane, broad, gibbous below the bafe, fwelled at the apex. Tongue flefhy. Legs naked. Feet webbed or finned.

GENERIC CHARACTER.

Bill ftrait, pointed; upper mandible longeft; edges bending inwards. Noftrils linear. Tongue, long, pointed, ferrated near the bafe. Legs thin and flat. Toes four, the exterior the longeft, back toe fmall, joined to the interior by a fmall membrane. Tail fhort; confifts of twenty feathers *.

SPECIFIC CHARACTER

AND

SYNONYMS.

Bill, Head, Neck, and upper fide black, marked with round fpots of white; a patch of white with black ftreaks under the chin, and another on each fide of the neck; fides of the breaft marked with fmall black ftreaks. Under fide white. Legs black.

COLYMBUS GLACIALIS. *Lin. Syft.* 1. *p.* 221. 5. *edit.* 12. *Holmiæ.* 1766.

---

* *Pennant, &c.*

E 2   Colymbus

- 293 -

PLATE LVIII.

Colymbus maximus stellatus nostras. *Sib. Hist. Scot.* 20. *tab.* 15.

Colymbus maximus caudatus. *Raii Syn. p.* 125. *A.* 4.

Northern Diver. *Penn. Br. Zool.* 2. 523. 237.

Lath. Gen. Syn. 5. 337.

Greatest speckled Diver, or Loon, *Will. Orn. p.* 341.

Albin. 3. *pl.* 93.

Le grand plongeon tacheté, *Brif. Orn.* 6. *p.* 120. 6. *pl.* 11. *fig.* 1.

L'Imbrim, *Buff. Oif.* 6. *p.* 258. *pl.* 22. *pl. enl.* 952.

Grosse Halb-Ente, Meer-Nœring. *Frifch* 2. 185. *A.*

This is a large bird, it measures more than three feet in length, in breadth four feet six inches: weight sixteen pounds.

It is far from common on our shores, and is entirely confined to the northern parts of the island, except in very severe winters. If we confider the authority of Albin worth quoting on this occafion, we may conclude it was fcarcely known as a native of this country in his time: he has given a figure of it in the third volume, plate 93. and fays, " It was brought from *Newfoundland*, and prefented to the *Right Honourable the Lord Ilay*, who was pleafed to lend it me to draw its picture." Willoughby mentions one being taken in the ifland of Jerfey; and Latham, in a note, fays, " One of thefe was caught alive near *Kefwick*, in *Cumberland*, in *July*, 1781. It was, as is fuppofed, making for the lake, but grew tired before it had power to reach it. *Dr. Heyfham*."

This laft circumftance is very remarkable, as it lives for the moft part on the open fea, and except in the breeding feafon feldom frequents frefh waters; nor are we certain whether it breeds on our coafts, as many water birds migrate to *Norway, Iceland, Greenland*, &c.

&c. to breed, and this fpecies is always found common in thofe northern regions. *Pennant* fays in *Scotland* it is called *Mur-buachaill*, or *Herdfman* of the fea, from its being fo much in that element.

The female lays two large pale brown or ftone-coloured eggs, in *June**. The fkins are tanned by fome northern nations with the down upon them, and are made into caps and other garments.

---

* *Latham.*

P L A T E

59

PLATE LIX.

FALCO CYANEUS.

HEN-HARRIER.

ACCIPITRES.

Birds of prey. Bill and claws ſtrong, hooked, an angle in each margin of the upper mandible. Body muſcular. Females larger, and more beautiful than the males.

GENERIC CHARACTER.

Bill arched from the baſe, which is covered with a wax-like membrane, or cere.

SPECIFIC CHARACTER

AND

SYNONYMS.

Bill black. Cere and irides yellow. General colour blue grey; back of the head white, with pale brown ſpots. Breaſt, belly and thighs white, with duſky ſtreaks. Two middle feathers of the tail grey. Legs yellow, long and ſlender.

FALCO CYANEUS. *Lin. Syſt.* 1. *p.* 126. N° 10.
Lanarius albus. *Aldr. av.* 1. 197.
Blue Hawk. *Edw.* 225. *male.*
Hen-harrier. *Pennant's Brit. Zool.* N° 58. *t.* 28.
　　　　　Lath. Gen. Syn. 1. *p.* 88. 74.

F　　　　　　　　　　　Le

PLATE LIX.

Le Lanier cendré. *Brif. orn.* 1. *p.* 365. N° 17.
Le Foucon a Collier. *Do.* 1. *p.* 345. N° 7. *male.*
L'Oiſſeau St. Martin. *Buff. oiſ.* 1. *p.* 212.—*Pl. enl.* 459.
Grau-weiſſe Geyer. *Friſch.* 1. 79. 80.
Rubetarius. *Turneri.*

Many authors have ſuppoſed the Ring-tail to be the female of the Hen-harrier, but Mr. Pennant does not ſubſcribe to this general opinion; he obſerves of the Ring-tail, " from ſome late obſervations by the infallible rule of diſſection, males have been found of this ſpecies." And Mr. Latham, after noticing the opinion of Pennant, ſays, " To this I may add my own obſervations; the Bird I now poſſeſs, as an *Engliſh* ſpecimen, being ſet down in my notes as a *male.*"

The Hen-harrier is very deſtructive to the young poultry, ſkims the ground when it flies, and does not perch on trees. Length ſeventeen inches, breadth three feet three inches, weight twelve ounces.

PLATE

60

PLATE LX.

MOTACILLA RUBETRA.

WHINCHAT.

PASSERES.

Bill conic, pointed. Noftrils oval, broad, naked.

GENERIC CHARACTER.

Bill ftrait, flender. Tongue jagged.

SPECIFIC CHARACTER

AND

SYNONYMS.

Above reddifh brown, with dark fpots. Beneath reddifh yellow, a white ftroke over the eye, and a broad one below it. Two white fpots on the wings. Upper half of the tail white, lower half black. Bill, mouth and legs black.

MOTACILLA RUBETRA. *Lin. Syft. Nat.* 1. 186. 18. *edit.* 10.
Scop. Ann. 1. N° 237.
Kram. el. p. 375. N° 5.
Whinchat. *Raii Syn. p.* 76. *A.* 3.
Will. Orn. p. 237.
Penn. Brit. Zool. 1. 158.
Lath. Gen. Syn. 4. 454. 54.

Le

PLATE LX.

Le grand Traquet, ou le Tarier. *Brif. orn.* 3. *p.* 432. N° 26.
pl. 24. *f.* 1.

     Buff. oif. 5. *p.* 224.—*Pl. enl.* 678. *f.* 2.

Geftettenfehlagar. *Kran.* 375.

Groffer Fliegenfuenger. *Frifch. f.* 22.

This is a common Bird in moft parts of Europe. In England it is feen in the North only in the fummer; but in the South it continues the whole year. It is frequently feen on the heaths with the Stone-chatter; but is not fo common as that Bird. It builds its neft among the furze: its food is chiefly infects.

The colours of the female are not fo beautiful as in the male. The white on the wing is lefs confpicuous, the breaft is of a plain colour, and inftead of the white and black ftreaks on the cheeks, it has only one broad ftreak of dull brown. Length about fix inches.

PLATE

61

PLATE LXI.

TURDUS TORQUATUS.

RING-OUZEL.

PASSERES.

Bill conic, pointed.   Noſtrils oval, broad, naked.

GENERIC CHARACTER.

Bill rather ſtraight : upper mandible notched, and bent near the apex : noſtrils naked, and half covered by a membrane : mouth ciliated with a few briſtles at the corners : tongue jagged.

SPECIFIC CHARACTER

AND

SYNONYMS.

Black-brown edges of the feathers white : a white creſcent on the breaſt.

TURDUS TORQUATUS : nigricans, torque albo, roſtro flaveſcente.
    Lin. Syſt. 1. *n.* 170. 13 *edit.* 10. *Fn. Suec.* 185.

Merula Torquata. *Geſn. av.* 607.

Merula congener, *Raii Syn. p.* 67. N° 12.

Ring-ouzel, or Amfel, *Raii Syn.* p. 65. A. 2.
    Albin. vol. 1. *pl.* 39.
    Pen. Br. Zool. 1. N° 110. *pl.* 46.
    —— *Arĉt. Zool.*
    Lath. Gen. Syn. 3. *p.* 46. 49.

Rock, or Mountain-ouzel. *Will. Orn. p.* 124.

Le Merle ou Collier. *Belon. av.* 318.

Le Merle à plaſtron bl. *Buff. oiſ. p.* 340. *pl.* 31.—*Pl. enl.* 516.

Le Merle a Collier, *Briſ. orn.* 2. 235.

     G   Ring-

PLATE LXI.

Ring lamfel. *Kram.* 360.

Ringel-Amfel. *Frifch.* 1. 30.

Mwyalchen y graig. *Camden. Brit.* 795.

The length of this bird is eleven inches; the breadth feventeen. The crefcent of white on the breaft is more obfcure in the female than in the male; in the former it is fometimes wholly wanting, and hence fome writers on birds have made the male and female two diftinct fpecies *. Mr. Latham mentions feveral varieties, one quite white; a fecond fpotted with white; and a third, which is bigger than the common one, fpotted with white, and without any crefcent on the breaft.

This appears to be a migratory fpecies in moft parts of Europe. It is met with in *Burgundy* in France, about the beginning of October, but ftays there only two or three weeks; it returns there again in April or May for a fhort time only; it is found as high as Lapmark, but not in *Ruffia* or *Siberia* †; it is alfo found in *Africa* and *Afia* according to Adanfon, and other authors.

It breeds in the North of Wales, in Cumberland and Scotland; but is very rarely feen in the Southern parts of this country, except during its migrations, at which time it flies in fmall flocks of five or fix. Mr. Pennant fays they are known to breed in Dartmoor, in Devonfhire, in banks on the fides of ftreams, and that they are very clamorous when difturbed; he further adds, thofe that breed in *Wales* and Scotland never quit thofe countries; in the laft they breed in the hills, but defcend to the lower parts to feed on the berries of the Mountain Afh.

---

* *Ring-Ouzel* and *Rock-Ouzel.* † *Pennant.*

PLATE

62

PLATE LXII.

HÆMATOPUS OSTRALEGUS.

SEA PIE, or OYSTER-CATCHER.

GRALLÆ.

Bill roundifh : tongue entire, flefhy : thighs naked : toes divided.

GENERIC CHARACTER.

Bill compreffed, obtufe, the tip equal cuneate: noftrils linear : tongue one third length of the bill : feet formed for running, toes three and cleft.

SPECIFIC CHARACTER

AND

SYNONYMS.

Bill orange : irides crimfon : head, neck, fhoulders, black: wings and tail black and white : beneath white : legs red.

HÆMATOPUS OSTRALEGUS. *Lin. Syft.* 1. 152. 81. *edit.* 10.
SEA PIE. *Raii Syn. p.* 105. 7.
         Will Orn. p. 297.
PIED OYSTER-CATCHER. *Br. Zool. p.* 213.
         Catefby Car. 1. *pl.* 85.
         Arct. Zool. N° 406.
Marfpitt, ftrandfkjura, *Faun. Suec. fp.* 192.
Tirma, Trilichan, *Martin's voyage, St. Kilda* 35.
Scolopax Pica. *Scop. Ann.* 1. *N°* 135.

L'Hutrier.

PLATE LXII.

L'Hutrier. *Brif. orn. 5. p. 38. pl. 3. fig. 2.*
Buff. oif. 8. p. 119. pl. 9.
——— *pl. enl. 929.*

The Oyfter-catcher is very common on fome of our fhores. It feeds on marine infects and fhell-fifh; chiefly on oyfters and limpets. When it finds an oyfter that gapes wide enough for the infertion of it's bill, it thrufts it in and eats the fifh; it's bill is well conftructed for this purpofe, it is flattened on the fides, for more than half it's length, and by forcing it into the fhell fideways, it anfwers the fame purpofe as a knife for opening it. In the winter thefe birds are feen in confiderable flocks, in fummer only in pairs; at this time they live in the neighbourhood of the fea and falt rivers. The female lays four or five eggs on the bare ground, above high water mark; they are of a whitifh brown hue, thinly fpotted and ftriped with black, according to Pennant; Latham fays they are of a greenifh grey blotched with black.

It is mentioned by many authors and navigators as an inhabitant of very diftant countries. It is common from *New York* to the *Bahama Iflands* *; found alfo in *New Holland* †, *New Zealand* ‡, *Japan* §, &c.

---

* *Arch. Zool. Catefby Car.* 1. 85.   † *Dampier voy.* 3. *pl. in* 123.
‡ *Hawkfworth voy.* 2. p. 333.   § *Kæmpfer, Jap.* 1. *p.* 113.

PLATE

63

PLATE LXIII.

FALCO TINNUNCULUS, *fem.*

FEMALE KESTRIL HAWK.

ACCIPITRES.

Birds of prey. Bill and claws ftrong, hooked. An angle in each margin of the upper mandible. Body mufcular. Females larger than the Males.

GENERIC CHARACTER.

Bill arched from the bafe, which is covered with a wax-like membrane, or cere.

SPECIFIC CHARACTER.

Cere and Feet yellow. Male : Head and Tail grey ; on the laft a black bar near the end ; tip white. Back bright brick colour, with black fpots. *Female :* Head pale red, back paler, with tranf-verfe ftripes of black. Tail pale brown, with feveral tranfverfe bars.

FALCO TINNUNCULUS. *Lin. Syft. p.* 127. *N°* 17. *edit.* 12.
Keftrel, Stannel, or Windhover. *Will. Orn. p.* 84, *t.* 5.

Albin. 1. *pl.* 7.

Keftrel. *Pen. Br. Zool. N°* 60.
Keftril. *Lath. Gen. Syn.* 1. *p.* 94. 79.
La Crefferelle. *Brif. Orn.* 1. *p.* 39. *N°* 27.

Buff. Oif. 1. *p.* 280. *t.* 18.

————— *Pl. enl.* 401. 471.

H

The

PLATE LXIII.

The female of this fpecies differs fo much from the male, that though a figure of the laft has been given in a plate of this work, it cannot be amifs to give a figure of the female at this time. The female exceeds the male in length, but confiderably more in bulk, weighing eleven ounces, the male only fix ounces and a half. They are both very fierce, and beautiful birds, and were formerly ufed in falconry ; but as that amufement is now much difregarded, they are feldom trained for that purpofe.

PLATE

64

PLATE LXIV.

ALCA TORDA.

RAZOR-BILL.

ANSERES.

Bill obtufe, covered with a thin membrane, broad, gibbous below the bafe, fwelled at the apex. Tongue flefhy. Legs naked. Feet webbed, or finned.

GENERIC CHARACTER.

Bill ftrong, compreffed on the fides. Noftril linear. No back toe.

SPECIFIC CHARACTER

AND

SYNONYMS.

Bill black; on the upper mandible four grooves or furrows. Head, Throat, Back, Tail, and Legs black. Tips of the Wings and Belly white. A white line from the eyes to the bill.

ALCA TORDA, *Lin. Syft. Nat.* 1. 130. 63. 1. *f.* 1. *edit.* 10.
    Scop. Ann. N° 94.
    Brun. N° 100.
    Muller, N° 16.
Razor-bill, Auk, or Murre, *Raii Syn. p.* 119. *a.* 3.
    Will. Orn. p. 323. *pl.* 64.
    Albin. 3. *pl.* 95.
    Edwards, pl. 358. *fig.* 2.

H 2

Penn.

PLATE LXIV.

Penn. Br. Zool. 2. N° 230. *pl.* 82.

—— *Arct. Zool.* N° 425.

Lath. Gen. Syn. 5. 319. 5.

—— *Suppl.* 265.

The Falk. *Martin's Voyage, St. Kilda* 33.

The Marrot. *Sib. Hist. Fife.* 48.

Le Pingoin. *Brif. Orn.* 6. *p.* 89. 2. *pl.* 8. *fig.* 1.

Buff. Oif. 9. *p.* 390. *pl.* 27.—*Pl. enl.* 1003, 1004.

Tord. Tordmule. *Faun. Suec. fp.* 139. *Nowegis* Klub-Alke, Klympe. *Danis,* Alke, *Brunnich.*

Length of this fpecies eighteen inches; weight twenty-two ounces.

Thefe birds breed in the ledges and cliffs of the moft ftupendous and craggy rocks on our coafts. They appear in the Britifh feas early in February, but do not inhabit their breeding places till May. The female lays only one egg; but that is of an extraordinary fize compared with the bird, being three inches long : it is of a pale fea green, irregularly fpotted with black ; fometimes the ground colour is white. They build no neft, but lay the egg on the bare rock, fo clofe to the verge of the precipice, that if it is the leaft difturbed, human ability can rarely place it on its former equilibrium. If the firft egg is deftroyed, or taken away, it lays a fecond, and fometimes a third, if the fecond is miffing.

Thefe eggs are fo eagerly fought after by the inhabitants of the fea coaft ; that they often brave the greateft dangers to find them ; and not unfrequently facrifice their lives in the attempt. The ufual method of taking them is for two perfons, having a rope tied round

I

the

PLATE LXIV.

the middle of each, to ftand clofe to the edge of the precipice, and one to lower the other down gradually, the perfon above holding the rope as faft as poffible whilft the other collects the eggs. It however often happens, in this perilous fituation, that the weight of the loweft overbalances the ftrength of his companion above, and both are forced down the precipice, where they muft inevitably be dafhed to pieces.

The Razor-bill is found very common in the north of Europe, in *Iceland*, *Greenland*, &c. They extend along the *White Sea* into the *Arctic Afiatic* fhores, and from thence to *Kamtfchatka* and the gulph of *Achotka* *.

Latham mentions, in his Supplement, the following curious particular of this fpecies. " The method this bird takes in fifhing is rather fingular, often diving and catching feveral fmall fifh, which it is obferved to range on each fide of the bill, with the head in the mouth, and the tails hanging out on each fide of the bill ; and when the mouth can hold no more, the bird retires to the rocks to fwallow them at leifure."

---

* Latham. Arct. Zool. &c.

H 3

PLATE

PLATE LXV.

MERGUS MERGANSER. (*fem.*)?

FEMALE GOOSANDER?

OR

DUN DIVER.

ANSERES.

Bill obtufe, covered with a thin membrane, broad, gibbous below the bafe, fwelled at the apex. Tongue flefhy. Legs naked. Feet webbed or finned.

GENERIC CHARACTER.

Bill long, roundifh, taper, ferrated and hooked at the apex. A creft on the head.

SPECIFIC CHARACTER

AND

SYNONYMS.

Female. Head and neck ferruginous. Chin and throat white. Back, wing coverts, fides of the body, tail, afh colour. Breaft and belly white.

MERGUS MERGANSER. *Faun. Suec. p.* 48. 8*vo.* 1761 ?
MERGUS CASTOR. *Lin. Syft.* 1. *p.* 209. 4.—*edit.* 12 ?
Anas rubricapilla. *Brun. N*° 93.
Mergus Gulo. *Scop. Ann.* 1. *N*° 88.

H 4

MERGUS

PLATE LXV.

Mergus Castor. Var. A. *Lath. Gen. Syn.* 6. 420. 2. ?
Dun-Diver, or Sparling fowl. *Raii Syn. p.* 134. *A.* 2.

> *Will. Orn. p.* 333. *pl.* 64. *(head)*
> *Albin.* 1. *pl.* 87.
> *Penn. Br. Zool.* 2. *p.* 557. *pl.* 92. *fig.* 2.
> ———— *Arct. Zool. N°* 465.
> *Lath. Gen. Syn. V.* 6. *p.* 420. 2.
> ———— *Suppl.* 270.

L'Harle cendré, ou le Bievre, *Brif. Orn.* 6. *p.* 254. *pl.* 25.
L'Harle femelle, *Brif. Orn.* 6. *p.* 236.

> *Buff. Oif.* 8. *p.* 272.
> ———— *Pl. enl.* 953.

After the account that has been given of this bird, in the defcription of the male, Mergus Merganfer, plate 49, we can add nothing material concerning it. In the Leverian Mufeum, the Dun-Diver and Goofander are placed together as male and female. Pennant has defcribed them as the two fexes of Mergus Merganfer ; and the authority of Linnæus may alfo be quoted to fanction this opinion. Yet Latham has endeavoured to prove, by the moft fatisfactory experiments, that they are diftinct fpecies, and that *Mergus Caftor* is only a variety of the Dun-Diver. Thus perplexed between fuch oppofite opinions, both of which are advanced by the moft refpectable naturalifts, we can fcarcely determine to which opinion we fhould incline ; but as the obfervation of Dr. Heyfham * moft evi-

---

* Vide Defcription of Mergus Merganfer, pl. 49.

dently

PLATE LXV.

dently tends to confirm the opinion of Mr. Latham, we will not hefitate to confider them as diftinct fpecies, and the Mergus Caftor only a variety. Our readers muft, notwithftanding, fee the neceffity of adopting the fynonyms as for the female of the Goofander.

Length twenty-three inches and a half.

PLATE

PLATE LXVI.

RECURVIROSTRA AVOCETTA.

SCOOPING AVOSETTA.

GRALLÆ.

Bill roundifh. Tongue entire flefhy. Thighs naked. Toes divided.

GENERIC CHARACTER.

Bill curved upwards, flexible at the point. Feet palmated: the webs deeply femilunated between each toe. Back toe very fmall.

SPECIFIC CHARACTER

AND

SYNONYMS.

White. Above marked with black. Legs blue and long.

RECURVIROSTRA AVOCETTA: albo nigroque varia. *Linn. Syft.*
        Nat. 1. 151. *8vo. edit.* 10.
        Amœn. Acad. 4. 591.
        Scop. Ann. 1. N° 129.
        Brun. N° 188.
        Mulleo. N° 214.
        Kram. el. p. 348.
Avofet. *Raii Syn. p.* 117. A. 1.
        Albin. 1. *pl.* 101.
        Penn. Br. Zool. 2. 228. *pl.* 80.
        —— *Arct. Zool. p.* 503. B.
        Lath. Gen. Syn. 5. 293. 1.
        —— *Suppl.* 263.

The

PLATE LXVI.

The Scooper. *Charl. ex.* 102.

The Crooked Bill. *Dale's Hiſt. Harwich.* 402.

L'Avocette. *Briſ. Orn.* 6. *p.* 538. *pl.* 47. *fig.* 2.

Buff. Oiſ. 8. *p.* 466. *pl.* 38.

—— *Pl. enl.* 353.

Avoſetta, Beccoſtorto, Becoroella, Spinzago d'acqua. *Aldr. av.* 3. 114.

Krumbſchnabl. *Kram.* 348.

Skerſlacka, Alſit. *Faun. Suec. ſp.* 191.

Danis. Klyde, Loufugl. Forkeert Regorſpove. *Br.* 188.

———

The length of this ſpecies is eighteen inches[*]. The body is ſmall, but the legs are remarkably long. The male differs very little from the female; and in the eſſential characteriſtic, the bill, they perfectly agree: this part, which is about three inches and a half long, is of a ſubſtance like whale-bone; it is very ſlender, and compreſſed, is flexible, and, unlike the bills of other birds, turns up towards the end, and tapers to a point.

This bird is common in winter on the eaſtern coaſts of this kingdom, particularly on thoſe of *Suffolk* and *Norfolk*; and ſometimes on the lakes of *Shropſhire*. They are found in great plenty, in the breeding ſeaſon, in the fens in *Foſsdike Waſh* in *Lincolnſhire*, and in the fens of *Cambridgeſhire* and *Suffolk*. They feed on worms and inſects, which they ſcoop out of the mud and ſand; and are ſometimes obſerved to wade or ſwim, but always cloſe to the ſhores.

———

[*] Sometimes twenty inches, or rather more.

They

PLATE LXVI.

They lay two eggs, about the fize of thofe of a pigeon. *Pennant* fays, they are white, tinged with green, and marked with large black fpots. In the defcription given by *Latham*, he obferves, they are of a cinereous grey, whimfically marked with deep brownifh black patches, of irregular fizes and fhapes, befides fome under markings of a dufky hue.

The Avofet is far more frequent in fome parts of Europe than in England. Albin fays, in *Rome* and *Venice* they are common. Salerne writes, in the breeding time they are fo plenty on the coafts of *Bas Poiĉou*, that the peafants take their eggs by thoufands. They are found alfo in *Ruffia* and *Siberia, Denmark, Sweden,* and other northern countries.

PLATE

67

PLATE LXVII.

CAPRIMULGUS EUROPÆUS.

EUROPEAN GOAT SUCKER.

PASSERES.

Bill conic, pointed. Noſtrils oval, broad, naked.

GENERIC CHARACTER.

Bill ſhort, bent at the end, briſtles round the baſe. Mouth very wide. Tail of ten feathers, not forked.

SPECIFIC CHARACTER

AND

SYNONYMS.

Plumage dark brown, black, white, aſh colour intermixed, and diſpoſed in ſpots, ſpecklings, &c.

CAPRIMULGUS EUROPÆUS. *Lin. Syſt. Nat.* 1. *p.* 134. *edit.* 12.
   Scop. Ann. 1. *N°* 254.
   Muller, p. 34. *N°* 291.
   Kram. el. p. 281.
   Georgi Reiſe. p. 174.
   Friſch. t. 101.

           Brun.

PLATE LXVII.

Brun. N° 293.

Faun. Arag. p. 91.

Sepp. Voy. pl. in p. 39.

Hirundo cauda æquabili. H. caprimulga.   *Klein. av.* 81.

Dorhawk accipiter Cantharo phagus.

Dorhawk, Night Jar, or Night Hawk.   *Charlton. ex.* 71. *N°* 8.

Caprimulgus, Fern-Owl, Churn-Owl, Goat-Sucker, or Goat-Owl.

Raii. Syn. p. 26.

Will. Orn. p. 107.

Albin. p. 10.

Borlafe's Hift. Corn. pl. 24. *f.* 13. 1758.

NOCTURNAL GOAT-SUCKER.   *Penn. Br. Zool. N°* 172. *pl.* 59.

—— *Arct. Zool.* 2. *p.* 437. *A.*

EUROPEAN GOAT-SUCKER.   *Lath. Gen. Syn.* 4. *p.* 593. *N°* 5.

—— *Suppl.* 194. 5.

Tette-chevre, ou Crapaud volant.   *Brif. Orn.* 2. *p.* 470. *N°* 1.

pl. 44.

L'Engoulevent.   *Buff. Oif.* 6. *p.* 512.

L'Effraye ou Frefaye.   *Belon. av.* 343.

Caprimulgus, Geiffmelcher.   *Gefn. av.* 241.

Calcobotto.   *Aldr. av.* 1. 288.

Covaterra.   Zinanni. 94.

Natfkrafa, Natfkarra, Quallknarren.   *Faun. Suec. fp.* 274.

Nat-Ravn, Nat-Skade, Aften-bakke.   *Brun.* 293.

Mucken ftecker.   Nach trabb.   *Kram.* 381.

It

PLATE LXVII.

It is difficult to defcribe the diverfified plumage of this beautiful bird. The colours are, throughout, of the plaineft kinds ; but they are fo exquifitely foftened, neatly fpeckled, and elegantly interfperfed and varied with ftreaks and waves of black, that no defcription can convey a juft idea of its beautiful appearance.

It has many characters of the Swallow tribe. Klein has placed it in that genus, and diftinguifhes it by its undivided tail from the other fpecies ; and Pennant fays, it may with juftice be called the *Nocturnal Swallow*, as it differs from the Swallows chiefly in the time of its flight, the latter being on the wing in the day, and the Goat-Sucker only in the evening. It agrees in feveral refpects alfo with the Owl tribe. Its manners are much the fame in moft countries in Europe : it retires into fome dark recefs in forefts, woods, or among rocks, and never ventures out in the day time but in very gloomy weather, or when difturbed. As it can fee beft in the twilight, it comes out in the dufk of the evening and morning, and collects its food ; this it does chiefly on the wing when it finds abundance of moths and other infects ftirring. In the month of July, it is faid to live entirely on the dorr beetle, or cock-chaffer * ; and from this circumftance Charlton has called it the Dorr-Hawk.

The notes of this bird are of two kinds : " the loudeft," fays Pennant, " fo much refembles that of a large fpinning-wheel, that the Welch call this bird *aderyny droell*, or the *Wheel Bird*." And he farther adds, " it begins its fong moft punctually on the clofe of

---

* Scarabæus Melontha,

I

day,

day, fitting ufually on a bare bough, with its head lower than the tail, the lower jaw quivering with the efforts. The noife is fo very violent, as to give a fenfible vibration to any little building it chances to alight on, and emit this fpecies of note. The other is a fharp fqueak, which it repeats often : this feems to be a note of love, as it is obferved to reiterate it when in purfuit of the female among the trees."

The male is diftinguifhed from the female by a large oval white fpot, fituated on the inner web of the firft three quill feathers, and another at the ends of the two exterior feathers of the tail.

The bill is alike in both male and female : it is fhort, but the gape is remarkably wide. It is, probably, from the ftructure of the mouth that the ancients fuppofed this bird fucked the teats of goats. In the days of Ariftotle, this ridiculous notion was generally preva-lent ; but among modern naturalifts, none except *Scopoli* feems in-clined to credit fuch an opinion.

The female makes no neft, but lays her eggs on the bare ground. They are ufually two in number, of a whitifh hue, and marbled with brown.

This is a very confined genus. Latham enumerates, including his fupplementary volume, but feventeen fpecies, and of thefe we find only our prefent fubject, mentioned as a native of Europe. It appears to be an inhabitant of every country on the continent, but is very fparingly diffufed in fome parts, and no where common : it is alfo faid to inhabit *Africa* and *Afia*. *Sonnerat* met with one on the coaft of Coromandel. With us it is a bird of paffage, and arrives about the latter end of May. It entirely difappears in the northern

parts

PLATE LXVII.

parts of the kingdom in Auguſt, but does not quit the ſouthern parts till September.

The ſize of this ſpecies is ten inches and a half, breadth twenty two inches and a half, weight two ounces and three quarters.

P L A T E

68

PLATE LXVIII.

PODICEPS CRISTATUS.

CRESTED GREBE.

ANSERES.

Bill obtufe, covered with a thin membrane, broad, gibbous below the bafe, fwelled at the apex. Tongue flefhy. Legs naked. Feet webbed, or finned.

GENERIC CHARACTER.

Bill ftrong, flender, fharp pointed. Noftril linear. Lore bafe of feathers. Body depreffed. Feathers thick, fmooth and gloffy. No Tail. Wings fhort. Legs placed far behind, much compreffed, and doubly at the back part. Toes furnifhed on each fide with a broad plain membrane *.

SPECIFIC CHARACTER

AND

SYNONYMS.

Above black brown. Beneath filvery white. Feathers of the Head long, and forming two ears above, and a ruff below.

---

* *Latham.*

I   COLYMBUS

PLATE LXVIII.

COLYMBUS CRISTATUS. Pedibus lobato-fiffis, capite rufo, collari nigro, remigibus fecundariis albis. *Fn. Sv.*— *Lin. Syft. Nat.* **1**. 135. 68. 2. *edit.* 10.

Colymbus major criftatus & cornutus. *Raii Syn. p.* 124. A.

Avis pugnax. *Aldr.* 169.

PODICEPS CRISTATUS. *Lath. Gen. Syn. vol.* 5. 281. 1.

GREAT CRESTED GREBE. *Penn. Br. Zool. N°* 223.

—— *Arct. Zool. p.* 498. A.

CRESTED GREBE. *Latham.*

Greater crefted and horned Ducker. *Will. orn. p.* 340. § 4. 5. *pl.* 61.

Plott. Hift. Staff. p. 229. *pl.* 22.

Albin. 1. *pl.* 81. 2. *pl.* 75.

Afh-coloured Loon. *Dr. Brown. Raii fyn. av.* 124.

The Cargoofe. *Charleton ex.* 107.

La Grebe huppée. *Brif. Orn.* 6. *p.* 38. 2. *pl.* 4.

Buff. Oif. 8. *p.* 233.

—— *Pl. enl.* 944.

La Grebe cornue. *Brif. Orn.* 6. *p.* 45. 4. *pl.* 5. *fig.* 1.

Buff. Oif. 8. *p.* 233. *pl.* 19.

—— *Pl. enl.* 400.

Grand Plongeon de riviere. *Belon. av.* 178.

Ducchel. *Gefner. av.* 138.

Smergo, Fifolo marino. *Zinan.* 107.

Danis Topped og Halfkraved Dykker,

Topped Haw Skiœre. *Brunnich.* 135.

Gehoernter Seehahn, Noerike. *Frifch.* 2. 183.

———————————

This is the largeft bird of the genus that inhabits Great Britain; its length is twenty-three inches, weight two pounds and a half.

The

PLATE LXVIII.

The extraordinary length of the neck, and remarkable ſtructure of the feet, give it a moſt aukward, yet ſingular appearance. The breaſt and belly is of a very beautiful ſilvery white, intermixed with ſhades of pale ferruginous colour, and has a gloſs like ſattin ; theſe ſkins are in much requeſt, and like thoſe of the Grebe of Geneva are made into muffs, tippets, &c. In February the ſkins loſe the bright colour, and in the breeding time the breaſt is almoſt bare.

Mr. Latham gives a minute account of the plumage of this ſpecies in ſeveral ſtages of life ; he ſays, " At firſt they are perfectly downy and ſtriped, eſpecially down the neck, with black : after this, when about half grown, the ſtripes on the neck are leſs diſtinct, being rather mottled than ſtriped, and the under parts, though white, is clouded with duſky ; at this period a fullneſs round the head is obſerved : as the bird advances ſtill further towards perfection, the brown and white appears clear and diſtinct, the head becomes much tufted, and the horns are a little elongated. But we have great reaſon to believe that the bird does not obtain the full and perfect creſt till the ſecond year at leaſt."

This ſpecies is common in ſome parts of this country. They breed in the meres of *Shropſhire* and *Cheſhire,* and in the *eaſt* fen of *Lincolnſhire* *. The female lays four white eggs, the ſize of thoſe of a pigeon ; the neſt in which they are depoſited, like others of the Grebe kind, are compoſed of different kinds of water plants, ſuch as the ſtalks of the *Water Lily, Pond Weed,* &c. careleſsly put together, and left floating on the water, among the flags and ruſhes. The old bird feeds on ſmall fiſh ; when the young brood is hatched,

---

* They are called by the country people of this part *Gaunts.*

they

PLATE LXVIII.

they are fed upon fmall eels.   This bird is rarely feen on the land, and though common, it is very difficult to be fhot, as it darts into the water on the leaft appearance of danger, and feldom flies farther than the end of the lake it frequents *.

---

* Pennant.

PLATE

69

PLATE LXIX.

PARUS MAJOR.

GREAT TITMOUSE.

PASSERES.

Bill conic, pointed. Noftrils oval, broad, naked.

GENERIC CHARACTER.

Bill fhort, ftrait, ftrong, fharp pointed. Noftrils covered with briftles. Tongue blunt; briftles at the end.

SPECIFIC CHARACTER

AND

SYNONYMS.

Bill, Head, Throat, black; Cheeks white. Above green. Beneath yellow with an irregular black line down the middle. Greater coverts of the wing tipped with white. Exterior fide of the outer feathers of the Tail white.

PARUS MAJOR: capite nigro, temporibus albis, nucha lutea.
> *Fn. Sv.—Lin. Syft. Nat.* 1. 189. 100. 2. *edit.* 12.

Parus Major, feu Fringillago. *Raii Syn. p.* 73. A. 1.
> *Gefn. av.* 640.

GREAT TITMOUSE, or OX EYE. *Will. Orn. p.* 240. *pl.* 43.
> *Albin.* 1. *pl.* 46.

I 3

GREAT

PLATE LXIX.

GREAT TITMOUSE. *Penn. Br. Zool.* 1. N° 162. *pl.* 57. *f.* 1.

———— *Arct. Zool.*

Lath. Gen. Syn. 4. *p.* 536. 1.

La Groffe Mefange, ou la Charbonniere, *Brif. Orn.* 3. *p.* 539. N° 1.

Buff. Oif. 5. *p.* 392. *pl.* 17.

———— *Pl. enl.* 3. *f.* 1.

Nonette ou Mefange. *Belon. av.* 376.

Spernuzzola, Paraffola. *Olina,* 28.

Snitza. *Scopoli,* N° 242.

Talg-oxe. *Faun. Suec.*

Mufvit. *Brunich,* 287.

Kohlmeife. *Kramer,* 378.

Frifch, 1. 13.

———————

This is a very common fpecies in this country, as well as in many other parts of Europe: it is alfo faid to inhabit the moft remote parts of Africa.

It frequents gardens, and does much injury to fruit trees in the fpring, by tearing off the young fhoots; but, it is alfo very beneficial, in deftroying the infects that infeft thofe trees. In confinement it prefers hemp feed to all others.

In its manner it very much refembles a Wood-pecker, it is continually running up and down the bodies of trees in queft of infects, and fuch as are concealed under the bark, it difcovers by founding with its bill. It is very alert and almoft always feen hanging by its legs, or running in a fufpended pofture. It is alfo very courageous, and will attack birds that are far more powerful than itfelf.

It

PLATE LXIX.

In the breeding-time it lives chiefly in the woods, and builds in hollow trees. The female lays from eight to twelve eggs; they are white, fpotted with ruft colour.

The length of this bird is five inches and three quarters, weight almoft an ounce.

PLATE

70

PLATE LXX.

FALCO HALIÆTUS.

OSPREY.

ACCIPITRES.

Birds of prey. Bill and claws ftrong, hooked. An angle in each margin of the upper mandible. Body mufcular. Females larger and more beautiful than the males.

GENERIC CHARACTER.

Bill arched from the bafe, which is covered with a wax-like membrane or cere.

SPECIFIC CHARACTER

AND

SYNONYMS.

Cere and Feet blue. Back brown. Head whitifh.

FALCO HALIÆTUS: cera pedibufque cæruleis, corpore fupra fufco fubtus albo, capite albido.—*Fn. Suec. Lin. Syft. Nat.* 1. *p.* 91. 21. *edit.* 10.

Haliætus, feu aquila marina. *Gefner av.* 804.

Falco cyanopus. *Klein Stem. Tab.* 8.

Auguifta piumbina, Aquilaftro, Haliætus, feu Morphnos. *Aldr. av. I.* 105. 114.

Falco Haliætus. *Georgi Reife. p.* 164.

Kolben Cape of Good Hope. 2. *p.* 137.?

OSPREY.

PLATE LXX.

Osprey. *Pen. Br. Zool.* 1. 174. 46.
     Lath. Gen. Syn. 1. 45. 26.

Bald Buzzard, or Sea Eagle. *Raii Syn. av.* 16.

Bald Buzzard. *Will. Orn. p.* 69. *t.* 6.

Fifhing Hawk. *Catefby's Carol.* 1. *Tab.* 2.

L'Aigle de Mer. *Brif. Orn.* 1. *p.* 440. *t.* 34. N° 10.

Le Balbuzard. *Buff. Oif.* 1. *p.* 103. *t.* 2.—*Pl. enl.* 414.

Une Orfraye. *Belon. av.* 96.

Balbufhardus. *Turneri.*

Blafot, Fifk-orn. *Faun. Suec.*—*Brunnich, p.* 5.

The length of this Bird is twenty-three inches; breadth five feet four inches. It is a very powerful creature, and is armed with long, hooked claws, and a remarkably ftrong bill; in the ftruǎure of it's feet it differs from all other birds of prey; the outer toe turns backwards, and the claw belonging to it is larger than that of the inner toe.

It frequents the fea-fhores, rivers and lakes. Some authors fay it feeds on water-fowl, but its chief food is fifh; and thefe it does not take by fwimming, but hovering in the air, with its eye direǎed into the water, it foon difcovers it's prey, when precipitating like lightning upon it, it brings the fifh up in its talons, and retires to a diftance to devour it. The Italians call it the *Leaden Eagle*, becaufe it defcends with fo much violence on its prey: and Latham has adopted with an (?) among his fynonyms the account of *Kolben*, of a bird he fuppofes the Ofprey. *Kolben* obferves, " That it is of all birds the moft deftruǎive to the *Flying-fifh*, taking them during their rife from the water."

Pennant

PLATE LXX.

Pennant fays it builds it's neft on the ground among reeds, and lays three or four white eggs of an elliptical form ; rather lefs than thofe of a hen.

Ancient writers have afferted that the left foot is *fubpalmated* ; many refpectable Naturalifts of late years have followed this opinion, and indeed the *authority* of *Linneus* had almoft finally fanctioned this error. *Pennant* contradicts this opinion, and gives a faithful defcription of the bird, and *Latham* has added a very curious and interefting note to his account of it *, which is certainly an apology in fome meafure for the miftake in the firft inftance ; fince a bird perfectly according with the defcription given by early writers may have occurred. We cannot fuppofe, without fome good authority in ancient writers, Linneus would have adopted their errors.

---

* " I do not believe," fays Mr. Latham, " that there is either bird or quadruped, in which each fide of the body does not correfpond in fize and fhape, in a natural ftate, though the contrary is fometimes feen in the infect tribe." He further adds, " In refpect to winged infects, it is obferved that even the marks of the wings exactly correfpond on each fide. Indeed a fingular circumftance occurs in one of the BLATTA or *Cockroach genus* (*Blatta theteroclita* *) which I believe is the only one recorded, at leaft obferved by me. In this fpecies, one of the elytra, or wing-cafes, is marked with four white fpots, and the other with three only ; which holds good in every fpecimen of it I have yet feen.

" As to *Lufus Naturæ*, they are far from being uncommon ; fuch as a duck without webs to the toes, which I have often feen ; a common fnail with the fpiral turns of the fhell reverfed, one of which was found in my garden a few years fince ; alfo a flounder having the eyes and lateral line on the left fide inftead of the right †, &c.

" Thefe, and an hundred fuch which might be mentioned, muft be reckoned as fingularities happening now and then, but by no means to be fet down for permanent diftinctions of fpecies."

---

* Linneus defcribes this Infect *Caffida petiveriana*,—*Caffida* 7 *guttata* nigra, coleoptris maculis feptem albis. *Syft. Ent.* 90. 26.—*Linn. Syft. Nat.* 2. 577. 19.——Fabricius, *Blatta petiveriana* nigra, elytris maculis quatuor flavefcentibus. *Syft. Ent.* 272. 11.—*Spec. Inf.* 1. 343. 13.

† Br. Zool. vol. 8. p. 293.

PLATE

71

PLATE LXXI.

ANAS TARDONA.

SHIELDRAKE.

ANSERES.

Bill obtufe, covered with a thin membrane, broad, gibbous below the bafe, fwelled at the apex. Tongue flefhy. Legs naked. Feet webbed or finned.

GENERIC CHARACTER.

Bill convex above, flat beneath, hooked at the apex, with mem branous teeth.

SPECIFIC CHARACTER

AND

SYNONYMS.

ANAS TARDONA roftro fimo, frente compreffa, corpore albo variegato.——Anas albo-variegata, pe&oris lateribus
ferrugineis abdomine longitudinaliter cinereo-
maculato.—*Fn. Sv.*—*Lin. Syft. Nat. vol.* I.
p. 122. *g.* 61. *fp.* 3. *edit.* 10.

Anas maritima. *Gefner av.* 803. 804.

SHIELDRAKE or Burrough Duck. *Raii Syn. p.* 140. A. I.
Will. Orn. p. 363. *pl.* 70. 71.
Albin. I. *pl.* 94.
Penn. Br. Zool. 2. N° 278.—*Ar&. Zool. p.* 572. D.
Lath. Gen. Syn. 6. *p.* 504. 51.—*Suppl.*

La

PLATE LXXI.

La Tardone. *Brif. Orn.* 6. *p.* 344. 9. *pl.* 33. *fig.* 2.
      Buff. Oif. 9. *p.* 205. *pl.* 14.—*Pl. enl.* 53.
      Belon. av. 172.

Vulpanfer Tardone. *Aldr. av.* 3. 71. 97.

Jugas. *Faun. Suec.*

Bergander. *Turneri.*

Danis, Brand-Gaas, Grav-Gaas. *Norvegis,* Ring-Gaas, Fager-
    Gaas, Ur-Gaas, Bodbelte. *Feroenfibus,* Hav-
    Simmer. *Iflandis,* Avekong.—*Br.* 47. *Pennant.*

Kracht-Ente. *Frifch.* 2. 166.

This is a very elegant Bird : length two feet, weight two pounds two ounces. The female differs very little from the male, except that her colours are not fo bright. It is found in vaft quantities on feveral of our fea-coafts, and particularly about the rivers and lakes in *Lancafhire* and *Effex,* where it finds abundance of fmall fifh, marine infects, &c. It breeds in holes that it digs in the earth, or in the deferted burrows of rabbits. The female lays from twelve to fixteen eggs, of a roundifh form and white colour. Thefe are depofited at the fartheft end of the hole, and are carefully covered with fine down, which the female fupplies from her breaft. The old Bird is very careful of her little brood, and ufes many cunning ftratagems to draw the attention of any difturber from her young : it is even faid that if fhe cannot favor their efcape from danger by that means, fhe will carry them away in her bill, or on her back. The time of fitting on the eggs is about thirty days.

Some have attempted to domefticate them by bringing them up under the common Duck ; but they do not thrive fo well as when

3

th_y

PLATE LXXI.

they are wild, nor will any such attempt be likely to succeed, unless it be made in the neighbourhood of the sea. The flesh is very rank, but their eggs are much esteemed.

This species remains with us all the year: in winter it collects in large flocks; leaves the Orknies in that season, and returns there again in the spring. It is also found very far to the North: in Asia about the *Caspian Sea**, and the salt lakes of the *Tartarian* and *Siberian Desarts* †, *Kamschatka* ‡ and the *Falkland Isles* ‖.

---

* *Latham.*   † *Decou. Russ.* 1. *p.* 472.   ‡ *Pen. Arct. Zool.*   ‖ *Penrose, p.* 34.

PLATE

72

PLATE LXXII.

SCOLOPAX PHÆOPUS.

WHIMBREL.

GRALLÆ.

Bill roundifh. Tongue entire, flefhy. Thighs naked. Toes divided.

GENERIC CHARACTER.

Bill long, flender, incurvated. Toes connected as far as the firſt joint by a ſtrong membrane.

SPECIFIC CHARACTER

AND

SYNONYMS.

Bill fhort. Above brown fpotted with black ; beneath whitifh. Tail croffed with black bars. Legs and feet bluifh or dull green.

SCOLOPAX PHÆOPUS roſtro arcuato, pedibus cærulefcentibus ma-
culis dorſalibus fufcis rhomboidalibus. *Linn,*
Syſt. Nat. 1. 146. 6. *edit.* 10. *Faun. Suec.*
Scop. Ann. 1. N° 132.
Kram. El. p. 350.
Georgi Reiſe. 171.

Phæopus altera, vel arquata minor. *Gefner av.* 499.
Numenius minor. *Briffon. Orn.* 5. *p.* 317. 2. *pl.* 27. *fig.* 1.
Numenius Phæopus. *Latham Suppl.*

K

WHIM-

PLATE LXXII.

WHIMBREL. *Raii Syn. p. 103. A. 2.*

Will. Orn. p. 294.

Edw. pl. 307.

Penn. Br. Zool. 2. Nº 177. p. 430.

Lath. Gen. Syn. 5. 123. 6.

Corlieu, ou petit Courly. *Briſ. Orn. 5. p. 317. 2. pl. 27. fig. 1.*

Buff. Oiſ. 8. p. 27.—Pl. enl. 842.

Tarango la Girardello. *Aldr. av. 3. 180.*

Windſpole, Spoſ. *Faun. Suec. p. 169.*

Kleiner Goiſſer. *Kram. 350.*

Kleine Art Brachvogel or Regenvogel. *Friſch. 2. 225.*

The length of this Bird is ſeventeen inches; breadth twenty-nine; weight fourteen ounces and a half. It is much leſs frequent in this country than the Curlew, to which it bears a great reſemblance in appearance, though not in ſize; the latter being uſually from twenty to twenty-five inches in length. In its manners it is alſo much like the Curlew.

It is met with in flocks from April to May on its paſſage to the North, where it is ſuppoſed they breed. *Mr. Pennant* ſays he received one from *Invercauld*, ſhot on the *Grampian Hills.* This ſpecies ſeems to vary in a great degree: the Bird deſcribed by *Mr. Latham* does not accord in ſeveral reſpects with that given by *Mr. Pennant.* A ſpecimen of this has been received from Sweden: *Mr. Latham* ſays it is alſo found in America.

INDEX

INDEX to VOL. III.

ARRANGEMENT

ACCORDING TO THE

SYSTEM of LINNÆUS.

ORDER I.

ACCIPITRES.

ORDER III.

ANSERES.

* Not defcribed by Linnæus.

L

ORDER

INDEX.

ORDER IV.

GRALLÆ.

ORDER VI.

PASSERES.

INDEX.

VOL. III.

ARRANGEMENT

ACCORDING TO

LATHAM's SYNOPSIS of BIRDS.

DIVISION I. LAND BIRDS.

ORDER I. RAPACIOUS.

GENUS II.

ORDER III. PASSERINE.

GENUS XXXI.

GENUS XXXV.

GENUS XLI.

L 2   GENUS

I N D E X.

D I V I S I O N II. WATER BIRDS.

O R D E R VII. WITH CLOVEN FEET.

O R D E R

INDEX.

VOL.

INDEX.

VOL. III.

ALPHABETICAL ARRANGEMENT.

THE

NATURAL HISTORY

OF

BRITISH BIRDS;

OR, A

SELECTION OF THE MOST RARE, BEAUTIFUL, AND INTERESTING

BIRDS

WHICH INHABIT THIS COUNTRY:

THE DESCRIPTIONS FROM THE

SYSTEMA NATURÆ

OF

LINNÆUS;

WITH

GENERAL OBSERVATIONS,

EITHER ORIGINAL, OR COLLECTED FROM THE LATEST
AND MOST ESTEEMED

ENGLISH ORNITHOLOGISTS;

AND EMBELLISHED WITH

FIGURES,

DRAWN, ENGRAVED, AND COLOURED FROM THE ORIGINAL SPECIMENS.

By E. DONOVAN.

VOL. IV.

LONDON:

PRINTED FOR THE AUTHOR; AND FOR F. AND C. RIVINGTON,
No. 62, ST. PAUL'S CHURCH-YARD. 1797.

73

PLATE LXXIII.

ARDEA MAJOR.

ARDEA CINEREA.

COMMON HERON.

Bill roundiſh. Tongue entire, fleſhy. Thighs naked. Toes divided.

GENERIC CHARACTER.

Bill long, ſtrong, pointed. Noſtrils linear. Tongue pointed. Toes connected as far as the firſt joint by a ſtrong membrane.

SPECIFIC CHARACTER

AND

SYNONYMS.

Head of the male creſted with long black feathers. Grey above. Breaſt white, marked with oblong black ſpots.

MALE.

ARDEA MAJOR. *Linn. Syſt. I. p.* 256. 12.
Scop. Ann. I. N° 117.
Kram. El. p. 346. *N°* 4.
Friſch. t. 199.

A 2

PLATE LXXIII.

Alia Ardea. *Gefner av.* 219.

Ardea Cinerea major feu pella. *Raii Syn. av.* 98.

Common Heron, or Heronfhaw. *Will. Orn.* 277.

Raii Syn. p. 98. A. 1.

The Heron, or Heronfhaw: Ardea cinerea major five pella.—
Albin. I. pl. 67.

COMMON HERON. *Penn. Br. Zool. N°* 173.

—— *Arct. Zool. N°* 343.

Lath. Gen. Syn. Vol. 5. *p.* 83. 50.

Heron cendrè. *Belon. av.* 182.

Le Heron hupé. *Brif. Orn.* 5. *p.* 396. 2. *pl.* 35.

Buff. Oif. 7. *p.* 342.

—— *Pl. Enl.* 755.

Garza cinerizia groffa. *Zinan.* 113.

Reyger. *Frifch.* 2. 199.

Blauer Rager. *Kram.* 346.

Hager. *Faun. Suec. fp.* 59.

FEMALE.

ARDEA CINEREA. *Linn. Syft. I. p.* 256. 11.

Scop. Ann. I. N° 117.

Kram. El. p. 346. *N°* 4.

Ardea pella five cinerea. *Gefn. av.* 211.

Ardea cinerea tertia. *Aldr. av.* 3. 159.

Common Heron. *Albin.* 3. *pl.* 78.

Pennant Br. Zool.

——— *Arct. Zool. N°* 343.

Le Heron. *Brif. Orn.* 5. *p.* 392. I. *pl.* 34.

Buff. Oif. 7. *p.* 342. *pl.* 19.

——— *pl. enl.* 787.

PLATE LXXIII.

Danis et *Norvegis* Heyre v. Hegre. *Cimbris* Skid-Heire, Skred heire. *Brunnich* 156.

The Heron is one of the moſt common birds that inhabit this kingdom: and there is ſcarcely any part of the globe that has been viſited by travellers, in which it has not been noticed. In Britain it was formerly held in high eſtimation, not only becauſe its fleſh was accounted a delicacy at the tables of the nobility*: but becauſe Heron Hawking was a favourite diverſion, inſomuch that laws were enacted for the preſervation of the ſpecies, and any perſon by deſtroying the eggs incurred a penalty of twenty ſhillings.

The plumage of the male bird is remarkable for its elegance ; perhaps we could with leſs propriety uſe the ſame expreſſion, if ſpeaking of its general appearance and proportions. Nature has not provided it with webbed feet to ſwim after its prey, which is almoſt wholly of the aquatic kind, but has furniſhed it with very long legs to wade after it, and theſe give it rather an aukward appearance when ſtanding on the land. The neck alſo is long and ſlender, but when it ſtands on the ſide of a ſtream or river waiting for the paſſing of a fiſh, its neck and head are drawn between the ſhoulders: in flying its neck is alſo crouched down, and the head almoſt concealed between the ſhoulders. The male is chiefly diſtinguiſhed from the female by having a fine creſt of black feathers; two in particular, in ſome ſpecimens, are eight

---

* It appears from a curious book, entitled *The Regulations of the Houſhold of the Fifth Earl of* NORTHUMBERLAND, *begun in* 1512, that Herons were valued at the ſame price as *Bytters* (Bitterns), *Feſſaunts* (Pheaſants), *Curlewes* (Curlews), and *Peacockes* (Peacocks).

" At PRINCIPAL FEASTS.—Item, it is thought in likewyze that HEARONSEWYS be bought for my *Lordes own mees* ; ſo that they be at xiid. a pece." The price of the Crane was 16d. at the ſame time, and the Goofe 3d. or at moſt 4d.—*Partridges* 2d. *Woodcocks* 1 or 1½.; and *Snipes* three for a penny.

inches

inches in length. Mr. Latham believes that this appendage is found only in males of a full age, or perhaps very old birds. Mr. Pennant fays that the long foft black feathers on the fides were ufed in old times as egrets for the hair, or ornaments to the caps of knights of the garter; and the crefts of the males are now ufed as ornaments in the Eaft.

The female has only a very fhort plume of dufky greyifh feathers, and the loofe feathers that hang over the breaft are very fhort, while thofe of the other fex are long. This has been generally fuppofed a diftinct kind: the accurate Linneus defcribed it as another fpecies, under the fpecific name *cinerea*, and many other naturalifts have been of the fame opinion, as appears by the fynonyms: Mr. Pennant obferves this was formerly fuppofed; " but later obfervations prove them to be the fame." Mr. Latham adopts precifely the opinion of Mr. Pennant, but in neither of their accounts can we find the authority on which that opinion is founded. It is worthy of remark that Albin, who lived at a time when Heronries were far more numerous than at prefent (though even now they are very common in fome parts *), in the firft volume of his Birds, has figured the male, and in the third volume the female, yet gives not the fmalleft reafon to conclude that he did not confider them as diftinct fpecies.

In the breeding feafon they unite in large focieties, and build on the higheft trees. The neft is made of fticks, and lined with rufhes, wool, feathers, &c. They lay four, five, or fix eggs, of a pale green colour. They defert the nefts in the winter, and are then found on the banks of rivers, or marfhy places.

The length is about three feet: breadth five feet; weight exceeds three pounds.

---

* " At *Creffi Hall*, near *Gofberton* in *Lincolnfhire*, I have counted eighty nefts in one tree." *Pennant.*

P L A T E

74

PLATE LXXIV.

STERNA FISSIPES.

BLACK TERN.

ANSERES.

Bill obtufe, covered with a thin membrane, broad, gibbous below the bafe, fwelled at the apex. Tongue flefhy. Legs naked. Feet webbed or finned.

GENERIC CHARACTER.

Bill ftrait, flender, pointed. Noftrils narrow. Tongue flender and fharp. Wings very long. Tail forked. A fmall back toe.

SPECIFIC CHARACTER
AND
SYNONYMS.

Head. Neck, breaft and belly as far as the vent black. Back and wings dark grey. Legs reddifh black.

STERNA FISSIPES. *Linn. Syft. I. p.* 228. 7. *edit.* 12. 1766.
Larus Niger (Meyvogelin) *Gefner av.* 558. *fig.* 589.
Larus Niger fidipedes. *Raii Syn. p.* 131. 4. A. 6.
Larus Merulinus. *Scop. Ann. I. N°* 108?
Sterna Nigra, *Sepp Vog. pl. in p.* 131.
BLACK TERN. *Penn. Br. Zool. N°* 256.
       —— *Arct. Zool. N°* 450.
       Lath. Gen. Syn. vol. 6. 366. *fp.* 22.
       —— *Suppl.* 267.

Scare-

PLATE LXXIV.

Scare-Crow. *Raii Syn. p.* **131**. A. 3.

Black cloven-footed Gulls. *Idem.* 1 3 2. N° 6.

Will. Orn. 354. §. 4. 6. *pl.* 78.

L'Hirondelle-de-Mer noire, ou l'Epouvantail. *Brif. Orn.* 6. *p.* 211.4.

Buff. Oif. 8. *p.* 341.

—— *Pl. enl.* 333.

Kleinote Moewe. *Frifch.* 2. 220.

Siælandis Glitter. *Brunnich,* 153.

The length of this fpecies is commonly about ten inches: breadth twenty-four: weight two ounces and a half. The male is known by a white fpot under the chin. Mr. Latham mentions a variety, (Var A) in which the lower part of the breaft, belly, thighs, under wing coverts and vent are white; and in fome fpecimens of the common kind the white at the vent is fpread towards the thighs. The webs of the feet are depreffed, and form a crefcent: the colour of the legs feem to vary; our bird had fcarcely any of the red tinge in the black colour.

Thefe birds frequent our fhores in fummer. Latham fays they are obferved on the coafts of Kent in a few days after the other terns; and, as they differ fomewhat in their manners, do not affociate. They are found during fpring and fummer in vaft numbers in the fens of Lincolnfhire. The eggs are three or four in number, of a greenifh or olive colour, fpotted with black, and have alfo a band of the fame colour about the middle. Thefe eggs are depofited among the reeds in fens and other marfhy places. The food is infects and fmall fifh, which it procures by hovering over the water, and darting on its prey in the fame manner as moft other birds of the fame genus.

It

PLATE LXXIV.

It is an inhabitant of moſt of the northern countries of Europe: very common in Siberia, and about the ſalt lakes of the deſarts of Tartary. It is alſo ſuppoſed to be the ſpecies which was ſeen by *Kalm* in vaſt flocks, beyond lat. 41. north, long. 47. W. He ſays, " It was rather darker than the common ſea-ſwallow ; the flocks confiſted of ſome hundreds, and ſometimes ſettled on the ſhip*."

---

* Vide *Kalm.* Travels in North America, &c. tranſlated by J. R. Forſter, 1770,

B

PLATE

75

PLATE LXXV.

SCOLOPAX ÆGOCEPHALA.

GODWIT.

GRALLÆ.

Bill roundifh. Tongue entire, flefhy. Thighs naked. Toes divided.

GENERIC CHARACTER.

Bill flender, ftrait, weak. Noftrils linear, in a furrow. Tongue pointed, flender. Toes divided, or flightly connected; back toe fmall.

SPECIFIC CHARACTER

AND

SYNONYMS.

Above pale reddifh brown; a dark mark down the middle of each feather. Beneath white. Tail barred with brown.

SCOLOPAX ÆGOCEPHALA roftro recto, pedibus virefcentibus, capite colloque rufefcentibus, remigibus tribus nigris bafi albis. *Linn. Syft. Nat. I. p.* 147. 77. 13. *edit.* 10.

C

Godwit,

PLATE LXXV.

Godwit, Yarwelp, Yarwip. *Raii Syn. p.* 105. A. 4.

Will. Orn. p. 292.

Albin. 2. *pl.* 70.

Penn. Br. Zool. 2. 439. 179.

—— *Arct. Zool. N°* 373.

Lath. Gen. Syn. 5. *p.* 144. 14.

—— *Suppl.* 245.

Le Grande Barge grife. *Brif. Orn.* 5. *p.* 272. 3. *pl.* 24. *fig.* 2.

——————— aboyeufe. *Buff. Oif.* 7. *p.* 501.

—— *Pl. Enl.* 876.

The length of this fpecies is commonly about fixteen inches; breadth twenty-feven, and weight twelve ounces; but they are liable to confiderable variation in weight and fize, as well as colour: fometimes they do not even exceed feven ounces.

It is very generally met with in *Europe,* and extends to *Afia* and *America.* Mr. Latham fays at *Hudfon's Bay* it is known by the name of *Wafawuckapefhew.* Thefe Birds are found in *England* in the fens amongft the Ruffs and Reeves one part of the year, but continues with us the whole winter, frequenting the open fands like the Curlew, and feeding on Infects.

PLATE

76

PLATE LXXVI.

ALAUDA OBSCURA.

DUSKY LARK.

PASSERES.

Bill conic, pointed. Noſtrils oval, broad, naked.

GENERIC CHARACTER.

Bill ſtrait, ſlender, bending a little towards the end. Noſtrils covered with feathers or briſtles. Tongue cloven. Toes divided to the origin, back claw very long.

ALAUDA OBSCURA, DUSKY LARK. *Lath. Ind. Orn.* 2. 494. *N°* 7.

BLACK LARK. *Albin. Vol.* 3. *pl.* 51.

L'Alouette noire. *Briſ. Orn.* 3. *p.* 34. B.

Buff. Oiſ. 5. *p.* 22.—*pl. enl.* 650. *f.* 1.

It ſeems undetermined whether we ought to conſider this as a diſtinct ſpecies, or only as a variety of the Sky Lark. Mr. Latham, in his Synopſis, conſiders it a variety, and obſerves that he is aware of this and other Birds becoming black by feeding on *Hemp-ſeed*, as was the caſe with a Goldfinch and Houſe Sparrow. It appears alſo liable to much variation of colours in different ſpecimens. One in

C 3

the

PLATE LXXVI.

the Britifh Mufeum is of a full deep black throughout, and that from which the figure in our plate is copied, is of a lighter colour in many parts than that feems to have been from which Albin engraved his plate. The account which this Author has given is curious: " This Lark," fays he, " was taken with a clap net by one of the Bird-catchers in a field near *Highgate*, and brought to me by *Mr. Davenport*, which I have taken care to draw exactly from the Bird, neither adding nor diminifhing in the draught or colouring. This being a curiofity, I was defired by one of my fubfcribers to make a plate of it."

The name Albin has given it is fcarcely juftified by this defcription which he has added. " The bill of this Bird was of a dufky yellow; the irides of the eyes yellowifh: it was all over of a dark reddifh brown, inclining to black, excepting the hind part of the head, on which was fome dufky yellowifh feathers; likewife fome feathers with whitifh edges on the belly."

Our fpecimen was fhot in Scotland by Mr. Agneau, Gardener to the late Duchefs of Portland, feveral years ago. Its length rather exceeds feven inches.

PLATE

77

PLATE LXXVII.

LARUS HYBERNUS.

WINTER GULL.

ANSERES.

Bill obtufe, covered with a thin membrane, broad, gibbous below the bafe, fwelled at the apex. Tongue flefhy. Feet webbed, or finned.

GENERIC CHARACTER.

Bill ftrong, ftrait, bending near the end; an angular prominency on the lower mandible. Noftrils linear. Tongue cloven. Leg and back toe fmall, naked above the knee.

SPECIFIC CHARACTER

AND

SYNONYMS.

General colour white. Head and neck marked with dufky fpots. Back grey. Scapulars grey, fpotted with brown. A black bar acrofs the end of the tail.

LARUS HYBERNUS. *Lath. Suppl. Gen. Syn. p.* 296.
Winter-Mew, or Coddy Moddy. *Raii Syn. p.* 130. *A* 14.
        Albin, 2. *pl.* 87.
        Will. Orn. p. 350. *pl.* 66.
Winter Gull. *Penn. Br. Zool.* 2. *pl.* 248. *p.* 537.
        Lath. Gen. Syn. 6. *p.* 384.

Gauca-

PLATE LXXVII.

Gauca-gaucu. *Raii Syn. p.* 130. 12.

Will. Orn. p. 352.

Gavia Hyberna, le Mouette d'hiver. *Brifon av.* 6. 189.

The length of this bird is eighteen inches : breadth three feet fix inches ; weight feventeen ounces. It is a common bird in England, and frequents the inland rivers, fens, and moift meadows many miles diftant from the fea fhore in winter.

Mr. Pennant obferves, that the gelatinous fubftance, known by the name of *Star Shot,* or *Star Gelly,* owes its origin to this bird, or fome of the kind ; being nothing but the half digefted remains of Earth-Worms, on which thefe birds feed, and often difcharge them from their ftomachs.

Mr. Morton in the *Nat. Hift. Northampt.* has given alfo the following curious obfervation :—" In the courfe of my correfpondence with the late Mr. *J. Platt* of *Oxford,* I recollect his having mentioned, that once meeting with a lump of this *ftar-jelly,* on examination he found the toes of a *Frog* or *Toad* ftill adhering, and undiffolved ; and from thence concluded it to be the remains of one of thefe, having been fwallowed whole by fome bird, and the indigeftible parts brought up in the condition he found it."

PLATE

78

PLATE LXXVIII.

COLYMBUS SEPTENTRIONALIS.

RED THROATED DIVER.

ANSERES.

Bill obtufe, covered with a thin membrane, broad, gibbous below the bafe, fwelled at the apex. Tongue flefhy. Legs naked. Feet webbed, or finned.

GENERIC CHARACTER.

Bill ftrait, pointed. Upper mandible longeft ; edges of each bending in. Noftrils linear. Tongue pointed, ferrated near the bafe. Legs thin, flat. Exterior toe longeft : back toe joined to the interior by a fmall membrane. Tail fhort, and confifts of twenty feathers.

SPECIFIC CHARACTER

AND

SYNONYMS.

Above dufky, marked with a few white fpots. Beneath white. Throat dull red.

COLYMBUS SEPTENTRIONALIS. *Linn. Syft.* 1. *p.* 220. 3.
Colymbus arcticus collo rufo. *Arct. Nidr. L. p.* 244. *b.* 2. *fig.* 2.

RED

PLATE LXXVIII.

Red Throated Diver. *Pen. Br. Zool. vol.* 2. *p.* 526. 240.—
Arct. Zool. N° 443.—*Lath. Gen. Syn.*
vol. 5. *p.* 344.

Red Throated Loon. *Edw. pl.* 97.

Le Plongeon à gorge rouge. *Brif. Orn.* 6. *p.* 111. 3. *pl.* 11. *fig.* 1.
——*Pl. Enl.* 308.

Iflandis & *Norvegis* Loom v. Lumme ; *Danis,* Lomm.
Brunnich, 132.

This fpecies breeds on the borders of lakes in the northern parts of *Scotland,* and very rarely migrates to the fouthward but in fevere winters. It is an inhabitant of many cold countries, fuch as *Ruffia, Siberia, Kamtfchatka, Iceland,* and *Greenland*; and is alfo found about the rivers in *Hudfon's Bay.* It breeds in *Greenland* in June. The neft is compofed of mofs and grafs, and is placed amongft the rufhes near the water: it contains two eggs of a more elongated form than thofe of the common Hen : they are of an afh colour, and are marked with a few black fpots.

Thefe birds are more frequent about frefh waters than thofe of the fea, and are injurious to the fifhermen by diving among the nets and devouring the fifh ; but they often entangle themfelves, and are by that means taken.

The weight of this Diver is three pounds, and the length two feet five inches.

PLATE

79

PLATE LXXIX.

PARUS ATER.

COLEMOUSE.

PASSERES.

Bill conic pointed. Noftrils oval, broad, naked.

GENERIC CHARACTER.

Bill ftrong, a little compreffed, fharp pointed. Briftles at the bafe. Tongue blunt and terminated by three or four briftles. Toes divided to the origin; back toe very large.

SPECIFIC CHARACTER

AND

SYNONYMS.

Head black. Breaft and Belly dirty, or white inclining to afh colour. Back and Wings greenifh.

PARUS ATER: capite nigro, dorfo cinereo, occipite pectoreque
albo. *Fn. Suec.* 241.—*Lin. Syft. Nat.* **1. 190.**
100. 5. *edit.* **10.**
Scop. Ann. **1.** *p.* **163.** Nº 245.
Kram. El. *p.* **379.** Nº 4.
Gefner av. **641.**

D

COLE-

PLATE LXXIX.

COLEMOUSE. *Raii Syn. p. 73. A. 2.*
>*Will. Orn. p. 241. t. 43.*
>*Penn. Br. Zool.* 1. N° 164. *pl. 57. f.* 3.
>―――*Arct. Zool.*
>*Lath. Gen. Syn.* 4. 540. 7.
>―――*Suppl.* 189. 8.

Parus Atricapillus.
La Mefange à tête noire. *Brif. Orn.* 3. *p.* 551. N° 5
La petite Charbonniere. *Buff. Oif.* 5. *p.* 400.
Quatriefme efpece de Mefange. *Belon. av.* 370.
Speermiefce, Creuzmeife. *Kram.* 379.
Tannen Meife, (Pine Titmoufe). *Frifch.* 1. 13.

It has been fuppofed by fome authors that the **Parus Ater,** and **Parus Paluftris** * of Linnæus were not diftinct fpecies, but merely the two fexes of one kind ; and others have thought the latter only a variety of the firft. Willughby is the firft author worthy of confideration who has noticed the precife difference between the two birds. He fays the Marfh Titmoufe differs from the Colemoufe in thefe particulars : 1ft, that it is bigger : 2d, that it wants the white fpot on the head : 3d, it has a larger tail : 4th, its under fide is white : 5th, it has lefs black under the chin : 6th, it wants the white fpot on the covert of the wings. Mr. Pennant obferves on this account given by Willughby, that the laft diftinction does not hold in general, as the fubject figured in the *Britifh Zoology* had thofe fpots ; yet wanted that on the hind part of the head.

---

* Marfh Titmoufe.

The

PLATE LXXIX.

The opinion of Mr. Latham in this particular deferves attention alfo; he fays, " it is much to be feared that the Marfh Titmoufe is not a diftinct fpecies; moft probably a mere variety of the Cole-moufe." *Gen. Syn. vol.* 4.

In the *Supplement* to the *General Synopfis of Birds,* Mr. Latham has added the following account under the head MARSH TITMOUSE. " In my *Synopfis* it has not appeared clear to me, whether the *Cole-moufe* and this were different fpecies. I find it to be the opinion of *Sepp,* that they form but one, being both figured in the fame plates as *male* and *female.* In one of them is a fpot of white on the hind head, and the fides of the head are white: the throat black. The other has the top of the head wholly black, and the black fpot of the throat wanting. The neft feems here compofed of fedge, mixed with large *cat's-tail,* lined with down and feathers: furnifhed with five white eggs, mottled with red brown.

The Colemoufe appears to be lefs injurious in gardens and orchards than others of the fame genus: it is alfo lefs numerous, and generally inhabits woods. The length is four inches. It is found throughout *Europe* and in *America.*

PLATE

80

PLATE LXXX.

CORVUS CARYOCATACTES.

NUTCRACKER.

PICÆ.

Bill compreffed, convex.

GENERIC CHARACTER.

Bill ftrong, conic, with briftles at its bafe reflected downwards. Tongue bifid.

SPECIFIC CHARACTER

AND

SYNONYMS.

Entirely dark brown, marked with triangular white fpots on every part, except the Wings and Tail.

CORVUS CARYOCATACTES: fufcus alboque punctatus, alis caudaque nigris: rectricibus apice albis: intermediis apice detritis.

Caryocactes, *Raii Syn. p.* 42.

　　　　　Will. Orn. p. 132. *pl.* 20.

NUTCRACKER, *Edwards, pl.* 240.

　　　　　Penn. Br. Zool. 2. *App. p.* 625. *pl.* 3.

E　　　　　　　　　　　　　*Lath.*

PLATE LXXX.

Lath. Gen. Syn. **1.** 400. 38.

—— *Suppl.* 82.

Ces Caffe noix, *Brif. Orn.* 2. *p.* 59. N° **1.** *pl.* 5. *fig.* **1.**

Buff. Oif. 3. *p.* 122. *pl.* 9.

—— *Pl. Enl.* 50.

Nicifraga, *Brif. Orn.*

Notwecka, Notkraka. *Faun. Suec. fp.* 19.

Tannen-Heher (Pine Jay). *Frifch.* 1. 56.

Danis Noddekrige. *Norvegis.* Not-kraake, *Brunnich*, 34.

Waldftarl, Steinheher, *Kram. el. p.* 334.

The Nutcracker is fo extremely rare in this Country, that Mr. Pennant has added it to his Britifh Zoology, in the third plate of the Appendix of Vol. II. And as we are indebted to his authority for afcertaining its being an Englifh fpecies, we have tranfcribed his account of it.

" The fpecimen we took our defcription from, is the only one we ever heard was fhot in thefe kingdoms: it was killed near *Moflyn, Flintfhire, Octoher* 5, 1753."

" It was fomewhat lefs than the Jackdaw: the bill ftrait, ftrong, black: the colour of the whole head and neck, breaft and body, was a rufty brown: the other parts marked with triangular white fpots: the wings black: the coverts fpotted in the fame manner as the body: the tail rounded at the end, black, tipt with white: the vent-feathers white: the legs dufky."

Mr. Latham mentions a fecond inftance: he faw the mutilated fkin of one that was fhot in Kent.

It

PLATE LXXX.

It appears from different authors, that this bird is moſt frequent in the pine foreſts in Ruſſia, Siberia, and Kamtſchatka : it is alſo found in Germany, where it is more common than in any other part of Europe, though it inhabits the mountainous parts of Sweden and Denmark : it ſometimes viſits France in flocks. Found alſo in North America.

In its manners it is ſaid to greatly reſemble the Jay. It feeds on the kernels of the pine, wild berries, and inſects; and makes its neſt in the holes of trees.

E 2

PLATE

PLATE LXXXI.

SITTA EUROPÆA.

EUROPEAN NUTHATCH.

PICÆ.

Bill compreſſed, convex.

GENERIC CHARACTER.

Bill ſtrait, triangular. Tongue ſhort, horny at the end.

SPECIFIC CHARACTER

AND

SYNONYMS.

Upper part of the Head, Back, and Wing coverts bluiſh grey. A black ſtroke through the Eye. Throat white. Breaſt and Belly dull orange. Tail black and white.

SITTA EUROPÆA: rectricibus nigris: lateralibus quatuor infra apicem albis. *Linn. Syſt. Nat.* 1. 115. 55. 1. *edit.* 10.

Picus cinereus, ſeu Sitta. *Gefner av.* 711.

NUTHATCH, or NUTJOBBER. *Will. Orn. p.* 142. *t.* 23.
   Raii Syn. Av. 47.
   Pen. Br. Zool. 1. N° 89. *pl.* 38.
   Lath. Gen. Syn. 2. 648. 1.—*Suppl.* 117.

E 3   NUT-

PLATE LXXXI.

NUTBREAKER. *Albin*. 2. *pl*. 28.

WOODCRACKER. *Plot's Oxfordſh*. *p*. 175.

Le grand Grimpereau, le Torchepot. *Belon. av*. 304.

La Sittelle, ou le Torchepot. *Buff. Oiſ*. 5. *p*. 460. *pl*. 20.

—— *Pl. Enl*. 623. *f*. 1.

Briſ. Orn. 3. *p*. 588. *No*. 1. *pl*. 29. *f*. 3.

Blau ſpecht. *Friſch. t*. 39.

Picchio grigio, Raparino. *Zinan*. 74.

Klener, Nuſſzhacker. *Kram*. 362.

Baileſs. *Scopoli, No*. 57.

Notwacka, Notpacka. *Faun. Suec. ſp*. 104.

Ziolo. *Aldr. av*. 1. 417.

The Nuthatch is a ſmall bird: it weighs about one ounce, and is five inches and three quarters in length. The female differs from the male only in ſize; the weight ſeldom exceeding five or ſix drams.

It breeds in the hollows of trees, and lays ſix or ſeven eggs, of a dirty white colour, dotted with rufous; theſe are depoſited on the rotten wood, mixed with a little moſs. If the entrance to the neſt is too large, it cloſes up part of it with clay, leaving only a very ſmall hole to creep through. The neſt of this bird is ſeldom diſturbed, or the eggs taken away, when the female is ſitting, for her hiſſing ſo nearly reſembles that of a ſnake, that few would venture to put their hands into the hole to ſearch for them; and it is ſaid, that the female will ſuffer her feathers to be plucked off rather than

2   deſert

PLATE LXXXI.

defert her eggs or young. The male alfo fhews the greateft tendernefs for them and its mate, during the time of incubation.

It feeds on all kinds of infects, as well as nuts: of the latter, it
lays up a confiderable hoard in the hollows of trees, and brings
them out when other food is fcarce. The manner of its cracking
the nut is curious, and has been noticed by feveral authors, and particularly *Willoughby*: he fays, " It is a pretty fight to fee her fetch
a nut out of her hoard, when, placing it faft in a chink, fhe ftands
above it, with the head downwards, and, ftriking it with all her
force, breaks the fhell, and catches up the kernel."

In its manners, it is not unlike the Woodpecker tribe. It is
not fuppofed to fleep perched; for, when confined in a cage, it
would creep into a corner at night to fleep. Dr. Plott fays, " this
bird, by putting its bill into a crack in the bough of a tree, can
make fuch a violent found as if it was rending afunder, fo that the
noife may be heard at leaft twelve fcore yards."

It is not migratory, but changes its fituation in winter. Pennant
obferves, that it makes a chattering noife in Autumn. Latham
fays, he has been informed, that it has, at times, a whiftle like
that of a man.

E 4

PLATE

82

PLATE LXXXII.

MOTACILLA PHOENICURUS.

REDSTART.

PASSERES.

Bill conic, pointed. Noftrils oval, broad, naked.

GENERIC CHARACTER.

Bill ftrait, flender. Tongue jagged.

SPECIFIC CHARACTER

AND

SYNONYMS.

Cheeks and throat black. Back bluifh grey. Wings brown. Breaft red.

MOTACILLA PHOENICURUS: gula nigra, abdomine rufo, capite dorfoque cano.—*Fn. Sv.* 224.—*Linn. Syft. Nat. I.* 187. 21. *edit.* 10.

Ruticilla, five Phœnicurus, (Sommerotele) *Gefn. av.* 731.

REDSTART   -   *Raii Syn. p.* 78. *A.* 5. *Will. Orn.* 218.

Albin,

PLATE LXXXII.

Albin, 1. *pl.* 50.

Penn. Br. Zool. 1. *No.* 146.

———— *Arch. Zool.*

Lath. Gen. Syn. 4. 421. 11.

Le Roffignol de Muraille, *Brif. Orn.* 3. *p.* 403. *No.* 15.

Buff. Oif. 5. *p.* 170.—*pl.* 6. *f.* 2.

— *Pl. enl.* 351. *fig.* 1. 2.

Codoroffo. *Olina,* 47.

Culo ranzo, Culo roffo. *Kinan.* 53.—*Scop. No.* 232.

Rodftjert. *Faun. Suec. fp.* 357.

Norvegis Blod-fugl. *Danis* Roed-ftiert. *Brunnich* 280.

Schwartzkehlein (Black-throat) *Frifch.* 1. 19.

Waldrothfchweiffl. *Kram.* 376.

This pretty fpecies is **very** common in the fummer. It is migratory; vifiting this country in the fpring, and departing again in autumn; but does not leave the warmer parts of Europe fo early. The neft is ufually made in the hollows of broken walls, or old trees: it is compofed of mofs, with a lining of hair and feathers; and contains four, fometimes five eggs, of a light blue colour; and in other refpects refembling thofe of the Hedge Sparrow, except that they are rather more elongated at the fmalleft end. This bird is fo very fhy that if the eggs are only touched it forfakes the neft entirely.

The Redftart is rather fmaller than the Redbreaft; meafuring about five inches. The male is known by the chin, cheeks, and throat being black: in the female the chin is white; and the red colour of the breaft is paler than in the male. It has one very peculiar habit,

when

when it fhakes its tail it does not move it up and down like the Wagtail, but horizontally, or fideways, like a Dog when he is fawning.
Its note is foft and pleafing; but it will not bear confinement in a
cage, unlefs when reared from neftlings, when it requires the fame
treatment as the Nightingale.   In the wild ftate it feeds on every
kind of Infects.

P L A T E

83

PLATE LXXXIII.

YUNX TORQUILLA.

COMMON WRYNECK.

Pic Æ.

Bill compreſſed, convex.

GENERIC CHARACTER.

Bill ſhort, roundiſh, pointed. Noſtrils concave, naked. Tongue very long, cylindrical. Two fore and two hind claws.

SPECIFIC CHARACTER

AND

S Y N O N Y M S.

Whole plumage fine grey, with ſpecklings and undulated marks of dark brown and black.

JYNX TORQUILLA. Cuculus ſubgriſeus maculatus, rectricibus nigris faciis undulatus. *Fn. Sv.* 78. *t.* 1. *f.* 78.
    Linn. Syſt. Nat. 1. 112. 53. 1. *edit.* 10.
Jynx ſive Torquilla. *Raii Syn. p.* 44. A. 8.
Jynx. *Geſner av.* 573.
The Wryneck. *Will. Orn. p.* 138. *t.* 32.
    Albin. 1. *pl.* 21.
    Pen. Br. Zool. N° 83.
    Lath. Gen. Syn. 2. 548.

The

PLATE LXXXIII.

The Emmet Hunter. *Charlton ex.* 93.

Le Torcol. *Brif. Orn.* 4. *p.* 4. *pl.* 1. *f.* 1.

Buff. Oif. 7. *p.* 84. *pl.* 3.

—— *Pl. enl.* 698.

Le Tercou, Torcou, ou Tarcot. *Belon av.* 306.

Dreh-hals. *Frifch. t.* 38.

Collotorto, Verticella. *Zinan.* 72.

Gjoktyta. *Faun. Suec. fp.* 97.

Bende-Hals. *Br.* 37.

Natterwindl, Wendhalfs. *Kramer, p.* 336.

Ifhudefch. *Scop. N°* 50.·

The Wryneck is the only fpecies of the genus (*Yunx*) yet defcribed by any author; and feems to have given Linnæus fome trouble to determine to what genus he fhould affign it; for though it has the tongue of the Woodpecker, as well as the fituation of the toes, the bill is too weak for that genus. Linnæus, in the former edition of the *Fauna Suecica*, placed it with the Cuckow; but it appears to be the opinion of later naturalifts that it fhould form a diftinct genus, his new genus having been generally adopted. The *Jyngi Congener* * of Aldravendus is certainly no other than a variety.

The colours are altogether very plain, but are fo beautifully varied and pencilled, that, as Mr. Pennant obferves, Nature has made ample amends for their want of fplendor. The colours are paler in the female than the male.

---

* Le Torcol rayé of *Briffon.*

This

PLATE LXXXIII.

This bird builds in hollow trees: *Latham* fays they make no neft, but lay the eggs on the bare rotten wood. *Pennant* fays it makes the neft of dry grafs. The eggs, according to *Buffon*, are as white as ivory; and *Pennant* adds, that they are fo thin that the yolk may be feen through them. The number of eggs feldom exceed nine.

The Wryneck is fuppofed to be a Bird of paffage, appearing in the fpring eight or ten days earlier than the Cuckow. It feeds on Infects, and feems particularly fond of Ants; thefe the extreme length of the tongue enables it to pick out of the cracks where they are concealed. It takes its name from a habit it has of turning its head back to the fhoulders when alarmed or terrified: it can alfo erect the feathers of the head like thofe of a Jay.—Weight of this Bird is one ounce and a quarter: length feven inches: breadth eleven.

This Bird is found throughout *Europe*, and in many other parts of the world.

PLATE

84

PLATE LXXXIV.

LANIUS RUFUS,

WOOD CHAT.

ACCIPITRES.

Birds of prey. Bill and claws ftrong, hooked. An angle in each margin of the upper mandible. Females larger and more beautiful than the males.

GENERIC CHARACTER.

Bill hooked towards the end, with a notch in the upper mandible. Tongue jagged.

SPECIFIC CHARACTER

AND

SYNONYMS.

Head and hind part of the neck bright bay. A black line through the eyes paffing round to the breaft. Wing brown : fcapulars white. Throat, breaft and belly dirty white. Tail dark brown ; two exterior feathers partly white.

Ampelis Dorfo grifeo, macula ad aures longitudinali. *Fn. Succ.* *edit.* 1ma. *No.* 180. *t.* 2. *fœm.*
Lanius minor cinerafcens, &c. *Raii Syn. p.* 19. *A.* 6.
Ampelis 3tia. *Kram. Elench. p.* 363.
Another fort of Butcher-bird. *Will. Orn. p.* 89. §. 4.

F

Wood

PLATE LXXXIV.

Wood Chat. *Penn. Br. Zool.* N° 73.

———————— *Lath. Gen. Syn.* 1. *p.* 169. 17.

Lanius rufus. *Suppl. p.* 282.

La Pie-griefche rouffe. *Brif. Orn.* 2. *p.* 147. N° 3.

    Buff. Oif. 1. *p.* 301.

    —— *Pl. enl.* 9. *f.* 2. *the male.* — 31. *f.* 1. *the female.*

Kleiner Neun-toder. *Frifch. pl.* 61. *male and female.*

———

We have only three fpecies of Butcher-birds, or Shrikes, in this country: the Great, Red-backed, and Wood Chat. The firft is very fcarce: the fecond is not common; and the laft is extremely rare: fo that we have little opportunity to notice the fingular manners of this tribe. Nature feems to have allotted more than an ordinary fhare of courage to thefe little creatures: they equal the eagle in the fiercenefs of their attacks on fmaller birds, and defend themfelves againft thofe they cannot overcome with the greateft vigour and refolution. Though in this refpect they imitate the larger carnivorous birds, they have not, like them, claws ftrong enough to tear their prey to pieces, but, to fupply this defect, they faften it on a thorn and pull it afunder with their bill. Even when confined in a cage, it is faid that they treat their food in a fimilar manner, fticking it againft the wires before they eat it.

We now fpeak of the manners generally peculiar to the tribe: how far thofe of the Wood Chat accord with them, we can only prefume from being of the fame genus; no Englifh naturalift having yet been fo fortunate as to meet with it fince *Willoughby* and *Ray.*

Pennant

PLATE LXXXIV.

Pennant has not given a figure of it in the Britifh Zoology; and it is a fact well known, that both Lewin and Walcot have given figures of it drawn from mere defcriptions: this is the more to be regretted as no fpecimen of it was either in the Leverian or Britifh Mufeums: and that eminent ornithologift, Mr. Latham, with his accuftomed candour acknowledges, in his account of it, that he has never feen it. " Mr. *Pennant*," fays he, " does not defcribe this bird from his *own* infpection; and I muft confefs that it has never come under *mine*."— He alfo, fays *Buffon*, does not fpeak of it as uncommon, but gives it, as his opinion, that the red-backed Shrike is a variety of this fpecies, as well as fome other kinds he mentions; and adds, that from his own obfervation he cannot deny the fact. We have com-pared them, and do not hefitate to give them as two diftinct fpecies.

It is only the male bird that we have in our poffeffion. We muft own ourfelves indebted for the following defcription of the female to Pennant and Latham; nor are we certain that our fpecimen was fhot in England, but rather fufpect that it came from Germany.—The length is feven inches and three quarters. The female differs from the male: the upper part of the head, neck and body are reddifh, ftriated tranfverfely with brown: the lower parts of the body are of a dirty white, rayed with brown, marked near the end with dufky, and tipped with red.

PLATE

85

PLATE LXXXV.

FRINGILLA MONTIFRINGILLA.

BRAMBLING.

PASSERES.

Bill conic, pointed. Noftrils oval, broad, naked.

GENERIC CHARACTER.

Bill ftrong, conic, ftrait, fharp.

SPECIFIC CHARACTER

AND

SYNONYMS.

Head, back of the neck, back, black, margins of the feathers rufous brown. Throat, forepart of the neck, and breaft, pale reddifh orange. Belly white. Wing: leffer coverts rufous pale. Quills brown with yellow edges.

MONTIFRINGILLA: alarum bafi fubtus flaviffima. *Fn. Sv.* 198.
　　　　　　t. 2. f. 198.—*Linn. fyft. Nat.* 1. 79. 3. *edit.* 10.
Montifringilla montana. *Gefner av.* 388.
Bramble or Brambling. *Will. Orn.* 254.

　　　　　　　　G　　　　　　　　　Mountain

PLATE LXXXV.

Mountain-finch. *Raii. fyn. av.* 88.

Brambling. *Penn. Br. Zool.* 126.

Lath. Gen. fyn. 3. 261.

Le Montain. *Belon. av.* 372.

Le Pinçon d'ardennes. *Brif. av.* 3. 155.

Pl. enl. 54. *f.* 2.

Pinofch. *Scop. N°* 218.

Quæker, Bofinkins, Hore-Unge, Akerlan. *Brunn.* 255.

Nioowitz; Mecker, Piencken. *Kram.* 367.

Bergfinck (Mountain finch). *Frifch.* 1. 3.

This is not a very common bird in England. It is of the migratory kind and never builds here: is fometimes feen in large flocks, or in company with the Chaffinches. The colours of the female are not fo bright as in the male: in fome fpecimens of the latter the throat is black. Length rather exceeds fix inches. They are found in vaft abundance in France according to *Buffon*; and are faid to breed about *Luxemburg*, making the neft on the tall fir-trees, compofed of long mofs without, and lined with wool and feathers within. The eggs are four or five in number, yellowifh, and fpotted: the young are fledged at the end of May. They are alfo found in the Pine forefts of *Ruffia* and *Sibcria*.

PLATE

86

PLATE LXXXVI.

MOTACILLA SYLVIA?

LESSER WHITE THROAT.

PASSERES.

Bill conic, pointed. Noftrils oval, broad, naked.

GENERIC CHARACTER.

Bill ftrong, conic, ftrait, fharp.

SPECIFIC CHARACTER

AND

SYNONYMS.

Above pale cinereous brown. Beneath white. Tail brown: out-fide feather half white, fecond white at the end.

MOTACILLA SYLVIA: fupra cinerea, fubtus alba, rectrice prima longitudinaliter dimidiato alba, fecunda apice alba. *Fn. Sv.* 228.—*Linn. Syft. Nat.* 1. 185. 9. *edit.* 10?

LESSER WHITE THROAT. *Lath. Suppl. n.* 186.

This fpecies has not been defcribed by Pennant in his Britifh Zoology, nor is it certain that any preceding author has noticed it as a Britifh fpecies. The Rev. Mr. Lightfoot found it near *Bulftrode,*

G 2 in

PLATE LXXXVI.

in *Buckinghamſhire*, in May and June, and it is from a ſpecimen found by him, and preſented to the late Ducheſs of Portland, that our figure is taken. The neſt on which the bird is placed is com-poſed of dry bents mixed with wool, and lined with a few hairs of ſome animal, probably of a Cow. There is at preſent only one egg in the neſt *; it is of a pale colour, with ſmall irregular ſpots of brown.

It was the opinion of Mr. Latham, to whoſe infpection Mr. Light-foot ſubmitted this bird, that it was perhaps the *Motacilla Sylvia* of Linnæus; or that certainly it differed very little from it. Mr. Pennant alſo feems undetermined whether the White Throat was the *M. Sylvia* of Linnæus; though Berkenhout † gives it as that ſpecies without heſitation. The opinion of Mr. Latham is of the moſt importance; and, if it does not poſitively confirm our bird being the true *M. Sylvia* of Linnæus, it proves, at leaſt, that the *White Throat* is not that bird as has been generally ſuppofed ‡.

The male and female are very much alike. The ſize is that of the Yellow Wren, length leſs than five inches.

---

* Mr. Latham ſays there were three in that which came under his infpection.

† Outlines of Nat. Hiſt.

‡ " That Linnæus's bird is not our *White Throat*, I believe is manifeſt, both from ſize and colours. That author expreſsly ſays, that the ſize ſcarcely exceeds that of the *Yellow Wren*, and that it bears great affinity to the *Sedge Bird*. But that the bird in queſtion is neither the *Yellow Wren*, nor *Sedge Bird*, I am clear, as I have all the three before me."

Lath. Suppl. Gen. Syn. ʀ. 186.

We

P L A T E LXXXVI.

We have quoted the fpecific definition of *M. Sylvia* for this bird with diffidence, for it clearly appears that the tail in our fpecimen does not exactly agree with his character of that bird.

PLATE

87

PLATE LXXXVII.

LANIUS EXCUBITOR.

GREAT SHRIKE.

ACCIPITRES.

Birds of prey. Bill and claws ftrong, hooked. An angle in each margin of the upper mandible. Body mufcular, Females larger and more beautiful than the males.

GENERIC CHARACTER.

Bill hooked at the end. A notch in the upper mandible. Tongue jagged.

SPECIFIC CHARACTER

AND

SYNONYMS.

Crown and back afh colour. Underfide white. A black mark paffes through the eye. Wings black with a white ftripe. Tail wedge fhaped, black in the middle, white on the fides.

LANIUS EXCUBITOR: cauda cuneiformi lateribus alba, dorfo cano, alis nigris macula alba. *Linn. Syft. Nat. I. p.* 94. 2. *edit.* 10.
> *Sepp. Vog. pl. in. p.* 121.
> *Faun. Arag. p.* 71.

G 4   Lanius

PLATE LXXXVII.

Lanius Cinereus. *Gefn. av.* 579.

Lanius Cinereus, Collurio major. *Aldr. av.* 1. 199.

GREAT SHRIKE. *Penn. Br. Zool.* 33. Nº 71.

Lath. Gen. Syn. 1. 160. 4.

—— *Suppl.* 51. 1.

Catefby Carolin. app. p. 36.

Greater Butcher Bird, or Mattagefs, *Raii. Syn. p.* 18. *A.* 3.

Will. Orn. p. 87. *pl.* 10.

Albin. 2. *pl.* 13.

Butcher Bird, Murdering Bird, or Shreek, *Mer. Pinax*, 170.

Night Jar. *Mort. Northampt.* 424.

La Pie-griefche grife, *Brif. Orn.* 2. *p.* 141. Nº 1.

Buff. Oif. p. 1. 296. *pl.* 20.

—— *Pl. enl.* 445.

Shrike Myn Murder. *Turneri.*

Caftrica, Ragaftola. *Olina*, 41.

Speralfter, Grigelalfter, Newntotder. *Kram.* 364.

Warfogel. *Faun. Suec.* 80.

Velch Skrakoper. *Scopoli*, Nº 18.

Berg-Aelfter, or Groffer Neuntodter. *Frifch.* 1. 59.

———————————

The Great Cinereous Shrike is uncommon in England. It is of the migratory kind, coming in May and departing in September. We learn from Buffon that it is not fcarce in France. It is found in Germany, Ruffia, and North America. In Ruffia it is trained to catch fmall birds; and the peafants value it becaufe they believe that it deftroys the rats, mice, and other vermin.

The

PLATE LXXXVII.

The favage peculiarities of this carnivorous tribe of birds have been already noticed in the account of the Wood Chat and Red-Back Shrike: it is only neceffary to add that this is the largeft kind found in this country, and is inferior to very few from foreign countries either in fize or courage. Its conflicts with larger birds are fome-times fevere; but the fmaller kinds it feizes by the throat and ftrangles: from this circumftance it is called, in Germany, the *Warchangel,* or Suffocating Angel.—The female differs from the male chiefly in the colour and markings of the breaft, that part being of a dufky white, and marked with a number of tranfverfe, femicircular brown lines. She lays fix eggs, of a dull olive colour, fpotted at the thickeft end with black. The neft is compofed of heath and mofs, lined with wool. In North America, at Hudfon's Bay, it is faid to build its neft half way up a Pine, or Juniper-tree, in April, and that the hen fits fifteen days *.

---

* *Lath. Suppl.* 51.

PLATE

88

PLATE LXXXVIII.

FRINGILLA MONTANA.

MOUNTAIN, or TREE SPARROW.

PASSERES.

Bill conic, pointed. Noſtrils oval, broad, naked.

GENERIC CHARACTER.

Bill ſtrong, conic, ſtrait, ſharp.

SPECIFIC CHARACTER

AND

SYNONYMS.

Head, back, wings, tail, brown. Underſide of the body white, two bars of white acroſs the wing ſide of the neck white. Spot under the throat black.

FRINGILLA MONTANA: remigibus reɛtricibuſque fuſcis, corpore grifeo nigroque, alarum faſcia alba gemina. *Lin. Syſt. Nat.* 1. *p.* 183. 28. *edit.* 10.

TREE SPARROW. *Pen. Br. Zool.* 1. 339. 128.
  Arɛt. Zool. 2. 246.

TREE FINCH. *Lath. Gen. Syn.* 3. *p.* 252. *N°* 2.

Le

PLATE LXXXVIII.

Le Moineau de Montagne, Paſſer Montanus. *Briſſ. Orn.* 3. 79.

Paſſere Montano. *Zinan.* 81.

Skov-Spurre. *Brun.* 267.

Feldſpatz, Rohrſpatz. *Kram.* 370.

Friſch. 1. 1.

Grabetz. *Scopoli*, Nº 220.

This is rather ſmaller than the Houſe Sparrow. The female is duller in colour than the male; and has not the black marks on the ſide of the head and throat. In ſome ſpecimens the brown colour is more intermixed with black, or dark ſhades, than our bird; this change of colour has been often noticed in the Houſe Sparrow, which is ſometimes quite black.

We muſt conſider this as a local ſpecies; common in *Lancaſhire, Lincolnſhire*, and *Yorkſhire* only, in this country. It is very common in many parts of Europe: in Siberia it is more common than the *Houſe Sparrow.* It frequents trees, and, according to *Sepp*, builds its neſt in a hollow. It is compoſed of *bents*, mixed with feathers; and contains five eggs of a pale brown, with ſpots of a darker colour.

Albin ſcarcely knew this ſpecies, from which we may infer, that it has ſeldom been taken near the metropolis; he ſays, " This bird delights in mountainous woody places, not frequented. It was ſhot by a gentleman in the country, and ſent in a letter to me, by the name of the Mountain Sparrow." Vol. III. p. 62.

PLATE

89

PLATE LXXXIX.

TETRAO UROGALLUS.

COCK OF THE WOOD,

OR

WOOD GROUS,

GALLINÆ.

Bill convex: the upper mandible arched. Toes connected by a membrane at the bottom. Tail feathers more than twelve.

GENERIC CHARACTER.

A bare scarlet spot above the eye.

SPECIFIC CHARACTER

AND

SYNONYMS.

Head and neck grey, with black lines. Breast green. Belly dark brown, with a few white spots. A white spot at the setting on of the wing. Legs feathered to the toes.

TETRAO UROGALLUS: pedibus hirfutis, rectricibus exterioribus
fubbrevioribus axillis albis. *Fn. Sv. Lin. Syft.
Nat.* 1. 159. 1.

Urogallus major (*male*). Grygallus major (*female*). *Gefn. av.* 490. 495.
Gallo.

PLATE LXXXIX.

Gallo cedrone, Urogallus five Tetrao. *Aldr. av.* 2. 29.

Gallo alpeftre, Tetrax. *Nemefiani (fem.) Aldr. av.* 2. 33.

* Pavo Sylveftris. *Girald. Topogr. Hibern.* 706.

Cock of the Wood. *Raii Syn. p.* 53.

Wood Grous. *A.* 1.

> *Will. Orn. p.* 172. *pl.* 30.
>
> *Penn. Br. Zool.* 1. 92. *pl.* 40. 41.
>
> —— *Arct. Zool.*
>
> *Lath. Gen. Syn.* 4. 729. 1.

Mountain Cock and Hen. *Albin.* 2. *pl.* 29. 30.

Le Coq ou le Tetras, *Buff. Oif.* 2. *p.* 191. *pl.* 5.

> —— *Pl. enl.* 73. 74.

Le Coq de Bruyère. *Brif. Orn.* 1. *p.* 182. 1.

Le Coc de bois ou Faifan bruyant. *Belon. av.* 249.

Kjader. *Faun. Suec. Sp.* 200.

Aurhan. *Kram.* 356.

Auerhahn. *Frifch.* 1. 107. 108.

Devi peteln. *Scopoli. N°* 169.

Capricalca. *Sib. Scot.* 16. *tab.* 14. 18.

———

This noble bird was formerly an inhabitant of the woody and mountainous parts of Scotland † and Ireland ‡. At this time it is fuppofed to be extinct in Ireland ; and in Scotland is found only in the Highlands north of Invernefs §. Mr. Latham fays, the laft bird

---

* Peacock of the Woods, from its fine fhining green breaft.

† *Boethicus.—Defcr. Regni Scutiæ.*

‡ *Giraldus Cambrienfis. Topogr. Hibern.* 706.

§ Rofsfhire. Sutherlandfhire.

of

PLATE LXXXIX.

of this kind found in Scotland was in *Chicholm*'s great foreft, in Strathglafs.

Albin gave a figure of it in his Ornithology, publifhed fixty years fince, and then he feems not to have known that it was a Britifh fpecies; he calls it the Mountain Cock from Mufcovy. From this we may infer that the breed was nearly extirpated half a century ago.

According to moft authors it thrives beft in cold countries. It is found in Italy, but only in the higher regions, where the air is bleak. In Ruffia it is not uncommon, particularly in the Afiatic part of that empire. There is a variety of this fpecies much fmaller than the common fort, which is probably owing to the fevere cold of the climate in which it lives, being found in Lapland and Norway, the fartheft extreme of Europe towards the Icy Sea.

The male of this fpecies is two feet nine inches in length, breadth four feet, and is as large as a Turkey *. The female is fmaller, twenty fix inches in length. The bill is dufky, throat red, neck and back marked with tranfverfe bars of red and black: a few white fpots on the breaft, the lower part of an orange colour, belly barred with pale orange and black, the tips of the feathers white; the feathers of the back and fcapulars black, the edges mottled with black and pale reddifh brown: the fcapulars tipped with white, the inner webs of the quills dufky; the exterior mottled with pale brown: the tail of a deep ruft colour, barred with black and tipped with white.

---

* In Scotland it is known by the name of *Capercalze*, *Aver-calze*, and in the old law books Caperkally: the laft fignifying the *horfe of the woods*. In Germany it is called Aur-han, or Urus, *Wild Ox Cock.*—Pennant.

Thefe

Thefe birds feed on many kinds of plants, and particularly on the tender fhoots and feeds of the Pine and Fir trees, which are in the greateft plenty in the forefts they frequent. The males never affociate with the females, except from the beginning of February till the end of March: when the male perches on a tree, with his tail fpread, the quills lowered to the feet, the neck protruded, and the feathers of the head ruffled. In this pofture it makes a loud and fhrill noife, like the whetting of a fcythe, and this it repeats till the females dif-cover its haunts. They lay from eight to fixteen eggs, of a white colour, fpotted with yellow, and larger than thofe of the common hen: thefe are depofited upon mofs, on the ground *. The females only, fit the whole time of incubation, and cover the eggs with dry leaves when fhe is compelled to leave them. The young run after the mother as foon as hatched.

The flefh of the Wood Grous is much efteemed, except when it feeds on the berries of Juniper which communicates a very unpleafant tafte to it.

---

* Mr. Latham fays of the laft bird of this kind fhot in Scotland, " I am well in-formed that the neft was placed on a Scotch Pine."

P L A T E

90

PLATE XC.

STRIX PASSERINA.

LITTLE OWL.

ACCIPITRES.

Birds of prey. Bill and claws ſtrong, hooked, an angle in each margin of the upper mandible. Body muſcular. Females larger and more beautiful than the males.

GENERIC CHARACTER.

Bill ſhort, hooked, without cere. Head large. A broad diſk ſurrounding each eye. Legs feathered to the toes. Tongue bifid. Fly by night.

SPECIFIC CHARACTER

AND

SYNONYMS.

Head ſmooth, circular feathers on the face white tipped with black. Head, back, and wing coverts, brown with white ſpots. Underſide white ſpotted with brown.

STRIX PASSERINA: capite lævi, remigibus albis: maculis quinque
ordinum. *Lyn. Syſt. Nat.* 1. *p.* 93. 11. *edit.* 10.

H   Noctua

PLATE XC.

Noctua Minor. *Raii. Syn. p.* 26. *N*° 6.

Little Owl. *Penn. Br. Zool. N*° 70.

Lath. Gen. Syn. 1. *p.* 150. 40.

—— *Suppl. p.* 48. 40.

Will. Orn. 105.

Edw. Glean. t. 28.

Albin. 2. *t.* 12.

La Cheveche. *Belon. av.* 140.

La petite Chouette. *Brif. Orn.* 1. *p.* 514. *N*° 5.

Buff. Oif. 1. *p.* 377. *t.* 28.

—— *Pl. enl.* 439.

Kleinfte Kautzlein. *Frifch. t.* 100.

Tfchiavitt. *Kram.* 324.

La Civetta. *Olina* 65.

Scop. N° 17.

Krak-Ugle. *Brunnich,* 20.

The Little Owl appears to be no where a common bird; in this country it is fcarce, and in France, Buffon informs us, it is alfo rare. It is very feldom found in the woods, which others of the fame tribe inhabit, but frequents ruined edifices, caverns in rocks, and other fuch gloomy and folitary places. It lays five eggs, fpotted with white and a yellowifh colour. In England it has been chiefly found in Yorkfhire and Flintfhire.

This bird feems liable to much variation in colour. Buffon mentions one from St. Domingo which had lefs white on the throat, and

brown

brown bands on the breaſt inſtead of longitudinal ſpots ; and another
variety from Germany, with the plumage darker than uſual and
black irides. _Friſch_ has a figure of this bird with irides of a dark
blue colour. The length of this ſpecies is eight inches. Albin's
figure is quoted in the Synonyms, but it is much more ſlender in
its form than our ſpecimen.

PLATE

91

PLATE XCI.

FALCO SUBBUTEO.

HOBBY.

ACCIPITRES.

Birds of prey. Bill and claws ftrong, hooked. An angle in each margin of the upper mandible. Body mufcular. Females larger and more beautiful than the males.

GENERIC CHARACTER.

Bill arched from the bafe, which is covered with a wax-like membrane, or cere.

SPECIFIC CHARACTER

AND

SYNONYMS.

Cere yellow. Back brown. Back of the head white. Body pale with oblong brown fpots. Legs yellow.

FALCO SUBBUTEO : cera pedibufque flavis, dorfo fufco, nucha albo, abdomine pallido maculis oblongis fufcis.—*Linn. Syft. Nat. I. p.* 89. 12. *edit.* 10.

Dendro Falco, *Raii. Syn. p.* 14. *No.* 8.

———— Subbuteo, *p.* 15. *No.* 14.

I   Æfalon.

PLATE XCI.

Æfalon, *Aldr. av. I.* 187.

The Hobby. *Will. Orn. p.* 83.

Penn. Br. Zool. No. 61.

Arct. Zool. 2. *p.* 227. o.

Lath. Gen. Syn. I. p. 103. 90.

—— *Suppl. p.* 28. 90.

Le Hobreau, *Brif. Orn.* 1. *p.* 375. *No.* 20.

Buff. Oif. 1. *p.* 277. *t.* 17.

Belon. av. 118.

Stein Falck. *Frifch. t.* 86.

Laerke-Falk. *Brunn.* 10. 11.

———————

The Hobby is found in the temperate parts of Europe. It was antiently ufed in falconry in this country, particularly in daring of Larks and other fmall birds. It is faid, that the Larks never venture to take their flight in fight of this bird; but that if it hovers over them, they will remain motionlefs on the ground while the fowler draws a net over them.

We find on comparing the different defcriptions given by authors of this bird, with fpecimens that have fallen under our obfervation, that few birds vary more, in the colours of their plumage, than the Hobby: in fome the back is reddifh, or deep brown; in others almoft black with a bluifh caft. Again, we find fome with the back throughout of a very deep lead colour; and others with the edges of the feathers of a pale yellow-brown. The breaft is generally of a pure white with dark fpots; but thefe alfo vary: the white is tinged with a faint dirty-brown in fome; and the fpots incline to brown, inftead of black, in others. The irides are brown of every fhade in different birds:

one

4

PLATE XCI.

one author * fays they are yellow. Some of thefe variations, we muft prefume, depend on the age of the bird; but one circumftance has been noticed in adult fpecimens that deferves particular notice, the vent and thighs, which are generally ferruginous or rufous, are fometimes white. Mr. Latham, fpeaking of this variation, fays he has a fpecimen, in which the thighs are dufky white, longitudinally marked with brown; and the vent of a plain white: one of our fpecimens alfo precifely agrees with this account.

The length of the male bird is twelve inches, breadth two feet and three inches, weight feven ounces: the female is larger.

* *M. Briffon.*

I 2

PLATE

92

PLATE XCII.

MOTACILLA RUBICOLA.

STONE-CHAT.

PASSERES.

Bill conic, pointed. Noſtrils oval, broad, naked.

GENERIC CHARACTER.

Bill ſtrait, ſlender. Tongue jagged.

SPECIFIC CHARACTER

AND

SYNONYMS.

Breaſt reddiſh. Head black. A broad white mark on the wing; and another on the ſide of the neck.

SYLVIA RUBICOLA: griſea ſubtus rufeſcens, jugulo faſcia alba, loris nigris, uropygio maculaque alarum alba. *Lath. Ind. Orn.* 2. *p.* 523. 49.

MOTACILLA RUBICOLA. *Linn. Syſt. Nat.* 1. *p.* 332. *No.* 17. *edit.* Kram. el. 375.

Scop. Ann. I. No. 236.

Rubetra. *Aldr. av.* 2. 325.

STONE-CHATTER. *Penn. Br. Zool.* 1. *No.* 159.

I 3

STONE-

PLATE XCII.

STONE-CHAT. *Lath. Gen. fyn.* 4. *p.* 448. 46.

Stone-Smith, Stone-Chatter, Moor-Titling, *Raii. Syn. p.* 76. *A.* 4.

Will. Orn. p. 235. *pl.* 41.

Albin. 1. *pl.* 52.

Le Traquet. *Briff. Orn.* 3. *p.* 428. *No.* 25. *pl.* 23. *f.* 1. *(male).*

Buff. Oif. 5. *p.* 215. *pl.* 13.

Pl. enl. 678. *f.* 1.

Le Traquet ou Groulard, *Belon. av.* 360.

Pontza. *Scopoli, No.* 236.

Occhio di bue. *Zinan.* 52.

Criftoffl. *Kram.* 375.

The Stone-Chat is a conftant inhabitant of this country. In Summer it frequents heaths and commons: in Winter it retires to the marfhes, being the only places in which its favourite food, Infects, is found in abundance.

The length of this bird is four inches and three quarters. The head of the female is ferruginous colour fpotted with black ; that of the male is entirely black : they differ very little in other refpects, except that the colours of the former are more obfcure than in the other fex.

Moft authors agree that this is a noify and reftlefs creature, inceffantly flying from bufh to bufh, and always carefully concealing the place where its neft is depofited ; never alighting on the fame fpot, but creeping to it on the ground in an artful manner. The neft is placed at the bottom of fome bufh, or under a ftone, and ufually contains five or fix eggs, of a pale greenifh colour with marks of rufous.

The

PLATE XCII.

The trivial Englifh name of this bird has been accounted for by Mr. Latham in a very curious manner; he fays, he cannot find it remarked any where for its having any fong. *Buffon* compares its note to the word *ouiftrata* frequently repeated; but he has ever thought it exactly imitated the clicking of two ftones together, one being held in each hand. If others, fays he, have thought the fame, it will eafily account for the reafon of its being called the *Stone-Chatter*.

I 4

PLATE

93

PLATE XCIII.

ANAS ÆGYPTIACA.

EGYPTIAN GOOSE.

ANSERES.

Bill obtufe, covered with a thin membrane, broad, gibbous below the bafe, fwelled at the apex. Tongue flefhy. Legs naked. Feet webbed or finned.

GENERIC CHARACTER.

Bill convex above, flat beneath, hooked at the apex, with membranous teeth.

SPECIFIC CHARACTER

AND

SYNONYMS.

Bill fomewhat cylindrical. Body waved and fpeckled with brown, Upper part of the wing white, lower part brown; a black ftripe acrofs the middle of the wing.

ANAS ÆGYPTIACA: roftro fubcylindrico, corpore undulato, vertice albo, fpeculo alari candido fafcia nigra. *Lath. Ind. Orn.* 2. *p.* 840. 21.

Egyptian Goofe. *Lath. Gen. Syn.* 6. *p.* 453. 16.

Gambo Goofe, *Will. Orn. pl.* 71.

The Ganfer, *Albin.* 2. *pl.* 93.

L'Oye d'Egypte, *Brif. Orn.* 6. *p.* 284. 9. *pl.* 27.

Buff. Oif. 9. *p.* 79. *pl.* 4.

—— *Pl. enl.* 379. 982. 983.

3

Mr.

PLATE XCIII.

Mr. Latham has given this fpecies a place in his lift of the Birds of Great Britain; and we cannot furely incur difapprobation by following his example. If the authority of Mr. Latham is unfupported by the opinion of Mr. Pennant, we muft recollect that the Zoology of the latter author appeared many years before the *Synopfis* of Mr. Latham; and probably the fpecies was not fo generally diffufed and domefticated in this country before the Britifh Zoology was publifhed, as fince that period. It is impoffible that we can account otherwife for what reafon he excluded it, fince he has given the *Peacock*, *Pheafant*, *Guinea Hen*, and other domefticated Birds of foreign extraction, which certainly had no better claim to his attention, in that work, than the beautiful Bird before us.

This fpecies is a native of Africa, particularly of the Cape of Good Hope: from the latter place vaft numbers have been brought to this country; and the climate favouring their increafe, the kind is not uncommon in many parts of the kingdom. It is rather an ornamental than ufeful fpecies, and is generally kept in Gentlemen's ponds for pleafure.

Albin publifhed a figure and defcription of this Bird from a fpecimen reared in this country fixty years ago: he fays it fed on grafs and corn like other Geefe, and thus concludes his obfervations on the two fexes, " The difference between the cock and hen could not be diftinguifhed neither by the colours or fhape, but only by the cock's running to the hen with open wings, clafping or embracing her round with them. I could not find any other name for them from the Poulterers but that of *Ganfer*. This bird comes neareft to Mr. *Willoughby's Gambo Goofe*, the fpan in the wings excepted, page 360. tab. 71."—*Vide Albin. vol. ii. p.* 84.

PLATE

94

PLATE XCIV.

FALCO ÆSALON.

MERLIN.

ACCIPITRES.

Birds of prey. Bill and claws ſtrong, hooked, an angle in each margin of the upper mandible. Body muſcular. Females larger and more beautiful than the males.

GENERIC CHARACTER.

Bill arched from the baſe, which is covered with a wax-like cere or membrane.

SPECIFIC CHARACTER

AND

SYNONYMS.

Cere yellow. Head ferruginous. Body, above bluiſh. Cinereous ſpotted and ſtriped with ferruginous : beneath yellowiſh, with oblong ſpots. Legs yellow.

FALCO ÆSALON : cera pedibuſque flavis, capite ferrugineo, corpore
ſupra cærules cente-cinereo maculis ſtriiſque ferrugineis, ſubtus flavicante-albo maculis oblongis.
Lath. Ind. Orn. 1. *p.* 49. 119.

Æſalon. *Bellon. & Aldr.*
Raii. Syn. p. 15. *No.* 15.
MERLIN. *Will. Orn. p.* 85. 63.
Penn. Br. Zool. No. 63.
Lath. Gen. Syn. V. I. p. 106. 93.

L'Emerillon

PLATE XCIV.

L'Emerillon, *Brif. Orn.* 1. *p.* 382. *No.* 23.

Belon. av. 118.

Kleinſte rothe-falck, *Friſch. t.* 89.

Mr. Pennant, and other writers on the Zoology of this country, ſuppoſed the Merlin never bred here, till Dr. Heyſham met with two neſts in *Cumberland :* they were placed on the ground like that of the Ringtail ; and in each were four young. The egg, formerly in the Portland Muſeum, was of an uniform purpliſh brown colour, round-iſh, and one inch and a quarter in length. The Merlin appears in England when the Hobby diſappears, which happens in October.

This ſpecies, like the Hobby, ſeems to vary exceedingly in colour in different ſpecimens ; in ſome the back and wings are bluiſh aſh-colour *, in others ferruginous : the bars of clay-colour and duſky on the tail, are from thirteen to fifteen in moſt birds ; but Mr. Pennant ſays, one he examined had only eight ; our ſpecimen has twelve bars acroſs.

The length of the Merlin is twelve inches : though ſmall, it was formerly trained for hawking, particularly for taking partridges, which ſome authors ſay, it could kill by a ſingle ſtroke on the neck. The Merlin flies low, frequents the ſides of roads, and ſkims from one part to another in ſearch of prey. It is deſcribed as a bird wanting neither cunning nor ſpirit. It is at this time very ſcarce in England, and ſeems to be uncommon in every part of Europe.

---

* A ſpecimen in the Leverian Muſeum anſwers to this deſcription.

PLATE

95

PLATE XCV.

CORVUS PICA.

MAGPIE.

Picæ.

Bill compreffed, convex.

GENERIC CHARACTER.

Bill ftrong, conic with briftles reflected from the bafe downwards.
Tongue bifid.

SPECIFIC CHARACTER

AND

SYNONYMS.

Varied with black and white. Tail fhaped like a wedge.

CORVUS PICA: albo nigroque varius, cauda cuneiformi. *Linn.*
   Syft. Nat. 1. *p.* 106. 48. *No.* 10. *edit.* 10.
   Scop. ann. 1. *p.* 38. *No.* 41.
Pica varia et caudata. *Gefn. av.* 695.
Magpie or Pianet. *Raii. Syn. p.* 41. *A.* 1.
   Will. Orn. p. 127. *p.* 19.
   Albin. 1, *pl.* 15.
   Pen. Br. Zool. 1. *No.* 78.
   Lath. Gen. Syn. V. I. p. 392. 29.

L2

PLATE XCV.

La Pie. *Briſ. Orn.* 2. *p.* 35. *No.* 1.

       Buff. Oiſ. 3. *p.* 85. *pl.* 7.

       —— *Pl. enl.* 488.

Aelſter. *Friſch. t.* 58.

       Kram. el. p. 335.

Guzza, Putta, *Zinan.* 66.

Skata, Skiura, Skara, *Faun. Suec. ſp.* 92.

Danis Skade, Huus Skade. *Norv.*

Skior. Tunfugl. *Brunnich,* 32.

Praka. *Scop. No.* 38.

The beautiful combination of vivid gloſſes with which the plumage of this common bird is enriched, has been ſo little attended to in paintings of it, that we are afraid we ſhall be accuſed of flattering its appearance, by a gaudy introduction of unnatural tints in the annexed figure: to avoid ſuch imputation we have ſtrictly obſerved, and accurately expreſſed the colours from a ſpecimen, in our collection, that had been taken in a ſtate of nature. We are aware that the colours will vary in different ſpecimens of every ſpecies, but by correctly repreſenting one that is perfect, it will convey a better idea of the bird than any of the mutilated creatures kept in cages for amuſement.

The tail of the Magpie is particularly remarkable: its colours are more ſplendid than any other part of the bird: its form is like a wedge; the two middle feathers eleven inches long, the reſt decreaſe gradually, the outermoſt being only five inches and an half. The principal colour produced by reflection on the black part of the body is fine blue, or purple; the firſt tint is very vivid on the wings. The

fineſt

PLATE XCV.

fineſt green is the predominant colour of the tail, which changes in the folds to reddiſh yellow, with a gilded hue, fine brown, blue, purple, and ſhades of gloſſy black throughout. Theſe colours cannot be ſeen at a diſtance, becauſe they are produced only by the light falling in a particular direction, and all the bright gloſſes aſſimilate with the blacker hues when the bird is removed far from the eye of the ſpectator.

It would be tedious to enlarge on the pecular habits of a bird, that is familiar to every ruſtic inhabitant in the kingdom. In all its actions it diſcovers a degree of inſtinct ſuperior to moſt birds. In many reſpects it reſembles the crow; like that creature, it feeds in-diſcriminately on every kind of food, Infects, grain, ſmall birds, or their eggs, carrion, and even young poultry. Sometimes it procures its food by ſtratagem, at others by annoying larger animals when it has no danger to apprehend from them. It is often ſeen perched on the back of a Sheep or a Cow, picking off the Infects that infeſt them. In this reſpect we allow them to be uſeful; but if they re-lieve the poor animals from their ſmaller enemies, they ſubject them to their own uncontroulable inſolence, and ſometimes pick out the eyes of animals that attempt to reſiſt them. In a domeſtic ſtate it is often taught to repeat words, or ſentences. It does not imitate the human voice with the ſame facility and propriety as the parrot, but ſufficiently diſtinct to be underſtood.

Its great ſhare of inſtinct is clearly demonſtrated by the ſituation and manner in which the neſt is built: it is placed conſpicuouſly on the top of ſome tree, or in a hawthorn buſh, but is always fenced below by brambles and other thick buſhes, that make it difficult of acceſs. The neſt is compoſed of thorny twigs well interwoven, and has the

thorns

PLATE XCV.

thorns fticking outwards: it is lined with wool, feathers and roots, and is plaiftered within with fine mud. It is defended above by a thorny covering, and has an entrance juft large enough to admit the bird. The Magpie lays fix or feven eggs of a greenifh colour, fpotted with black.

PLATE

96

PLATE XCVI.

STERNA MINUTA.

LESSER TERN.

ANSERES.

Bill obtufe covered with a thin membrane, broad, gibbous below the bafe, fwelled at the apex. Tongue flefhy. Legs naked. Feet webbed or finned.

GENERIC CHARACTER.

Bill ftrait, flender, pointed. Noftrils narrow, on the bafe of the bill. Tail forked. Feet webbed.

SPECIFIC CHARACTER

AND

SYNONYMS.

Tail forked, body white, back grey. Head black, front white, the white continued in a band over each eye.

STERNA MINUTA : cauda forficata, corpore albo, dorfo cano, fronte
    fuperciliifque albis. *Lath. Ind. Orn.* 2. *p.* 809. 19.

Larus Pifcator. *Gefn. av.* 587. *fig.* 588.

LESSER TERN. *Penn. Br. Zool. No.* 155. *pl.* 90.

    Arct. Zool. No. 449.

    Lath. Gen. Syn. v. 6. *p.* 364. 18.

K   Leffer

PLATE XCVI.

Leſſer Sea Swallow.　*Raii. Syn. p.* **131.** *A.* 2.
　　　　　　　　　　Will. Orn. p. 353. *pl.* 68.
　　　　　　　　　　Albin. 2. *pl.* 90.
La petite Hirondelle-de-Mer.　*Briſ. Orn.* 6. *p.* 206. 2. *pl.* **19.** *fig.* 2.
　　　　　　　　　　Buff. Oiſ. 8. *p.* 337.
　　　　　　　　　　——— *Pl. enl.* 996.
Larus Piſcator.　*Geſn. av.* 587. *fig.* 588.
Hætting Tærne.　*Brun.* 152.

———————

Five ſpecies of this tribe of birds are natives of this country, if we include the Brown Tern mentioned by Mr. Latham as a doubtful kind, and not noticed by Mr. Pennant. The three Terns deſcribed in the Britiſh Zoology are the Greater or Common*, Black †, and Leſſer Terns: the two former being in the early part of this work we deem it unneceſſary to deſcribe them in this place: the latter, which is the ſubjeƈt repreſented in the annexed plate, is rather ſmaller than the Brown Tern: the body is conſiderably leſs; but the wings are nearly as long, meaſuring between the tips when expanded twenty inches: the length is about eight inches and an half.

The haunts and manners of this bird are nearly the ſame as thoſe of the Common Tern; it feeds on ſmall Fiſh and Inſeƈts, lives on the ſides of rivers, or on the ſea coaſt, and breeds amongſt the ruſhes. The egg is about an inch and a half in length, of an olive colour with reddiſh blotches. They leave their breeding-places at the approach

———————

* *Sterna Hirundo.*　　　　† *Sterna Fiſſipes.*

4

of

PLATE XCVI.

of winter. This bird is found alfo in the fouthern parts of Ruffia, and in America.

Albin has given this fpecies and the Black Tern, or a variety of it, as male and female, in Plate 89 and 90, Vol. II.

PLATE

97

PLATE XCVII.

TETRAO TETRIX.

BLACK GROUS.

GALLINÆ.

Bill convex: the upper Mandible arched. Toes connected by a membrane at the bottom. Tail feathers more than twelve.

GENERIC CHARACTER.

A bare fcarlet fpot above the eyes. Legs feathered to the feet.

SPECIFIC CHARACTER

AND

SYNONYMS.

Blue black. Tail forked. Lower half of the fcondary feathers of the wings white.

TETRAO TETRIX. *Linn. Syft. Nat.*

TETRAO TETRIX: nigro-violacea, cauda bifurca, remigibus fecundariis verfus bafin albis. *Lath. Ind. Orn.* 2. *p.* 635. 3.
Scop. Ann. I. No. 169.
Kram. el. p. 356. 2.
Gmel. Syft. I. p. 748.

L   Urogallus

PLATE XCVII.

Urogallus minor, *Raii. Syn. p.* 53. *A.* 2.

      Will. Orn. pl. 124. *t.* 41.

Black Cock, Black Grous, *Albin. v.* 1. *pl.* 22.

      Penn. Br. Zool. 1. *No.* 93. *pl.* 42.—*Arct.*

      Zool.

      Lath. Gen. Syn. 4. *p.* 733. 3.

Le Coq de bruyeres à queue fourchue. *Buff.* 2. *p.* 210. *t.* 6.——

      Pl. enl. 172, 173.

Birckhahn, *Gunth. Nest. u. Ey. t.* 34.

Orre, *Faun. Suec. sp.* 102.

Berkhan Schildhan, *Kram.* 356.

Gallo sforcello, *Scopoli. No.* 169.

The Black Cock, like the Cock of the Wood, is seldom found, except in northern countries ; in those near the south, which it sometimes inhabits, it prefers the coldest situations amongst woods and mountains : it feeds on the birch trees and mountain fruits. In Russia and Siberia, they are very abundant, as they were formerly in Scotland, Wales, and the north of England ; at present they are much diminished in this country, and, perhaps, may become as scarce as the Cock of the Wood is at present, the flesh being much esteemed, and therefore eagerly sought for.

It seems to partake greatly of the habits of the Cock of the Wood ; it frequents the same situations, and subsists on the same kind of food. It never pairs with the females ; but, in the spring, the male ascends some eminence, crows, and claps his wings, and the females, attentive to his note, resort to the spot.

The

PLATE XCVII.

The female is much fmaller than the male. Its length is eighteen inches, weight two pounds; the colours are red, black, and dufky white, which are difpofed in alternate bars and fpots, in different directions. The moft remarkable part of the male bird is the tail, which confifts of fixteen feathers; the exterior ones curve very much outwards, and give it a forked appearance; but when the tail is expanded, it refembles a large fan. Length of the male is twenty-one inches.

The female lay fix or eight eggs, of a yellowifh colour, fpeckled with ferruginous, and blotched at the fmall end with the fame colour. The young males leave the female parent in the beginning of winter, and keep in flocks, of fix or eight, till fpring. They are very quarrelfome, and fight like game-cocks.

L 2   PLATE

98

PLATE XCVIII.

ARDEA GARZETTA.

EGRET.

GRALLÆ.

Bill roundifh. Tongue entire, flefhy. Thighs naked. Toes divided.

GENERIC CHARACTER.

Bill ftrait, long, acute. Toes four.

SPECIFIC CHARACTER

AND

SYNONYMS.

Bill black. Back of the head crefted. Body white. Lore and feet greenifh.

ARDEA GARZETTA. *Linn. Syft. Nat.*
ARDEA GARZETTA. Occipite criftato, corpore albo, roftro nigro, loris pedibufque virefcentibus. *Lath. Ind. Orn.* 2. *p.* 694. 64.
Ardea Alba minor feu Garzetta. *Raii. Syn. av.* 99.
Will. Orn. p. 280.
Egret. *Pen. Br. Zool. Appen. pl.* 7.—*Arct. Zool. No.* 347.
Little Egret, *Lath. Gen. Syn.* 5. *p.* 90. 59.

L 3   Dwarf

PLATE XCVIII.

Dwarf Heron, *Barbot* 29.
L'Aigrette, *Buff. Oif.* 7. *p.* 372. *t.* 20.—*Pl. enl.* 901.
Kleiner Weiſſer Rager. *Kram.* 345.

Amongſt the number of curious and elegant ſpecies that have been extirpated in this country, the Engliſh Naturaliſt will moſt regret the loſs of this bird. It was formerly very common, and its fleſh much admired. It formed a part of many of the old Engliſh feaſts; and, amongſt others, that recorded by Leland, which was given by George Nevell, archbiſhop of York, in the reign of Edward the Fourth, alone included " *one thouſand Egrittes.*" At this time it is conſidered ſuch a rarity, that Mr. Pennant obſerves, in his Appendix to the Britiſh Zoology, " We once received out of *Angleſea* the feathers of a bird ſhot there, which we ſuſpeɔt to be the Egret; this is the only inſtance, perhaps, of its being found in our country." The ſame author adds, in another part, " We have never met with this bird, or the Crane, in England, but formed our deſcriptions from ſpecimens in the elegant cabinet of Dr. Mauduit, in Paris."

In ſome foreign countries it is ſtill very common: is found in ſeveral parts of Europe and Aſia: it is alſo ſaid to be found in Africa, and on the American continent.

The weight of this bird is one pound, the length about eleven inches; the appendage of looſe feathers, which is ſituated on the back, and hang over the rump, were anciently uſed to decorate caps, or head pieces; and hence the ornament to a cap, in later times, was called an aigrette.

PLATE

99

PLATE XCIX.

COLYMBUS IMMER.

IMBER DIVER.

ANSERES.

Bill obtufe, covered with a thin membrane, broad, gibbous below the bafe, fwelled at the apex. Tongue flefhy. Legs naked. Feet webbed or finned.

GENERIC CHARACTER.

Bill ftraight, flender, pointed. Noftrils linear, at the bafe of the bill. Legs near the tail. Feet webbed.

SPECIFIC CHARACTER

AND

SYNONYMS.

Above dufky. Beneath white.

COLYMBUS IMMER. *Linn. Syft. Nat.*

COLYMBUS IMMER : corpore fupra nigricante albo undulato fubtus
        toto albo. *Lath. Ind. Orn.* 2. 800. 2.
        Gmel. Syft. I. 588.

EMBER GOOSE, *Sibbald Scot.* 21.—*Wallace Orkney* 16.—*Debes Ferroe
        Ifles* 138.

Gefner's Greater Doucker. *Will. Orn.* 342.
        Raii. Syn. av. 126. *No.* 8.

L 4

Imber

PLATE XCIX.

Imber Diver. *Br. Zool. No.* 238.

Lath. Gen. Syn. 5. 340. 2.

Le Grand Plongeon. *Briſſon* 6. 105. *Tab.* 10.

Buff. Oiſ. 8. *p.* 251.—*Pl. Enl.* 914.

The Imber Diver inhabits the ſeas about the Orkney and Ferroe Iſlands, and never viſits the ſouthern parts of Great Britain, except in ſevere winters. Living chiefly at ſea, it is taken with much difficulty. If purſued when ſwimming, it dives under the water, and does not appear again till it is at a conſiderable diſtance from its purſuers. It is often caught under water by a hook, baited with ſmall fiſh. Willoughby ſays, they are ſometimes taken in this manner ſixty feet under water.

Being rarely ſeen on land, it has been believed that it never quitted the water, and that it hatched its young in a hole formed by nature under the wing. Naturaliſts have diſcovered its neſt among *reeds* and *flags* in the water, where it is kept continually wet, as in ſome of the Grebe genus.

This ſpecies is larger than the common gooſe : the length is about twenty-five inches. The male is ſaid to be diſtinguiſhed by a few brown ſpecks on the ſide of the neck, and by having the colours throughout more defined than in the female : ſome authors have, however, conſidered the ſuppoſed females as birds not in an adult ſtate.

PLATE

100

PLATE C.

ALCEDO ISPIDA.

KINGSFISHER.

Picæ.

Bill compreffed, convex.

GENERIC CHARACTER.

Bill triangular, thick, ftrait, long. Tongue fhort, fharp.

SPECIFIC CHARACTER

AND

SYNONYMS.

Back bright blue. Beneath rufous. Lore brown. Chin whitifh.

ALCEDO ISPIDA. Brachyura, fupra cœrulea, fubtus fulva. *Linn. Syft. Nat. edit.* 10.

ALCEDO ISPIDA. Brach. Suboriftata cœrulea, fubtus rufa, loris fulvis, vertice nigro undulato, macula aurium gulaque albis. *Lath. Ind. Orn. I. p.* 252. 20. *Gmel. Syft. I. p.* 448. *Faun. Arag. p.* 73.

Scop.

PLATE C.

Scop. an. 1. *p.* 55. *No.* 64.

Raii. Syn. p. 48. *No. A. I.*

Kingsfisher. *Lath. Gen. Syn.* 2. *p.* 626.—*Suppl.* 113.

Kingsfisher. *Will. Orn. p.* 146. *t.* 24.

Albin I. pl. 54.

Pennant. Br. Zool. I. 246. *pl.* 38.

Le Martin-pêscheur, *Brif. Orn.* 4. *p.* 471. *No. I.*

Le Martin-pêscheur, ou l'Alcyon. *Buff. Oif.* 7. *p.* 164. *Pl. enl.* 77.

Piombino, Martino pescatore.

Pescatore del re. *Zinan.* 116.

Isfogel. *Muf. Fr. ad.* 16. *Scopoli, No.* 64.

Eifvogel. *Frifch.* 2. 223.

Meerfchwalbe. *Kram.* 337.

The Kingsfisher is feven inches in length; its weight is one ounce and a quarter. It is almoft needlefs to remark, that this bird is efteemed the moft beautiful of the feathered race that inhabits the fouth of Europe. In its form it is rather inelegant; but its colours are fine throughout: the azure of its back is exceedingly bright; and when the creature is hovering in the air, in a fine day, it appears refplendent in the higheft degree.

The abfurd fictions that poets, in the vigour of their imagination, have formed concerning this bird, have particularly inclined naturalifts to examine its manners of life with attention. The poets placed it in a floating neft, during the time of incubation, and endowed it with power to calm the adverfe winds and feas. *Ariftotle* and *Pliny* tell us, that this bird is moft common in the feas of Sicily: that it fat only a few days, and thofe in the depth of

winter,

PLATE C.

winter, and during that period the mariner might fail in full fecurity, for which reafon they were ftyled *Halcyon days* *. Among the moderns, its flefh has been thought unperifhable, and capable of preferving woollen and other veftments from decay ; and it has alfo been fuppofed to turn its breaft to the north when hung up dead.

Specimens of this bird are brought from almoft every part of the world : in England it is not uncommon : it frequents the fides of running ftreams, and takes its prey, which confifts entirely of fifh, by darting on it in the water. It makes no neft, but lays feven or more, beautiful tranfparent white eggs, in a large hole in the bank of a river or ftream.

---

* Pennant.

INDEX to VOL. IV,

ARRANGEMENT

ACCORDING TO THE

SYSTEM of LINNÆUS.

ORDER I.

ACCIPITRES.

ORDER II.

PICÆ.

ORDER

INDEX.

ORDER III.

ANSERES.

ORDER IV.

GRALLÆ.

ORDER V.

GALLINÆ.

ORDER VI.

PASSERES.

FRIN-

I N D E X.

V O L.

I N D E X.

V O L. IV.

A R R A N G E M E N T

ACCORDING TO

LATHAM's SYNOPSIS of BIRDS.

GENUS

INDEX.

M   DIVI-

INDEX.

VOL.

INDEX.

VOL. IV.

ALPHABETICAL ARRANGEMENT.

T H E,

NATURAL HISTORY

O F

BRITISH BIRDS;

OR, A

SELECTION of the MOST RARE, BEAUTIFUL, and INTERESTING

B I R D S

WHICH INHABIT THIS COUNTRY:

THE DESCRIPTIONS FROM THE

S Y S T E M A   N A T U R Æ

O F

L I N N Æ U S;

WITH

GENERAL OBSERVATIONS,

EITHER ORIGINAL, OR COLLECTED FROM THE LATEST
AND MOST ESTEEMED

ENGLISH ORNITHOLOGISTS;

AND EMBELLISHED WITH

F I G U R E S,

DRAWN, ENGRAVED, AND COLOURED FROM THE ORIGINAL SPECIMENS,

By E. DONOVAN.

V O L. V.

LONDON:

PRINTED FOR THE AUTHOR; AND FOR F. AND C. RIVINGTON,
No. 62, ST. PAUL'S CHURCH-YARD. 1798.

101

PLATE CI.

PHASIANUS COLCHICUS.

COMMON PHEASANT.

GALLINÆ.

Bill convex: the upper Mandible arched. Toes connected by a membrane at the bottom. Tail feathers more than twelve.

GENERIC CHARACTER.

Bill convex, fhort, ftrong. Head carunculated with bare flefh on the fides. Legs (moftly) furnifhed with fpurs behind.

SPECIFIC CHARACTER

AND

SYNONYMS.

General colour reddifh. Head blue. Tail long, wedge fhaped. Membrane of the cheek warted, and of a bright red colour.

PHASIANUS COLCHICUS: rufus, capite cæruleo, cauda cuneata, genis papillofis. *Lath. Ind. Orn.* 2. 629. 4.

      Gmel. Syft. I. p. 741.

Phafianus Colchicus. *Linn. Syft. I. p.* 271. 3.

      Brun. Orn. 58.

      Frifch. pl. 123.

      Olin. uc. p. 49.

     A 2   Pheafant.

PLATE CI.

That mind which is inclined to admire the wonders and beauties of creation, will pause to examine with more than ordinary attention, a bird, in which nature has displayed an elegance and variety of colours, sufficient to arrest the admiration of ancient philosophers; and furnish them with the happiest simile to abash human ostentation *.

Perhaps, there are few tribes of birds in which nature has been more profuse of her amplest colouring than that of the Pheasant. The common species as we now consider it in this country, notwithstanding its beauty, is inferior in that respect to two others that are also found at large in some of our woods. The Ring and painted Pheasants are far more richly decorated, and these may probably be as abundant in future generations as the Common Pheasant is at present. The variegated Pheasant is beautiful, and the scarcely exampled delicacy of the White kind renders it an interesting variety.

---

* " When Crœsus, king of Lydia, was seated on his throne, adorned with royal
" magnificence, and all the barbarous pomp of eastern splendour, he asked Solon if
" he had ever beheld any thing so fine! The Greek Philosopher, no way moved by
" the objects before him, or taking a pride in his native simplicity, replied, that after
" having seen the beautiful plumage of the Pheasant, he could be astonished at no
" other finery."
                                                 BUFFON.

The

PLATE CI.

The beauty of all thefe varieties are, however, eclipfed by the Argus Pheafant; and probably, were we better acquainted with the Phafianus fuperbus, and fome other gigantic Chinefe fpecies*, we might place them among the moft brilliant of the feathered race.

At what period of time the Pheafant was introduced into this Country, it is impoffible now to afcertain. They have, in all probability, been long naturalized in this Country. Some of our domeftic fowls, it is fuppofed, were introduced more than two thoufand years ago, *Cæfar* noticing them. Whether this circumftance may affift conjecture, concerning the introduction of the Pheafant, we dare not prefume to determine. Pheafants were firft brought into Europe from the banks of the Phafis, a river of ancient Colchis, in Afia Minor: at prefent it is found throughout Europe, in a wild ftate. It has not hitherto been difcovered in America.

The female is fmaller than the male; the general colour, brown, variegated with other obfcure colours, the tail is fhorter than in the male; and the fpace round the eye, which is bare in that fex, is covered with feathers in the female.

They breed like the Partridge, on the ground. Lay from twelve to fifteen eggs, fmaller than thofe of the hen, and of a paler colour than thofe of the Partridge. The young follow the females like Chickens.

Several authors have noticed a circumftance of this bird, which furprifing as it may appear, is by no means peculiar to the Pheafant

---

* Colonel Davies has a drawing of the tail feather of one of the Chinefe fpecies of Pheafants, which is fix feet in length.

A 3   only.

PLATE CI.

only. After the hen has done laying and fetting, the plumage of the female becomes like that of the male, and fhe is then entirely neglected by him. *Salerne*, *Edwards*, and others, have mentioned this of the Pheafant, Guinea Hen, Rock Manakin, &c. and Mr. J. Hunter had a paper in the Philofophical Tranfactions on that fubject. Latham obferves, that it does not always require *mature age* to give the hen Pheafant the appearance of the male.

PLATE

102

PLATE CII.

ANAS ALBIFRONS.

WHITE FRONTED GOOSE.

ANSERES.

Bill obtufe, covered with a thin membrane, broad, gibbous below the bafe, fwelled at the apex. Tongue flefhy. Legs naked. Feet webbed or finned.

GENERIC CHARACTER.

Bill convex above : flat beneath : hooked at the apex ; and befet with membranous teeth.

SPECIFIC CHARACTER

AND

SYNONYMS.

Afh coloured, front white.

ANAS ALBIFRONS. Cinerea, fronte alba,
Gmel. Syft. I. p. 509.
Lath. Ind. Orn. 2. 842. 27.
Anas Erythropus. *Lin. Syft.—Faun. Suec.* 166. *(fem.)*
Georgi Reife. p. 166.
Anas Septentrionalis fylveftris. *Brif.* 6. *p.* 269. 3.
Laughing Goofe. *Edw. pl.* 153.

A 4

White

PLATE CII.

White Fronted Goofe. *Br. Zool. No.* 268. *pl.* 94. *I.* (*the head*)
Arct. Zool. No. 476.
Lath. Gen. Syn. Vol. 6. *p.* 463. 22.
L'Oye Sauvagè du nord. *Brif. Av.* 6. 269. 3.
L'Oye rieufe. *Buff. Oif.* 9. *p.* 81.
Polnifche Ganfs. *Kram.* 339.
Vild Gaas. *Brunnich.*

The length of the White Fronted Goofe exceeds two feet; the weight is about five pounds. It has neither beauty of colours or elegance of form to render it an interefting fpecies. The white fpace on the forehead is the moft ftriking peculiarity of the bird, and its name is fufficiently characteriftic of that part, to diftinguifh it from every other Britifh fpecies of the Duck tribe. It is found in the fens in fmall flocks, during winter, and migrates in March. In England it is rather uncommon.

Linnæus confidered the White Fronted Goofe as the female of the Bernacle Goofe, of which credulity has reported fo much, and we may think naturalifts have faid too little; for it feems yet, but doubtful with fome Ornithologifts whether the opinion of Linnæus be wholly unfounded in truth or not, though they have ventured to feparate them into diftinct fpecies.

As many kinds of the Duck tribe inhabit the lakes and forefts of Lapland, and other arctic regions, during the breeding feafon, it is difficult in fome inftances, to diftinguifh the mere differences of fex or age from fpecific diftinctions. The bernacle of which the white fronted Goofe has been fuppofed the female, were believed about two hundred years to be bred on the coaft of Scotland; but thofe

who

PLATE CII.

who afferted this, declared alfo, that they were generated out of *decayed wood*, or were hatched in the fhell of the Lepas Antifera, a marine production very common in thofe parts. It has feveral membranous branches or arms, and at the end of each, is fituated a multivalve fhell. The feathered beard of the fifh hanging out of the fhell, were the fuppofed feathers or limbs of the young *Tree Geefe*, as they were called by the projectors of this whimfical hypothefis.

PLATE

103

PLATE CIII.

FRINGILLA CARDUELIS.

GOLDFINCH.

PASSERES.

Bill ftrong, pointed.　Noftrils oval, broad, naked.

GENERIC CHARACTER.

Bill ftrong, conic, ftraight, fharp.

SPECIFIC CHARACTER

AND

SYNONYMS.

Wings marked in the middle with yellow : the tips white.　Tail black : moft of the feathers marked with a white fpot near the end.

FRINGILLA CARDUELIS.　remigibus antrorfum luteis, extima immaculata, rectricibus duabus extimis medio reliquifque apice albis.　*Lath.*
Ind. Orn. I. 449. 58.
Lin. Syft. I. p. 318. 7.
Gmel. Syft. I. p. 903.
Klein. p. 365. *I.*
Schæf. El. Orn. t. 24.

Frifch.

PLATE CIII.

Frisch. t. I.
Muller, No. 258.
Faun. Arag. p. 87.
Merian Inf. t. 173.

GOLDFINCH, OR THISTLEFINCH. *Br. Zool.* 2. *No.* 124.
Arct. Zool. 2. *p.* 283.
Lath. Gen. Syn. 3. *p.* 281.
Russel. Alep. p. 70.
Will. Orn. p. 246. *t.* 46.
Albin. I. t. 64.
Raii. Syn. av. 89.

Le Chardonneret. *Buff. Ois.* 4. *p.* 187. *t.* 10.
Pl. Enl. t. 4. *f. I.*
Brif. av. 3. 53.

Cardellino, *Zinnan. Vov. p.* 58. *t.* 8. 47.

Cardelli, *Olina.* 10.

Stiglitza, *Faun. Suec. sp.* 236.

Stiglitz, *Kram.* 365.
Scopoli. 211.

Stigeliz, *Wirsing. Vog. t.* 9. *and* 30.

Amongst the common birds that inhabit this Country, the Goldfinch claims a decided preference to our attention. It would be superfluous to expatiate on the beauty of a bird so well known, and difficult to add any information to its general history, that has escaped the notice of ornithologists.

The Goldfinch is found throughout Europe, and in many parts of Africa and Asia. The varieties of it are numerous. Latham mentions

PLATE CIII.

tions no lefs than eight kinds. One of thefe is like the common fort, except the fore part of the head, which is red, and about the eyes white. Another, fuppofed to be a mixed breed with the Lark, has a flefh-coloured bill, irides yellowifh; head, throat, and neck black, fpotted with red near the bill; breaft, back, fcapulars, and rump yellowifh brown; belly, fides, thighs, and under tail coverts, white.

Inftances of Goldfinches wholly white fometimes occur; one fpecimen of that kind is preferved in the Leverian Mufeum, and another, in which thofe parts only, which are red in the common fort, have a gloffy tinge of that colour. A third fort in the fame collection, is white except the crown of the head, which is mottled with red, and a crefcent of the fame colour under the throat; the wings are yellowifh.

Goldfinches of the oppofite extreme of colour are not uncommon. Some are entirely black with a flight trace of red about the head, in others even this trace is obliterated. Birds that are fed on hemp feed, of which the Goldfinch will eat freely, often become entirely black. Buffon mentions one, in which the head only was of that colour. Willughby and Ray defcribe a variety that had no red on the head, but a faffron-coloured ring furrounded the bill. Brown has another, with the head ftriped alternately with red and yellow. Buffon and Briffon have a fort, in which the wings and tail are brownifh afh-colour: and that part dingy, which in the common fort is yellow. In young birds of the common fort the head is grey.

The neft of the Goldfinch is curioufly conftructed of mofs, liver-wort, thiftle-down, &c. and lined with wool, hair, and the down of

the

PLATE CIII.

the fallow. It lays five eggs, of a whitiſh colour, and marked with deep purple ſpots.

Theſe birds breed twice in a year. In winter they aſſemble in flocks. Generally frequent places where thiſtles grow in abundance, being particularly fond of the ſeeds of thoſe plants.

PLATE

104

P L A T E CIV.

RALLUS AQUATICUS.

WATER RAIL.

GRALLÆ.

Bill roundifh. Tongue entire, flefhy. Thighs naked. Toes divided.

GENERIC CHARACTER.

Bill compreffed, incurvated. Tongue jagged at the end. Body compreffed. Tail fhort. Toes four divided to the bafe.

SPECIFIC CHARACTER

AND

SYNONYMS.

Wings olive-brown with black fpots. Sides of the lower part of the belly marked with white.

RALLUS AQUATICUS. Alis grifeis fufco maculatis, hypochondriis albo maculatis, roftro fubtus fulvo. *Lath. Ind. Orn.* 2. 755. *1.*

RALLUS AQUATICUS. *Linn. Syft. I. p.* 262. 2.
Gmel. Syft. I. p. 712.

Schæff.

PLATE CIV.

Schæff. El. t. 60.

Muller, No. 219.

Scop. Ann. I. No. 155.

Klein. av. p. 103. 2.

Gallinula aquatica, *Mars. Dan. v. p.* 68. *t.* 32.

Gallina ferica Gefneri, *Raii. Syn. p.* 114. *A.* 4.

Ralla aquatica. *Aldr. av.* 3. 179.

Gallina Cinerea. *Gefner. av.* 515.

Water Rail, Bilcock.

Brook-Ouzel. *Br. Zool.* 2. *No.* 214. *t.* 75.

Albin. I. t. 77.

Will. Orn. p. 314.

Lè Rale d'eau, *Buff.* 8. *p.* 154. *t.* 13.

——— *pl. enl.* 749.

Brif. av. 151. *tab.* 12. *fig.* 2.

Waffer hennl. *Kram.* 348.

Jord-Koene. *Brunnich,* 193.

This is the only fpecies of its genus we have in Britain. Briffon and Linnæus place it with the Land Rail or Crake, and Ray with the Water Hens. Pennant obferving the difference between the effential characters of the two latter tribes, and that of the Water Rail, conftitutes a new genus of our fpecies. The Water Rail is diftinguifhed by its flender, compreffed and incurvated bill. The Crake (Gallinule) by the bafe of the upper mandible reaching far upon the forehead, and being membranaceous: the bill is alfo thick at the bafe, and floping to the point.

The

PLATE CIV.

The length of the Water Rail is twelve inches, breadth fixteen inches, weight four ounces. It frequents the rufhy and fheltered fides of rivulets and ponds, among which it can conceal itfelf from danger. It is a very fhy bird. Flies indifferently, but walks with great celerity, and has been feen to run on the furface of the water when there has been any weeds to bear it up *.

The Egg is more than an inch and an half in length ; of a pale yellowifh colour, marked with dufky fpots.

* Latham.

B

PLATE

105

PLATE CV.

FALCO OSSIFRAGUS.

SEA EAGLE.

ACCIPITRES.

Birds of prey. Bill and claws ftrong, hooked, an angle in each margin of the upper mandible. Body mufcular. Females larger than the males.

GENERIC CHARACTER.

Bill arched from the bafe, which is covered with a wax-like membrane or cere.

SPECIFIC CHARACTER

AND

SYNONYMS.

Cere and legs yellow; the latter feathered half-way down. Body brown. Tail marked on interior webs with white.

FALCO OSSIFRAGUS: *Lin. Syft. Nat.*

FALCO OSSIFRAGUS: cere lutea pedibufque femilanatis, corpore
ferrugineo, rectricibus latere interiore albis.
Lath. Ind. Orn. 1. *p.* 12. 7.

C

Haliætus

PLATE CV.

Haliætus feu Offifraga. *Raii. Syn. p.* 7. *No.* 3.

Haliætos. *Turneri.*

Sea Eagle. *Will. Orn. p.* 59. *t.* 1.

Br. Zool. 1. *p.* 167. *t.* 17.

Lath. Gen. Syn.

Bone-breaker, *Kolb. Cap.* 2. *p.* 137.

Le Grand Aigle de Mer. *Brif. Orn.* 1. *p.* 437. *No.* 9.

L'Orfraie, *Buff. Oif.* 1. *p.* 112. *t.* 3.

Le Grand Aigle de Mer. *Pl. enl.* 415. *(fem.)*

Gaas Orn. *Brunnich.* 13.

Bein-brecher, *Offifraga.*

Meeradler, Fifch-arn, Haliætos. *Gefn. av.* 201. 203.

This fpecies is little inferior in fize to the Golden Eagle. The length is three feet fix inches; it is a ftout bird, and is armed with formidable talons: it may be diftinguifhed from the Golden Eagle by the legs, which are, for half their length, bare of feathers in the Sea Eagle: the legs of the Golden Eagle, on the contrary, are feathered to the toes.

It inhabits moft parts of Europe. In thefe kingdoms it is found, in Scotland and Ireland; and fometimes, though rarely, in England. Our fpecimen was fhot in the *Hebrides.* It is obferved of this fpecies, that it grows much larger in North America than in Europe. In Ruffia and Siberia it is very common.

This Bird lives chiefly on Fifh, which it takes in the fame manner as the Ofprey. It is fuppofed that the Eagle mentioned by *Kolben,*

is

4

PLATE CV.

is this fpecies; he fays, at the *Cape of Good Hope* it feeds on the Land Tortoifes, which it carries into the air to a confiderable height, and, by letting it fall on fome rock, dafhes the fhell in pieces, that it may more eafily pick out the flefh.

C 2

PLATE

106

PLATE CVI.

FULICA ATRA.

COMMON COOT.

GRALLÆ.

Bill roundifh. Tongue entire, flefhy. Thighs naked. Toes divided.

GENERIC CHARACTER.

Bill fhort; from this a callus extends up the forehead. Noftrils narrow. Toes furnifhed with a broad fcalloped membrane.

SPECIFIC CHARACTER

AND

SYNONYMS.

A thin fkin covers the fore-part of the fkull. Body black. Feet lobed or fcalloped.

Fulica atra, fronte calva, corpore nigro, digitis lobatis. *Lin. Syft. Nat.*

Fulica atra, fronte incarnata, armillis luteis, corpore nigricante. *Lath. Ind. Orn.* 2. 777. 1.

Fulica recentiorum. *Gcfner. av.* 390.

C 3

Com-

PLATE CVI.

Common Coot.   *Raii. Syn. p.* 116. *A.* 1.

           Will. Orn. p. 319. *pl.* 59.

           Albin. I. pl. 83.

           Penn. Br. Zool. No. 220. *pl.* 77.

           Arct. Zool. No. 416.

           Lath. Gen. Syn. 5. 275. 1.

           — *Suppl.* 259.

La Foulque, ou Morelle, *Brif. Orn.* 6. *p.* 23. 1. *pl.* 2. *fig.* 2.

           Buff. Oif. 8. *p.* 211. *pl.* 18.—*Pl. enl.* 197.

Folago o Polon.   *Zinan.* 108.

Blas-klacka.   *Faun. Succ. fp.* 193.

Lifka.   *Scopoli. No.* 149.

Kleiner Blœfsling, *Gunth. Neft. u. eg. t.* 29.

Danis Vand-Hoene,

Bles-Hoene.   *Brun.* 190.

Thefe Birds are common in the fummer throughout England, and are fometimes met with in the winter: it frequents feveral northern countries, fuch as Sweden, Norway, Ruffia, Siberia, Greenland, &c. It is alfo found in Jamaica, in Carolina, and other parts of North America.

It frequents the borders of ponds and lakes, and makes its neft among the reeds, grafs, &c. The neft is large, and contains fourteen or fifteen, fome fay twenty eggs, two inches and a quarter in length, of a pale brownifh white, fprinkled with minute chocolate-coloured fpots, in a very regular manner.

                                                 The

PLATE CVI.

The food confifts of fmall Fifh and water Infects, grain, roots of plants, &c. which it takes partly by diving into the water. The adult birds are as large as a fmall fowl. The colour of the fkin on the forehead, Briffon fays, is of a full red; Latham fays it is white, except in the feafon of incubation, when it is not of a full red, though it is tinged with that colour.

C 4

PLATE

107

PLATE CVII.

COLUMBA OENAS.

STOCK PIGEON.

PASSERES.

Bill conic, pointed. Noftrils oval, broad, naked.

GENERIC CHARACTER.

Bill foft, ftrait. Noftrils half covered by a naked fkin. Toes divided to their origin.

SPECIFIC CHARACTER

AND

SYNONYMS.

Bluifh. Back of the neck fhining green, changeable. Two fmall black bars acrofs the wings. End of the tail black.

COLUMBA ŒNAS. *Linn. Syft. Nat.*

Columba Œnas: cærulefcens, cervice viridi-nitente, dorfo poftico cinereafcente, fafcia alarum duplici apiceque caudæ nigricante. *Lath. Ind. Orn.* 2. 589. 1.

Columba, lignorum proprie, *Klein. av. p.* 119. 8.—*Id. ov. p.* 33.

Stock Pigeon, or Stock Dove. *Br. Zool.—Arct. Zool.*

Albin.

PLATE CVII.

Albin. 2. *t.* 46.

Lath. Gen. Syn. 4. *p.* 604. 1.—*Suppl.*
p. 197.

Le Pigeon fauvage, *Brif. Orn. I. p.* 86. *No.* 5.

It is the opinion of Pennant, that all the beautiful varieties of Pigeons, fo highly efteemed by *Pigeon-fanciers,* are defcended originally from one fpecies, the *Stock Dove.* Latham has, with confiderable induftry, arranged thefe *fancy* varieties under their Linnean fpecific names, which Pennant has omitted. Briffon imagines that the Roman Pigeon, *Columba Hifpanica* of Linneus, has given birth to all thefe varieties.

That kind called the *Carrier,* is much celebrated for its particular attachment to its native place. It was anciently ufed in many eaftern countries to convey letters with expedition ; at prefent the cuftom is not fo general. *Joinville* fpeaks of them in the crufade of St. Louis to Paleftine ; and *Taffo* in the fiege of Jerufalem, &c. The cuftom of conveying letters by means of thefe Birds, may be traced to a very early period. It is not only related of them, but of Swallows, by *Pliny* and *Ælian,* that they were employed on fuch fervice ; and the earlieft poets, who generally intermingled fome truth with their allegory, made the Dove the meffenger of the lover, and emblem of innocence,

" Gentle Dove,
" Whither fly'ft thou from above ?"

" From *Anacreon,* friend, I rove,
Bearing *mandates* to his love."

Anacreon, Ode 9, *to Bathyllus.*

All

PLATE CVII.

All writers on Egypt mention the vaſt number of Pigeons that are bred in that part of the world, where they proverbially conſtitute a great portion of the poor huſbandman's eſtate. In Perſia alſo they are bred in immenſe numbers. They are altogether a pleaſing and uſeful ſpecies in whatever country they are domeſticated *.

In a wild ſtate theſe Birds have two broods in a year; in a ſtate of confinement, ſometimes three. They uſually lay two eggs at a time, and ſit from fourteen to ſeventeen days before the young are hatched. They migrate in vaſt multitudes into the ſouth of England at the approach of winter, and return again in ſpring. They frequent woody places, and commonly build in the hollows of decayed trees.

---

* Their dung is uſed for tanning leather, is a valuable manure for the land, and is employed in medicine: formerly ſaltpetre was collected from it. It is uſed for many other purpoſes by diſtant nations.—*Vide Lath.*

PLATE

PLATE CVIII.

MOTACILLA LUSCINIA.

NIGHTINGALE.

PASSERES.

Bill conic, pointed.   Noftrils oval, broad, naked.

GENERIC CHARACTER.

Bill flender, weak.   Noftrils fmall.   Exterior toe joined at the under part, to the bafe of the middle one.

SPECIFIC CHARACTER

AND

SYNONYMS.

Reddifh above, beneath dirty white.   Tail red-brown.

MOTACILLA LUSCINIA.   *Linn. Syft.*

SYLVIA LUSCINIA: rufo-cinerea fubtus cinereo-alba, rectricibus
fufco-rufis, armillis cinereis.   *Lath. Ind. Orn.*
2. 506. 1.

Sylvia Lufcinia.   *Scop. ann.* 1. *No.* 227.

Nightingale.   *Raii Syn. p.* 78.
Will. Orn. p. 220. *pl.* 41.
Albin. 3. *pl.* 53.

Ruff.

- 523 -

PLATE CVIII.

Ruff. Alep. p. 7.

Penn. Br. Zool. No. 1. 145.

—— Arct. Zool. 2. p. 416. A.

Lath. Gen. Syn. 4. 408. 1.—Suppl. 180.

Le Roffignol. Belon. av. 335.

Buff. 5. p. 81. t. 6. f. 1.—Pl. enl. 615. 2.

Slauz. Scopoli. No. 227.

Nachtergahl. Faun. Suec. Sp. 244.

Nattergale. Brun.

Nachtigall. Frifch. 1. 21.

Au-vogel, Auen-nachtigall. Kram. 376.

Rufignulo. Zinan. 54.

The Nightingale is very common in England, except in the northern parts, where it is never feen. It comes in the beginning of April, and leaves us in Auguft. It is found in Sweden, Germany, France, Italy, Greece, China, and Japan. The female makes her neft in a low coppice or quickfet hedge, that is thickly cloathed with foliage. It is compofed of hay, and reeds, intermingled with oak leaves, &c. She lays four or five eggs of a greenifh brown colour. The male fcarcely differs from the female.

In the evening the Nightingale begins a fong that continues till morning. Concealed in fome thicket, this charming fongfter pours forth thofe melodious ftrains, whofe harmony, fweetnefs and variety, combine to fill the mind with foft emotions of fenfibility, and endear folitude to the contemplative man. The Nightingale is the favourite of every rural poet, and the loftieft *genius* has conftantly noticed it in his folemn defcriptions of evening, or of night.

" The

PLATE CVIII.

" —— The wakeful bird
Sings darkling, and in fhadieft covert hid
Tunes her nocturnal note."

" —— The amorous bird of night
Sung fpoufal, and bid hafte the evening ftar
On his hill-top to light the bridal lamp."

Milton's Paradife Loft.

The Nightingale is fometimes kept in cages. Thofe reared from the neft are better than fuch as are caught in a wild ftate, becaufe they fing throughout the year, except in the time of moulting. Thofe which are caught begin to fing about fix or eight days after. Mr. Latham fays, that neither this nor the *Blackcap* is found in Ireland. Mr. Pennant fays it does not inhabit Scotland, though *Sibbald* places it in his lift of the Birds of that country.

PLATE

109

PLATE CIX.

PAVO CRISTATUS.

CRESTED PEACOCK.

GALLINÆ.

Bill convex: the upper Mandible arched. Toes connected by a membrane at the bottom. Tail feathers more than twelve.

GENERIC CHARACTER.

Bill ſtrong, convex. Noſtrils large. Head ſmall, creſted. Spurs on the legs. Feathers above the tail very long, broad, expanſible, conſiſting of ranges of feathers, adorned at their ends with rich ocellated ſpots.

SPECIFIC CHARACTER

AND

SYNONYMS.

An erect creſt on the head.

PAVO CRISTATUS: capite criſta erecta. *Linn. Syſt. Nat. — Faun.*
Suec. 197.
Scop. Ann. 1. *No.* 162.
Brun. p. 58.
Friſch. pl. 118.
Kram. el. p. 355.

D   Peacock.

PLATE CIX.

Peacock. *Raii Syn. p.* 51. *A.* 2. *p.* 183. 18.

        Will. Orn. p. 158. *pl.* 217.

        Sloan. Jam. p. 302. *No.* 23.

        Brown. Jam. p. 470.

Le Paon. *Brif. Orn.* 1. *p.* 281. *pl.* 27.

        Buff. Oif. 2. *p.* 288. *pl.* 10. — *Pl. enl.* 433. *(male)*. 434

        (female).

Pfau, *Gunth. neft u. Ey. t.* 22.

Pavone, *Zinnan. Uov. p.* 25. *t.* 1. *No.* 1.

In compliance with preceding writers on the ornithology of Great Britain, we have added the *Peacock* to our work: the propriety of placing that magnificent Indian fpecies, with the humble and fimple-coloured birds of this country, we prefume not to defend. We may proudly claim it for a Britifh domefticated fpecies, but other nations of Europe have the fame privilege. Of the beauty of a bird fo well known, and which has excited admiration in all ages, we can fay little: language would but feebly exprefs the variety and brilliance of colours that profufely adorn the plumes of this majeftic creature.

Peacocks were known three thoufand years ago. In the days of Solomon, the *Tarfhifh fleet* of that monarch brought them to Jerufalem. In Greece they were alfo known very early: at Athens they were highly prized. It is fuppofed they were carried hence to *Samos*, where they were preferved near the temple of Juno. The epicures of Greece thought them a delicacy; and the young Pea-fowl is efteemed among us. It is faid, that when Alexander was in India he found vaft numbers on the banks of the *Hyarftis*, and was fo de-

5   lighted

PLATE CIX.

lighted with their beauty, that he appointed a punifhment for thofe who fhould kill any of them.—It is alfo found in Africa.

In our climate the Peacock does not come to its full plumage till the third year. The female lays five or fix greyifh eggs, the fize of thofe of a Turkey : the time of fitting is from twenty-feven to thirty days. The young are fed on barley-meal, chopped leeks, and curd; the old ones on *wheat, barley,* &c.

It is an Italian proverb, that the Peacock has the appearance of an angel, and voice of the devil ; for its cry is exceedingly inharmonious. In India, it is related, that they are taken by carrying lights to the trees where they rooft, and having painted reprefentations of the bird prefented to them at the fame time : when they put out the neck to look at the figure, the fportfman flips the noofe over their heads and fecures them. It is faid alfo, that the inhabitants of the mountains on both fides of the *Ganges* take them with a kind of *bird-lime,* made from oils and the juices of certain trees.

As we believe a figure of that fuperb variety of the common Peacock, *Le Paon Panaché* of Buffon, would be more acceptable to the reader than the fort which fo frequently occurs, we have preferred it for this work. It is a moft fuperb and elegantly variegated fpeci-men : and we only regret that the limits of our plate, and imper-fect ftate of that imitative art, colouring, will not permit us to do juftice to its incomparable beauty. The original meafures feven feet, of which the train forms a confiderable part. This appendage, which is ufually miftaken for the tail itfelf, rifes from the back and defcends to a vaft length. The tail is not more than one foot and an half in length, and confifts of eighteen brownifh-grey feathers. The female has a very fhort train, and the fpurs are generally wanting.

PLATE

110

PLATE CX.

FULICA CHLOROPUS.

WATER HEN, or COMMON GALLINULE.

GRALLÆ.

Bill roundiſh. Tongue entire, fleſhy. Thighs naked. Toes
divided.

GENERIC CHARACTER.

Bill ſhort, thick, convex. Forehead bare. Toes finned.

SPECIFIC CHARACTER

AND

SYNONYMS.

Callus and garters red. Above deep olive, beneath cinereous.
Outer edge of the wing, and tail coverts white.

FULICA CHLOROPUS. *Linn. Syſt.* 1. *p.* 258. 4.—*Gmel. Syſt.* 1.
  p. 698.
Fulica chloropus major. *Raii Syn. p.* 113. *A.* 1.—*Will. Orn.*
  p. 233. 5. 58.
Chloropus major noſtra. *Aldr. av.* 3. 177.

E   Gallinella

PLATE CX.

Gallinella aquatica. *Zinan.* 109.

GALLINULA CHLOROPUS. *Lath. Ind. Orn.* 2. *p.* 770. 13.

Common Water-Hen, or Moor-Hen. *Will. Orn.*

Raii. Syn.

Albin. II. pl. 72. 3. *pl.* 91.

Common Gallinule. *Br. Zool. No.* 217. *pl.* 77.

Lath. Gen. Syn. 5. *p.* 258. *ſp.* 12.

Poule d'eau. *Buff.* 8. *p.* 171. *t.* 15.—*Pl. enl.* 877.

Briſ. Orn. 6. *p.* 3. 1. *pl.* 1. *fig.* 1, 2.

Waſſerhennl. *Kram.* 358.

Length of this ſpecies fourteen inches, breadth twenty-one, weight fifteen ounces. It is a common bird in this country, frequents the ſides of rivers and ponds, and is ſuppoſed to feed on ſmall fiſh, and on plants.

It has two or three broods in the ſummer; the neſt is uſually placed on a low ſtump near the water, and contains ſeven eggs of a dirty white colour, ſparingly ſpeckled and ſpotted with ruſt colour, and nearly two inches in length. The female is ſmaller than the male, the colours are throughout much paler, and the throat is ſometimes white; in ſome birds it is grey, in others the colour of the reſt of the neck.

PLATE

111

PLATE CXI.

ANAS GLACIALIS.

LONG-TAILED DUCK.

ANSERES.

Bill obtufe, covered with a thin membrane, broad, gibbous below the bafe, fwelled at the apex. Tongue flefhy. Legs naked. Feet webbed or finned.

GENERIC CHARACTER.

Bill broad, depreffed, hooked at the apex, with membranous teeth. Noftrils oval, fmall. Tongue broad, edges fringed near the bafe. Feet, — middle toe longeft.

SPECIFIC CHARACTER

AND

SYNONYMS.

Tail long, pointed. Body black, beneath white.

ANAS GLACIALIS. *Lin. Syft. Nat. p.* 203. 20.
ANAS HYEMALIS. *Lin. Syft.* 1. *p.* 202, 29.
ANAS GLACIALIS, cauda acuminata elongata, corpore nigro fubtus
        albo. (*mas adultus.*) *Lath. Ind. Orn.* 2. *p.* 864.
        82.

E 2   LONG-

PLATE CXI.

Long-tailed Duck. *Edw. v.* 280.

Penn. Br. Zool. 283.

Lath. Gen. Syn. 6. p. 528. 73.

Swallow-tailed Shieldrake. *Will. Orn. p.* 364.

Le Canard a longue queue d'Iflande. *Brif. Orn.* 6. 379.

Canard de Miclon. *Buff. Oif.—Pl. Enl.* 1008.

This fpecies varies exceedingly in the colours of the plumage. In fome the principal colour is a kind of chocolate brown, in others deep black. It varies no lefs alfo in the difpofition of the white fpaces on the head, neck, and body in different birds. In the male, the fore part and fides of the head are of a reddifh grey, with an oval black fpot on each fide of the neck, a little below the head; the remainder of the neck white. The female has only the fides of the head white (except the belly), the neck being of a dufky black in general; though fpecimens have been feen, that much refemble the adult male bird. The colour of the legs vary much in different fpecimens alfo.

Linnæus divided the fuppofed male and female into two fpecies; the firft he called Anas Glacialis, the other Anas Hyemalis. Later authors, who have had more opportunity of obferving their manners of life, have fuppofed them only the two fexes of one fpecies.

Mr. Pennant, in his *Arctic Zoology, appen.* defcribes the two Linnæan fpecies as the two fexes of Anas Glacialis; and Mr. Latham fince, in his *Index Ornithologicus*, places Anas Glacialis as

the

PLATE CXI.

the adult male; this is the bird which Buffon calls, *canard à longue queue*; that which Ray terms *Anas caudacuta, Havelda,* Mr. Latham defcribes as the *young male. (β)* The female had been defcribed by Mr. Pennant in the Arctic Zoology*: Mr. Latham makes it the *(γ.)* of Anas Glacialis:—the fame author has added further, the Querquedula ferroenfis, of Briffon, or *Sarcelle de Ferroe,* of Buffon, as the *(δ.)* of the fame fpecies.

Thefe Birds frequent the more northern parts of the world, fuch as *Sweden, Lapland, Greenland,* &c. and only vifit the Englifh coafts in very rigorous winters. In the Orknies they are feen in flocks from October to April. Thefe Birds living chiefly on the water, dive and fwim well, and fubfift on fmall fhell-fifh. They build their nefts among the grafs, &c. on the fea fhore. The variety mentioned by Mr. Latham, which is called O'Edel by the inhabitants of the Ferroe ifles, has the black ftreak down the middle of the crown to the hind head, as in our fpecimen, from which we have been led to confider our Bird a variety alfo.

The length of our Bird is twenty-one inches.

---

* Vol. II. p. 76.

E 3   PLATE

112

PLATE CXII.

SCOLOPAX CALIDRIS.

RED SHANK.

GRALLÆ.

Bill roundifh. Tongue entire, flefhy. Thighs naked. Toes divided.

GENERIC CHARACTER.

Bill long, flender, ftrait, weak. Noftrils linear. Tongue pointed. Toes divided, back toes fmall.

SPECIFIC CHARACTER

AND

SYNONYMS.

Beak and feet red. Body afh-colour. Secondaries of the wings tipped with white.

SCOLOPAX CALIDRIS. *Lin. Syft. Nat.*

SCOLOPAX CALIDRIS, roftro rubro, pedibus coccineis, corpore cinereo, remigibus fecundariis albis. *Lath. Ind. Orn.* 2. 722. 25.

E 4　　　　　　　Scolopax

PLATE CXII.

Scolopax Totanus. *Faun. Suec. No.* 167.

* Totanus. *Aldr. av.* 3. 171.

Red Shank or Pool-Snipe. *Will. Orn.* 299.

Raii. Syn. av. 107.

Albin. 3. *t.* 87.

Br. Zool. 2. *No.* 184. *t.* 65.

Lath. Gen. Syn. Vol. 5. 150. 20.

—— *Suppl.* 245. 20.

Chevalier aux pieds rouges. *Buff.* 7. *p.* 513. *t.* 28.

Le Chevalier. *Brif. Orn.* 5. *p.* 188. 4. *pl.* 17. *fig.* 1.

Glareola. *Klein. av. p.* 101. 1.

Rothfufstler. *Kram.* 353.

Gallinula erythropus. *Gefner. av.* 504.

The length of this Bird is twelve inches, its breadth twenty-one inches, and its weight exceeds five ounces. In this country thefe birds are not uncommon, except in the northern parts. They frequent fens and marfhes in the breeding feafon, but feparate and conceal themfelves in the winter.

Its noife is fimilar to that of the lapwing, whofe manners it alfo imitates when it is difturbed. It lays four eggs, of a whitifh olive

* Albin fays, " This bird is not the *Totanus* of *Aldrovandus*;" he adds, " This is much lefs, has a fhorter bill and feet, and differs in the dufky colour of its back, and the red colour of its legs and feet."—*Alb. p.* 82. *Vol.* 3.

colour,

colour, marked with irregular black fpots. In winter the colours of thefe Birds become paler.

The Redfhank is found in moft parts of Europe and America. Mr. Latham defcribes a variety of this fpecies from the drawings of the late Dr. Fothergill, which inhabits the marfhes of China.

PLATE

113

PLATE CXIII.

STRIX FLAMMEA.

WHITE OWL.

ACCIPITRES.

Birds of prey. Bill and claws ſtrong, hooked, an angle in the margin of the upper mandible. Body muſcular. Females larger and more beautiful than the males. *

GENERIC CHARACTER.

Bill hooked, without cere. Noſtrils covered with briſtly feathers. Head large. Eyes and ears very large. Tongue bifid.

SPECIFIC CHARACTER

AND

SYNONYMS.

Head ſmooth. Body yellowiſh, with white ſpots. Beneath white, with pale black ſpots.

---

* Mr. Latham has a note on this character of the owls in page 46, Supplement, in which he ſays, that in ſuch of the owl genus as he has ſeen, the male was larger than the female, and therefore the owls differ in that reſpect from all other birds of prey.

STRIX

PLATE CXIII.

Strix Flammea : *Lin. Syſt. Nat.*

Strix Flammea : capite lævi, corpore luteo punctis albis, ſubtus
albido punctis nigricantibus. *Lath. Ind. Orn.*
T. 1. p. 6o. 28.

Aluco minor. *Aldr. av. 1. 272.*

White Owl. *Br. Zool. 1. No. 67.*
Lath. Gen. Syn. 1. p. 138. No. 26.
———— *Suppl. p. 46. 26.*

Common Owl, Howlet, Madge, Gillihowſter, &c. *Will. Orn. 104.*
Raii Syn. av. 25.
Alb.Vol.11.pl.11.

L'Effraie, ou Freſaie. *Buff. 1. p. 366. t. 26.*
Pl. Enl. 440.

Le petit Chat-haunt. *Briſ. av. 1. 503.*

Perl Eule. *Friſch. 1. 97.*
Kramer El. p. 324. 5.

Alloco *Zinnan. 99.*

———————

It need ſcarcely be ſaid that the white owl is common in every
part of England, and is generally found throughout the continent of
Europe. It inhabits North and South America, and is very frequent
in ſome parts of Aſia. Except in the breeding ſeaſon, it lives in
barns and out-houſes, where it is uſeful in deſtroying the mice that
infeſt ſuch places. In the breeding ſeaſon it retires to holes in
lofty buildings, or the hollows of trees.

Some

P L A T E CXIII.

Some fpecies of owls fee well in the day-time: the white owl has not that faculty; it is only in the twilight in mornings and evenings, or in moonlight nights, that it can fee clearly to take its prey. While the young are in the neft, the male and female go alternately in queft of food, make a circuit round the fields, drop on their prey inftantly, and return with it in their claws.

Thefe birds caft up the fur or feather of the creatures they devour in the form of fmall pellets, like thofe of the hawk tribe.

PLATE

114

PLATE CXIV.

FRINGILLA LINARIA.

LESSER RED-HEADED LINNET,

OR

REDPOLE.

PASSERES.

Bill conic, pointed. Noftrils oval, broad, naked.

GENERIC CHARACTER.

Bill conic, flender towards the end, and fharp pointed.

SPECIFIC CHARACTER

AND

SYNONYMS.

Dufky and reddifh brown, varied with black. Belly whitifh. Two whitifh bars on the wing coverts. Pole of the head, and the breaft red.

FRINGILLA LINARIA. *Lin. Syft.* 1. *p.* 322. 29. *Fn. Sv.* 241.

FRINGILLA LINARIA, fufco grifeoque varia, fubtus albo-rufefcens, fafcia alarum duplici albida, vertice pecto-reque rubris. *Lath. Ind. Orn.* 1. *p.* 458. 83.

Linaria

Linaria rubra minor. *Raii Syn. p.* 91. *A.* 3. *Will. p.* 191. *t.* 46.

Leſſer red-headed Linnet, or Redpole. *Br. Zool. No.* 132. *t.* 54.

Lath. Gen. Syn. 3. 305. 75.

Suppl. pl. 167.

Le petite Linotte de Vignes. *Brif. Orn.* 3. *p.* 138. 31.

Le Sizerin. *Buf. Oiſ.* 4. *p.* 216.

Grafiſka. *Faun. Suec. ſp.* 241.

Grafel, Meerzeiſel.

Tſchotſcherl. *Kram.* 369.

Rothplattige Stænfling. *Friſch.* 1. 10.

All the Finches except the Siſkin are ſuppoſed to continue in theſe kingdoms throughout the year; but ſhift to different parts according to the ſeaſons. The Redpole is known to breed in the mountains of Wales *. Mr. Pennant ſaw the neſt of this ſpecies on an alder ſtump near a brook, about two or three feet from the ground. The outſide was compoſed of dried ſtalks of grafs, and other plants, intermixed with a little wool: it was lined with hair and a few feathers. The eggs, four in number, were of a pale bluiſh green, thickly ſprinkled near the blunt end with ſmall reddiſh ſpots.

In October and November theſe birds arrive near London in vaſt numbers. The colours of the female are generally paler than thoſe of the male: the ſpot on the forehead is ſaffron colour.

---

* Barrington Miſcel. p. 217.

PLATE

115

P·L A T E CXV.

LOXIA PYRRHULA.

BULFINCH.

PASSERES.

Bill conic, pointed. Noſtrils oval, broad, naked.

GENERIC CHARACTER.

Bill ſtrong, convex above and below, and thick at the baſe. Noſtrils ſmall and round. Tongue truncated at the end. Toes placed three before, and one behind.

SPECIFIC CHARACTER

AND

SYNONYMS.

·Head, wings, tail, black.   Upper tail, coverts, and vent white.

LOXIA PYRRHULA   artubus nigris, tectricibus caudæ remigumque
poſticarum albis.   *Linn. Syſt.—Fn. Sv.* 178.
Scop. Ann. 1. *No.* 202.
Faun. Arag. p. 86.
Sepp. Vog. t. p. 133.
Schæff. Elem. Orn. t. 59.

F   Rubicilla,

PLATE CXV.

Rubicilla, five pyrrhula. *Gefner. av.* 733.

Coccothrauftes fanguinea. *Klein. Av. p.* 95. 5.—*Id. Stem. p.* 19.
t. 19. *f.* 13. *a. b.*—*Georgi Reife, p.* 174.

Bulfinch, Alp, or Nope, *Will. Orn.* 247.

Raii Syn. p. 86. *A.*

Albin. I. pl. 59. 60.

Br. Zool. 1. *No.* 116.

Lath. Gen. Syn. 3. *p.* 143. 51.

—— *Suppl. p.* 152. 51.

—— *Ind. Orn.* 1. *p.* 387.

Le Bouvreuil. *Briffon. av.* 3. 308.

Buff. Oif. 4. *p.* 372. *pl.* 17.

Monachino, Sufolotto. *Zinan.* 58.

Domherre. *Faun. Suec. fp.* 225.

Gumpel. *Gunth. Neft. u. Ey. t.* 54.

Gumpl. *Kramer.* 365.

Gimpl. *Scopoli,* N*o.* 202.

Cuifolotto. *Olina,* 40.

Blutfinck. *Frifch.* 1. 2.

Le Pivoine. *Belon av.* 3. 59.

The male of this common but beautiful Bird is diftinguifhed from the female by the rich black colour on the crown of the head, and the crimfon on the cheeks, breaft, belly, and throat, thofe parts being of an obfcure hue in the female.

In the winter and fpring it frequents gardens, and does much injury to the fruit-trees. In fummer it retires into the woods to breed. It forms a neft chiefly of mofs, about fix feet from the

ground,

ground, and depofits five or fix bluifh eggs, marked with dark fpots. The wild note of this Bird is not admired, though they may be taught to whiftle any tune in a tame ftate, and are then much valued.

There are feveral varieties of this fpecies, one of them in particular, is entirely black. Inftances are recorded of thefe birds, after being taken in full feather, in the courfe of three or four years becoming jet black, and then again recovering their former colours.

PLATE

116

PLATE CXVI.

GALLINULA CREX.

CRAKE GALLINULE.

GRALLÆ.

Bill roundiſh. Tongue entire, fleſhy. Thighs naked. Toes divided.

GENERIC CHARACTER.

Bill thick at the baſe, ſloping towards the point: baſe of the upper mandible reaching far on the forehead, callous. Body compreſſed. Wings ſhort and concave. Tail ſhort. Toes divided to their origin.

SPECIFIC CHARACTER

AND

SYNONYMS.

Above, greyiſh brown, middle of each feather black. Wings rufous brown. Beneath, reddiſh white.

RALLUS CREX. *Lin. Syſt. Nat.*

GALLINULA CREX: griſea pennis medio nigricantibus, alis rufo-
               ferrugineis, corpore ſubtus albo-rufeſcente.
               Lath. Ind. Orn. 2. 766. 1.

G   Porphyrio

PLATE CXVI.

Porphyrio rufefcens, *Brif.* 5. *p.* 533.

Darker Hen, or Rail, *Raii Syn.* *p.* 58. *A.* 8.

Will. Orn. *p.* 170. *pl.* 29.

Albin. 1. *pl.* 32.

Corn-crek. *Sib. Scot.* 16.

Land Hen. *Will. Orn.* *p.* 316.

Crake Gallinule. *Br. Zool. No.* 216. *pl.* 75.—*Arct. Zool. No.* 412.

Lath. Gen. Syn. 5. *p.* 250. 1.

Le Râle de Genet, ou Roi des Cailles, *Brif. Orn.* 5. *p.* 159. *pl.* 13?

f. 2.

La Poule-Sultane rouffatre, *Brif. Orn.* v. *p.* 533. 5.

Re delle Quaglia. *Zinan.*

Wiefen Schnarre, Wachtel Koenig, *Gunth. Neft. u. Ey. t.* 45.

Wachtel-konig. *Kram.* 349.

Roftz. *Scopoli, No.* 154.

This Bird is common in many parts of Great-Britain in fummer, and departs before winter, except in Ireland, where it is fuppofed they remain throughout the year. Being conftantly found in company with the Quails, the Crake has been called in many countries, their king, or leader.

Independent of a ftriking generical difference, the manners of thefe birds are altogether diftinct from thofe of the Water Rail, with which it has been fometimes confounded. It is found among *corn, grafs, broom,* or *furze* on heaths, and never in watery places. It lays ten or twelve eggs of a reddifh white colour, marked with ferruginous blotches: feeds on all kinds of infects, and on grain; and its flefh is much efteemed.

Length

PLATE CXVI.

Length of this fpecies is nine inches and a half; weight, from fix to eight ounces. Its note has been compared to the word *Crek* often repeated; and hence its name, Crake.

117

PLATE CXVII.

CORVUS CORNIX.

HOODED CROW.

PICÆ.

Bill compreffed, convex.

GENERIC CHARACTER.

Bill ftrong conic, with briftles at the bafe. Tongue cleft at the end.

SPECIFIC CHARACTER

AND

SYNONYMS.

Afh colour. Head, throat, wings, and tail, black.

CORVUS CORNIX. *Lin. Syft. Nat.*

CORVUS CORNIX: cinerafcens, capite jugulo alis caudaque nigris.
Lath. Ind. Orn. 1. 153. 7.

Cornix cinerea, *Brif. Orn.* 2. *p.* 19. 4.

Cornix nigra Monedula, *Ger. Orn.* 2. *p.* 35. *t.* 146, 147.

G 3   ROYSTON

PLATE CXVII.

Royston Crow. *Albin.* 2. *t.* 23.

                           Raii Syn. p. 39. *A.* 4.

                           Will. Orn. p. 124. *pl.* 18. 77.

Hooded Crow. *Br. Zool.* 1. *No.* 77.

                           Arct. Zool. 2. *p.* 251. *D.*

                           Lath. Gen. Syn. 1. *p.* 374. 5.

                          ————— *Suppl. p.* 77.

La Corneille mantelée. *Buff. Oif.* 3. *p.* 61. *t.* 4.

Mulacchia cinerizia, Monachia. *Zinan.*

Kraka. *Faun. Succ. Sp.* 88.

Grave Kran, Kranveitl. *Kramer* 333.

Urana *Scopoli, No.* 37.

———————————

The Hooded Crow is not uncommon during winter in many parts of England. With us it is a bird of paſſage. In thoſe countries where it breeds, it retreats to the mountains for that time, and deſcends into the plains as the winter approaches. In many parts of Scotland it is the only ſpecies of crow known, and in the northern iſlands and mountains, it is ſaid to remain the whole year.

Like the rook and crow, this Bird feeds on carrion and the offals of animals; and alſo on ſhell-fiſh, which they find on the banks of rivers; at other times, on ſeeds and grain, and mountain-berries. They are more elegant and varied in their plumage than the common crow, but not leſs miſchievous to young birds, or any

                                      wounded

PLATE CXVII.

wounded or defencelefs animals. Their nefts are built in trees, and commonly contain fix eggs. The length of this Bird is twenty-one inches.

PLATE

118

P L A T E CXVIII.

T A N T A L U S I G N E U S.

G L O S S Y I B I S.

GRALLÆ.

Bill roundifh. Tongue entire, flefhy. Thighs naked. Toes divided.

GENERIC CHARACTER.

Bill long, thick at the bafe, incurvated. Face naked, noftrils linear. Tongue fhort. Toes connected by a membrane.

SPECIFIC CHARACTER

AND

S Y N O N Y M S.

Very gloffy, general colour blackifh ; variegated with red, blue, and green. Head and neck black, tips of the feathers whitifh.

TANTALUS

PLATE CXVIII.

TANTALUS IGNEUS: corpore nigricante cœruleo viridi et vinaceo variegato-nitente, capite colloque nigris pennis albido fimbriatus. *Lath. Ind. Orn.* 2. *p.* 708. 16.

Tantalus igneus, *Gmel. Syst.* 1. *p.* 649.

Numenius igneus, N. C. *Petr.* 15. *p.* 460. *t.* 18.

Gloffy Ibis. *Lath. Gen. Syn.* 5. *p.* 115. 14.

The Gloffy Ibis is extremely rare in this country. It is not noticed by Mr. Pennant in the laft editions of the Britifh Zoology. Dr. Latham has placed it in his lift of Britifh Birds, but mentions only one inftance of its being found in England. " In the *Leverian Mufeum* is one of thefe, which was fhot in Cornwall." *Lath. Gen. Syn. v.* 5.

Our fpecimen of this fpecies, we are informed, was alfo fhot in England, and on diffection proved to be a male. Whether the fpecimen from which Dr. Latham's defcription is taken, differed in fex from this, we are unable to determine. The defcription does not exactly agree with our bird ; and the fpecimen referred to, is removed from the Leverian collection. In the defcription, the eyes are placed in a white fpace : the eyes in our Bird are furrounded with black. The legs are alfo defcribed of an olive colour in the dead bird, and green when living; thofe parts appear reddifh in our preferved fpecimen : we had no opportunity of obferving the true colour in the living bird ; but it muft certainly have been more of

a red,

PLATE CXVIII.

a red, or at leaſt reddiſh brown, than olive. Neither of theſe differences, however, affect the eſſential character of the ſpecies; and
we conſider our bird, beyond diſpute, the *Gloſſy Ibis* of Dr. Latham.
The length of this Bird exceeds twenty-two inches.

PLATE

119

PLATE CXIX.

EMBERIZA CITRINELLA.

YELLOW HAMMER.

Passeres.

Bill conic, pointed. Noftrils oval, broad, naked.

GENERIC CHARACTER.

Bill conic, the fides of each mandible bending inwards; a hard knob in the roof of the upper mandible.

SPECIFIC CHARACTER

AND

SYNONYMS.

Crown of the head, throat, and belly, yellow. Tail dark: feathers edged with olive, the two outmoft with white, and a white fpot on the interior fides, at the tip of each.

EMBERIZA CITRINELLA : rectricibus nigricantibus, extimis duobus
latere interiore macula alba acuta. *Lin.*
Syft. Nat.
Lath. Ind. Orn. 1. 400. 7.

EMBERIZA FLAVA. *Gefner. av.* 653.
Klein. av. p. 92. 5.
Brif. 3. 258. 1.

Yellow

PLATE CXIX.

Yellow Hammer, *Raii Syn. p.* 93. *A.* 2.

                 Will. Orn. p. 268. *pl.* 40.

                 Albin. 1. *pl.* 66.

                 Br. Zool. 1. *No.* 119. *pl.* 50.

                 Arƈt. Zool.

YELLOW BUNTING. *Lath. Gen. Syn. Vol.* 3. 170. 7.

Le Bruant, *Brif. Orn.* 3. *p.* 258. 1.

                 Buff. Oif. 4. *p.* 342. *pl.* 8.

                 —— *Pl. Enl.* 30. 1.

Cia pagglia riccia, Luteæ alterum genus. *Aldr. av.* 2. 372.

Sternardt. *Scopoli. No.* 209.

Zivolo, Zigolo. *Olin. uc. t. p.* 50.

Ammering, Goldammering. *Kram.* 370.

Groning, Goldfpink. *Faun. Suec. Sp.* 230.

This is a very abundant fpecies throughout Europe. In England, its manners of life are fo well known, that we avoid entering into a minute detail of them. The colours of the female are dull ; and it has fcarcely any yellow about the head.—The male is a very pretty bird; liable, however, to variation in different fpecimens. In fome, the head is brown, in others yellow, marked with brown ; in birds of fine plumage it is of a beautiful yellow ; the colours throughout very full, with the yellow delicately blended into the olive, and have a fine effeƈt.

Thefe Birds feed on grain and infeƈts. They make a large neft of hay and ftraw, mixed with mofs, and dried leaves, and lined

                                                             with

with hair and wool. The neft is generally placed on the ground, or in a low bufh. It lays five or fix eggs, veined irregularly with purplifh or brown colour, and fparingly blotched with the fame. There is more than one brood of this fpecies in the year.

PLATE

120

PLATE CXX.

STERNA SANDVICENCIS.

SANDWICH TERN.

ANSERES.

Bill obtufe, covered with a thin membrane, broad, gibbous below the bafe, fwelled at the apex. Tongue flefhy. Legs naked. Feet webbed or finned.

GENERIC CHARACTER.

Bill ftrait, flender, pointed. Noftrils linear, Tongue flender and fharp. Wings very long. A fmall back toe. Tail forked.

SPECIFIC CHARACTER

AND

SYNONYMS.

White. Back and wings pale, hoary, lead colour. Upper part of the head black; front fpeckled and white. Outer margins of the quill feathers black. Shafts white.

STERNA SANDVICENSIS. *Lath. Suppl. Gen. Syn.*

STERNA BOYSII : alba, dorfo alifque canis, pileo nigro, fronte maculis albis, remigibus nigricantibus fcapo albo.
Lath. Ind. Orn. 2. 806. 10.

H   Sterna

PLATE CXX.

Sterna Cantiaca. *Gmel. Syst.* 1. *p.* 6c6.

Sandwich Tern. *Lath. Syn.* 6. *p.* 356. 9.—*Boy's Sandwich.*

Young *bird*, or variety β.

Sterna Nivæa, cauda emarginato, corpore variegato, macula aurium

nigra. *Linn.*

Rallus lariformis. *Scop. Ann.* 1. *No.* 156.

La Guifette. *Buff. Oif.* 8. *p.* 339.—*Pl. enl.* 924.

This is certainly the new fpecies of Tern, which has been found on the Sandwich coaft, and defcribed under the name *Sandvicenfis* * by Mr. Latham: Our fpecimen does not precifely agree with the defcription given by that author; but the following confiderations feem to juftify cur conclufion.

The plumage of many Birds differ confiderably in colours and markings in the various ftages of their growth; and even in the adult ftate, we frequently obferve varieties which it is difficult to refer to their true fpecies.—From the general appearance of the Bird before us, it has been thought the young of the common Tern, and it ftands for fuch in the Leverian Mufeum: It cannot however be the young Bird of that fpecies; in the *contour* it is obvioufly different, and the form of the beak, with other ftriking peculiarities, muft certainly remove it from that fpecies.

---

* Catalogue of Britifh Birds. Vide Supplement of Synopfis. Altered to Sterna Boyfii in *Index Orn.*

Mr.

6

PLATE CXX.

Mr. Latham, in defcribing the Sandwich Tern, fays, the back and wings are a pale hoary lead colour, and in the young Birds are much clouded with brown : he fays alfo the head is much dotted with white in fome fpecimens ; but obferves, that all Terns with black heads are liable to the fame variation *. Thus far our fpecimen may be confidered as the Sandwich Tern ; but the colour of the legs and claws of that Bird is uniformly faid to be black, while in ours they are orange : this is however accidental, and by no means a permanent character ; the orange-coloured legs and feet is ftriking, but not invariable in the Common Tern, as is proved by the variety β. with black feet, defcribed by Mr. Latham in the Index Ornithologicus † ; we alfo find a fpecimen of the Sandwich Tern in the Britifh Mufeum, which has the legs and feet of a dull yellowifh or orange colour, and differs from the fpecimen we have figured only in the form of the tail, which is not forked as in the adult Birds.

The *Sterna Nævia* of Linnæus, and *La Guiffette* of Buffon, is confidered by Mr. Latham as the young Bird of the Sandwich Tern. Our fpecimen differs very little from the defcriptions given by thefe authors.

Length of our Bird fourteen inches. It was fhot in the Chelfea road.

---

* In the adult common Tern, the black of the head extends to the bafe of the bill ; in the Sandwich Tern the forehead is white.

† Sterna Hirundo. β. *var.* pedibus nigris, rectricibus extimis toto albis. *Lath. Ind. Orn.* 2. 808. 15.—*Phil. Tranf.* lxii. *p.* 421. *Forfter.*

PLATE

121

P L A T E   CXXI.

S T R I X   S T R I D U L A.

T A W N Y   O W L.

Accipitres.

Birds of prey. Bill and Claws ſtrong hooked, an angle in the margin of the upper mandible. Body muſcular. Females larger and more beautiful than the males.

GENERIC CHARACTER.

Bill hooked, without cere. Noſtrils covered with briſtly feathers. Head large. Eyes and ears very large. Tongue bifid.

SPECIFIC CHARACTER

AND

SYNONYMS.

Head ſmooth. Body reddiſh, or tawny brown. White ſpots on the wing.

Strix Stridula: capite lævi, corpore ferrugineo, remige tertia
longiore. *Lin. Syſt. Nat.*—*Fn. Suec.* 55.

I

Strix.

PLATE CXXI.

Strix. *Aldr. av.* 1. 285.

Strix Orientalis. *Haffelquiff, Itin.* 233.

Noctua Major, *Frifch.*

COMMON BROWN, OR IVY OWL. *Will. Orn.* 102. *t.* 14.

—*Albin. I. t.* 9.

Raii Syn. av. 25

TAWNY OWL. *Br. Zool. No.* 68.

Lath. Gen. Syn. 1. 139. 27.

—*Ind. Orn.* 1. *p.* 58. 25.

Le Chathaunt. *Brif. Orn.* 1. *p.* 500. *No.* 1.

Buff. Oif. 1. *p.* 362. *t.* 25.

—— *Pl. Enl.* 437.

Braune, oder ftock Eule. *Frifch. t.* 96. (*maf.*)

Gelblicke, oder brand Eule. *Frifch. t.* 95. (*fem.*)

Strige. Zinnan. *Uov. p.* 100. *t.* 16. *f.* 89.

Skrik uggla. *Faun. Suec.*

Nacht Eule, Gemeine. *Kram.* 324.

Nat Ugle. *Brun.* 18.

———————————————

The length of this bird is fourteen inches; breadth thirty-two inches. The male is darker in colour than the female. This kind of owls inhabits woods: in England they remain the whole year. The fpecies is found throughout Europe and America.

La Chouette, ou Grande Chevêche, of Buffon, is fuppofed to be the female of the Tawny Owl; but the defcription differs in fome particulars from thofe fpecimens we have examined.

PLATE

122

PLATE CXXII.

RALLUS PORZANA.

SMALL SPOTTED GALLINULE, or WATER-HEN.

GRALLÆ.

Bill roundifh. Tongue entire, flefhy. Thighs naked. Toes divided.

GENERIC CHARACTER.

Bill compreffed, fharp. Noftrils oval. Toes four. Body com-preffed.

SPECIFIC CHARACTER

AND

SYNONYMS.

Above olive brown, variegated with fpots and dafhes of black and white: beneath afh-colour, with white marks.

Rallus Porzana. *Lin. Syft. Nat.* 1. *p.* 262. 3.—*Gmel. Syft.* 1. *p.* 712.

Gallinula Porzana: fufco-olivacea nigro albidoque variegata et maculata, fubtus cinerea albido varia, rectricibus

K duabus

PLATE CXXII.

duabus intermediis albo marginatis. *Lath. Ind. Orn.* 2. 772. 19.

Rallus aquaticus minor, five Mauetta. *Brif.* 5. *p.* 155. 2. *t.* 13. *f.* 1.

Gallinula ochra Gefneri. *Raii Syn. p.* 115. 7.

Spotted Gallinule. *Br. Zool.* 11. *No.* 215. *Lath. Gen. Syn.* 5. *p.* 264. 18.

Petit Rale d'eau, ou le Marouette. *Buff. Oif.* 8. *p.* 157.

Kleines gefprenkeltes Wafferhuhn. *Frifch.* 2. 211.

This elegant fpecies is fcarce in Great-Britain. It is of the migratory kind; but is known to breed here. Mr. Latham fays in Cumberland.

This is a folitary creature, living entirely among reeds in marfhy places. The neft is very fingular, and is built on the water; it is compofed of rufhes matted together in form of a boat, and is faftened by one end to a reed that it may float in fecurity on the water while the female fits on the eggs.

The length of this bird is nine inches.

PLATE

123

PLATE CXXIII.

MOTACILLA RUBECULA.

RED-BREAST.

PASSERES.

Bill conic, pointed. Noftrils oval, broad, naked.

GENERIC CHARACTER.

Bill ftrait, flender. Tongue jagged.

SPECIFIC CHARACTER

AND

SYNONYMS.

Greyifh. Throat and breaft ferruginous orange.

MOTACILLA RUBECULA, grifea, gula pectoreque ferrugineis,
   Lin. Syft. Nat.

Sylvia Rubecula. *Lath. Ind. Orn.* 2. 520. 42.

Robin Red-breaft, or Ruddock. *Will. Orn.* 219.

Red-breaft. *Br. Zool. Raii Syn. av.* 78. 1. 147.
   Lath. Gen. Syn. 4.

Le Rouge-gorge. *Brif. av.* 3. 418.

Pettiroffo. *Olin. uc. t. p.* 16.

Rotgel. *Faun. Suec. fp.* 260.

K 2

Roed-

PLATE CXXIII.

Roed-Finke, Roed-Kielke. *Brun.* 283.

Rothkehlein. *Frifch.* 1. 19.

Rothkropfl. *Kram.* 376.

Smarnza, Tafchtza. *Scop. No.* 231.

The manners and œconomy of this little creature are familiar to every one. It frequents inhabited places in the winter; in fummer it retires into thickets or decayed buildings to breed. The neft is compofed of dried leaves mixed with hair and mofs, and lined with feathers: it contains from five to feven eggs of a dufky white colour, fprinkled with irregular reddifh fpots. The young birds are very unlike the adults, being fpotted with white.

The Robin has been chofen by our earlieft poets to pourtray inftinctive affection towards man. An artlefs tale *: a pathetic appeal to the tendereft feelings, pleads its behalf to the infant mind, and maturer age rather cherifhes than difcards its firft impreffions. Hence the Robin, through fucceffive ages, has become an object of fondnefs, and fuperftitious refpect; and, as if confcious of our pity and protection, it boldly vifits our dwellings in winter, and claims that fubfiftence the inclement feafon denies.

> ——" The RED-BREAST, facred to the houfehold gods,
> Wifely regardful of th' embroiling fky,
> In joylefs fields and thorny thickets leaves
> His fhivering mates, and pays to trufted man
> His annual vifit. Half afraid, he firft

------

* An ancient and fimple ballad, *The Babes in the Wood.*

Againft

PLATE CXXIII.

Againſt the windows beats: then briſk alights
On the warm hearth; then hopping o'er the floor,
Eyes all the family aſkance,
And pecks, and ſtarts, and wonders where he is,
Till more familiar grown, the table crumbs
Attraċt his ſlender feet."

THOMSON.

Diveſted of the pleaſing poetic ſimiles that are interwoven with the hiſtory of the Robin, it is a ſavage little animal, and in perpetual warfare with its own ſpecies, and every other tribe of ſmall birds. It feeds on infeċts and feeds, or when preſſed by hunger, on many other kinds of food. The note is fine and ſoft.

K 3

PLATE

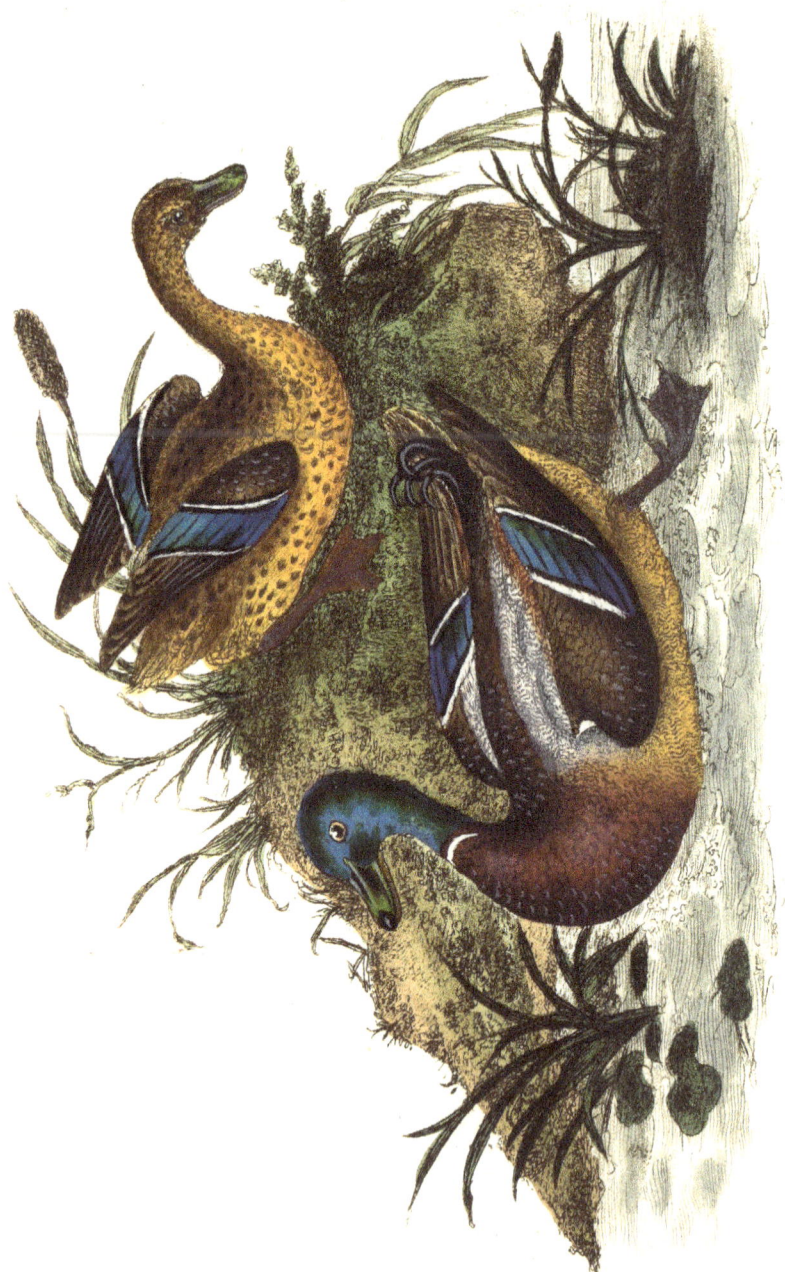

124

PLATE CXXIV.

ANAS BOSCHAS.

MALLARD.

ANSERES.

Bill obtufe, covered with a thin membrane, broad, gibbous below the bafe, fwelled at the apex. Tongue flefhy. Legs naked. Feet webbed or finned.

GENERIC CHARACTER.

Bill convex above, flat beneath, fwelled at the apex, with membranous teeth.

SPECIFIC CHARACTER

AND

SYNONYMS.

Afh colour, middle tail feathers of the male recurved. Bill ftrait. An incomplete white collar on the front of the neck.

ANAS BOSCHAS cinerea, rectricibus intermediis (maris) recurvatis, roftro recto, torque alba. *Lin. Syft. Nat.*—
Lath. Ind. Orn. 2. 850. 49.
Anas domeftica. *Gefner av.* 113. 96.

K 4

Common

PLATE CXXIV.

Common Wild Duck and Mallard.   *Will. Orn.* 371. 380.

Raii Syn. p. 145. *A.* 1. 150. 1.

Albin. 2. *pl.* 10.—1. *pl.* 99.

Br. Zool. 2. *p.* 279. *pl.* 97.

Arct. Zool. No. 494.

Lath. Gen. Syn. 6. *p.* 489. 43.

Le Canard Sauvage.   *Brif. Orn.* 6. *p.* 318. 4

Buff. Oif. 9. *p.* 115. *pl.* 7. 8.

—*Pl. Enl.* 776. 777.

Gras-And, Blanacke.   *Faun. Succ. Sp.* 131.

Wolde Ento.   *Frifch.* a1. 158—159.

Ratza.   *Scopoli.*

Einheimifche ent. Stock ent.   *Kram.* 341.

This is the parent ftock of our domefticated or common Duck. The varieties in a tame ftate are endlefs, but they uniformly pre-ferve one character by which we can trace them to this wild origin; this is the fhort curled tail feathers of the Drakes, which are con-ftant in all its varieties of plumage, and the form of the bill in both fexes, whether in a wild or domeftic ftate.

Thefe birds are fo well known, that we beftow little attention on their beauty; or we fhould confider the Mallard Drake the moft beautiful of the web-footed birds that inhabit this country. The plumage throughout is of fingular richnefs, and the various dotted ftreaks and lines on the plainer colours are uncommonly elegant. The colours of the female, as in other inftances, are more fimple.

Wild

PLATE CXXIV.

Wild Ducks abound in this country, but are no where more plenty than in the fens of Lincolnſhire*. The means of taking theſe birds are various, and have been deſcribed with minuteneſs by many authors. The method in common uſe in England is to ſet large decoy nets in the places they frequent, and by means of a trained bird entice them into its labyrinths †. The inhabitants of other countries have alſo various and peculiar contrivances to entrap theſe, and others of the water fowl. Theſe birds breed in marſhy places, and lay from ten to ſixteen eggs. Dr. Latham ſays they are ſometimes known to lay the eggs in a high tree, in a deſerted Magpie, or Crow's neſt, and mentions an inſtance of one being found at *Etchingham* in Suſſex, ſitting upon nine eggs, in an Oak, twenty-five feet from the ground: the eggs were ſupported by twigs laid crofsways ‡.

Length of this ſpecies near two feet; weight, two pounds and an half.

---

* In only ten decoys, in the neighbourhood of *Wainfleet*, thirty-one thouſand two hundred have been taken in one ſeaſon.—*Britiſh Zoology.*

† The decoy nets are generally placed on a piece of water nearly ſurrounded with wood, that the birds may not be frightened or diſturbed. They are ſo contrived, that different pipes lead to it from ſeveral directions; theſe pipes are ſo many avenues of net-work ſupported by hoops, which become gradually narrower from the opening, and lead up a ditch, at the end of which the funnel net is placed. Along theſe pipes, at proper intervals, are ſcreens of reeds, behind which the decoy man conceals himſelf from the birds. The Ducks trained for decoys are fed on hemp ſeed, which being light, floats on the ſurface of the water. When the evening ſets in, the decoy man throws ſome of the ſeeds from behind one of the ſcreens to the trained bird; this entices him into the pipe, and the wild fowl follow. When they arrive at a certain part of the avenue, the decoy bird dives under the water, and the reſt paſs on till they enter the purſe net, where they are taken.—*Brit. Zool. &c. &c.*

‡ *Lath. Gen. Syn.*

INDEX

INDEX TO VOL. V

ARRANGEMENT

ACCORDING TO THE

SYSTEM OF LINNÆUS.

ORDER I.

ACCIPITRES.

ORDER II.

PICÆ.

ORDER III.

ANSERES.

ORDER

INDEX.

ORDER IV.
GRALLÆ.

ORDER V.
GALLINÆ.

ORDER VI.
PASSERES.

VOL.

INDEX.

VOL. V.

ARRANGEMENT

ACCORDING TO

LATHAM's SYNOPSIS of BIRDS.

ORDER

INDEX.

DIVI-

INDEX.

VOL.

I N D E X.

V O L. V.

A R R A N G E M E N T

A C C O R D I N G T O

PENNANT's BRITISH ZOOLOGY.

INDEX.

L INDEX.

INDEX.

VOL. V.

ALPHABETICAL ARRANGEMENT.

FINIS.

THE

NATURAL HISTORY

OF

BRITISH BIRDS;

OR, A

SELECTION OF THE MOST RARE, BEAUTIFUL, AND INTERESTING

BIRDS

WHICH INHABIT THIS COUNTRY:

THE DESCRIPTIONS FROM THE

SYSTEMA NATURÆ

OF

LINNÆUS;

WITH

GENERAL OBSERVATIONS,

EITHER ORIGINAL, OR COLLECTED FROM THE LATEST
AND MOST ESTEEMED

ENGLISH ORNITHOLOGISTS;

AND EMBELLISHED WITH

FIGURES,

DRAWN, ENGRAVED, AND COLOURED FROM THE ORIGINAL SPECIMENS.

By E. DONOVAN.

VOL. VI.

LONDON:

PRINTED FOR THE AUTHOR; AND FOR F. C. AND J. RIVINGTON,
No. 62, ST. PAUL'S CHURCH-YARD. 1816.

Law and Gilbert, Printers, St. John's, Square, London.

125

PLATE CXXV.

COLUMBA TURTUR.

TURTLE DOVE.

PASSERES.

GENERIC CHARACTER.

Bill ſtraight, defcending towards the tip : noſtrils oblong, and half covered with a foft tumid membrane.

SPECIFIC CHARACTER

AND

SYNONYMS.

Tail feathers tipped with white : back grey : breaſt fleſh colour : each fide the neck a fpot of black feathers with white tips.

COLUMBA TURTUR : rectricibus apice albis, dorſo grifeo, pectore incarnato, macula laterali colli nigra lineolis albis. *Scop. Ann.* 1. *n.* 181.

Turtur. *Olin. ucc.* 34.—*Briſſ. av.* 1. *p.* 92. *n.* 7.—*Norzem. nederl. Vogel. t.* 6.

Tourterelle. *Buff. Hiſt. Oiſ.* 2. *p.* 545. *t.* 25.—*Pl. Enl. n.* 394.

Turtle, or Turtle-dove. *Brit. Zool.* 1. *n.* 10° *t.* 45.—*Albin. av.* 2. *t.* 47.

A 2

The

PLATE CXXV.

Turtle Doves inhabit Europe, and fome parts of Afia, being met with in China, and India. In Britain they are not uncommon during the fummer feafon, arriving however later in the fpring, and leaving the country earlier in autumn than any other of the pigeon tribe.

The conftancy of the Turtle-dove to its mate is proverbial. They build in general in the thickeft woods, forming their nefts on the branches of the loftieft trees: like the reft of the pigeons, they lay only two eggs; and, it is concluded, breed only once in a feafon. Thefe birds are of a fhy and timid difpofition, and feldom venture from their woody retreats in the day-time, unlefs in flocks of about twenty together. They fubfift on berries, fruits, and various kinds of vegetables, and are efpecially partial to peas, the fields of which they vifit as foon as the crop begins to ripen, and oftentimes commit vaft depredations. There are feveral fuppofed varieties of this fpecies, fome of which may prove hereafter to be diftinct kinds.

This is a bird of extremely beautiful plumage, and meafures in length about twelve inches.

PLATE

126

PLATE CXXVI.

HIRUNDO RIPARIA.

SAND MARTIN.

PASSERES.

GENERIC CHARACTER.

Bill fmall, weak, incurvated, fubulate, and depreffed at the bafe : gape larger than the head : tongue fhort, broad, and cleft : wings long : tail generally furcated.

SPECIFIC CHARACTER

AND

SYNONYMS.

Cinereous, throat and abdomen white.

HIRUNDO RIPARIA : cinerea, gula abdomineque albis. *Linn. Fn. Suec.* 273.
    Kram. el. p. 381. *n.* 4.
    Müll. Zool. p. 34. *n.* 289.

HIRUNDO RIPARIA. *Gefn. av.* 656.
    Aldr. orn. 2. *p.* 694. *t.* 695.
    Ray av. p. 71. *n.* 3.

Hirondelle de rivage. *Buff. Hift. Nat.* 6. *b.* 632.—*Pl. enl. n.* 643. *f.* 2.

          Uferfchwalbe.

PLATE CXXVI.

Uferfchwalbe. *Frifch. av. t.* 18.

Georg. it. p. 175.

Sand-Martin, or Shore-bird. *Arct. Zool.* 2. *p.* 430. *t.* 332.

Will. orn. p. 213. *t.* 39.

A fmall bird of very delicate appearance, meafuring rather lefs than five inches in length, and having the upper part of the plumage brown, the lower white. This little fpecies inhabits various parts of Europe and America: in Britain, it may be confidered, if not a rare, at leaft a very local fpecies.

The Sand-Martin, as its name implies, is in a great meafure peculiar to fandy places. Their ufual haunts are the fides of fand-banks contiguous to lakes and rivers, where they live in deep openings, dug in an horizontal courfe, and at the furtheft end of which receffes the neft is depofited. Sometimes the neft of the Sand-Martin is built in cavities of rocks, fand-pits, or hollow trees; almoft invariably, however, on the banks of lakes or rivers, the old birds fubfifting, for the moft part, on the infects which it takes on the wing when fkimming the furface of the water.

The neft confifts of fibres of grafs, intermixed with ftraw, and a few feathers: the eggs, about fix in number, are of a beautiful white colour, and tranfparent. It cannot be abfolutely afferted whether the Sand-Martin has two, or only a fingle brood in the year.

PLATE

127

PLATE CXXVII.

MEROPS APIASTER.

COMMON BEE-EATER.

PICÆ.

GENERIC CHARACTER.

Bill curved, quadrangular, compreffed, carinate, pointed : noftrils fmall, at the bafe of the bill : tongue flender, the tip moftly jagged : feet grefforial.

SPECIFIC CHARACTER
AND
SYNONYMS.

Back ferruginous : abdomen and tail green-blue : two tail feathers longeft : throat yellow.

MEROPS APIASTER : dorfo ferrugineo, abdomine caudaque viridi-cœrulefcente, reftricibus duabus longioribus, gula lutea. *Gmel. Syft.* 1. *p.* 460.—*Linn. Syft.* 1. *p.* 182. 1.

Merops galilæus. *Haffelq. It.* 247.—*Fn. Arab. p.* 1.

Bienenfraas. *Wirfing. Vog. t.* 27.

La Merope. *Cet. uc. Sard. t. p.* 93.

Ifpida cauda molli. *Kram. El. p.* 337.

BEE-

PLATE CXXVII.

BEE-EATER. *Will. p.* 147.

Albin. 2. *t.* 44.

Linn. Tranf. v. 3. *p.* 333.

The **Common Bee-eater** is a bird of very fplendid plumage, the male in particular, the colours in that fex being of a more vivid caft than in the female. It has been denominated the European Bee-eater; a term not, perhaps, fufficiently difcriminate, fince another fpecies of the fame genus, the Merops Congenor, is alfo found in Europe. Neither is the Merops Apiafter peculiar to Europe: it extends both to Africa and Afia: in Europe it is confined to fouthern countries, and in Afia to more temperate regions.

Till within the laft few years, this fpecies was altogether unknown as a native of Britain. The firft account we have of its being met with in this country, is recorded in the third volume of the Linnæan Tranfaftions, from which it appears, that a folitary example of the fpecies was fhot in the year 1794, and communicated to Dr. Smith.

This bird meafures about ten inches, from the tip of the bill to the end of the tail. Its food confifts of infefts of various kinds, more efpecially thofe of the bee tribe, in reference to which it bears the name of Bee-eater. Their nefts are faid to be compofed of mofs, and to be depofited, like that of the fand-martin, in deep holes in the banks of rivers. The eggs are fmaller than thofe of a blackbird, of a white colour, and from five to feven in number.

PLATE

128

PLATE CXXVIII.

PERDIX COTURNIX.

COMMON QUAIL.

GALLINÆ.

GENERIC CHARACTER.

Bill convex and ftrong : noftrils with a prominent margin : orbits papillous : legs naked, and moftly armed with a fpur.

SPECIFIC CHARACTER

AND

SYNONYMS.

Legs unarmed : body grey, fpotted : eye-brows white : tail feather with ferruginous edge and crefcent.

PERDIX COTURNIX : mutica, corpore grifeo maculato, fuperciliis albis, rectricibus margine lunulaque ferruginea. *Lath. Ind. Orn. t. 2. p.* 651. *n.* 28.

TETRAO COTURNIX. *Linn. Syft.* 1. *p.* 278. 20.—*Fn. Suec.* No. 206.—*Gmel. Syft.* 1. *p.* 765.

Quaglia. *Zinnan. Uov. p.* 36. *t.* 5. *f.* 19.

Wachtel. *Gunth. Neft. u. Ey. t.* 35.

La Caille. *Buff.* 2. *p.* 449. *t.* 16.—*Pl. enl.* 170.

Quail. *Arct. Zool.* 2. *p.* 320. B.

Albin, 1. *t.* 80.

Br. Zool. 1. *No.* 97.

VOL. VI.   B   The

PLATE CXXVIII.

The Quail is a fmall bird, about feven inches and a half in length; the female differs from the male, in being deftitute of the black fpots on the fore part of the neck, breaft, and lateral feathers, and in the rufous fpace being paler.

This bird occurs in the greateft plenty throughout Ruffia, Tartary, China, and other parts of India, changing its fituation in immenfe flocks according to the feafons, in the fpring proceeding northward, and returning fouthward in autumn. In the iflands of the Archipelago, and the fouth of Europe, they appear at certain times in immenfe myriads. Twice in the year, it is faid, they come in fuch vaft abundance into the ifland of Capri, that the bifhop derives the chief part of his revenue from them ; and on the weft coaft of the kingdom of Naples, within the fpace of four or five miles, an hundred thoufand have been taken in a day. The Quail is not a common bird in this country, although it breeds with us : during the fummer it is found in the northern counties, and migrates fouth in autumn. Thefe birds feed on corn, and other grain.

PLATE

129

PLATE CXXIX.

ANAS ACUTA.

PIN-TAIL DUCK.

ANSERES.

GENERIC CHARACTER.

Bill convex, obtufe, the edges divided into lamellate teetn : tongue fringed and obtufe : three fore toes folitary.

SPECIFIC CHARACTER

AND

SYNONYMS.

Tail pointed, elongated, and black beneath ; hind head on each fide with a white line : back waved cinereous.

ANAS ACUTA: cauda acuminata elongata fubtus nigra, occipite utrinque linea alba, dorfo cinereo undulato. *Linn. Fn. Suec.* 126.

Anas cauda forcipata pedibus longiore, macula alarum cupreo-fufca. *Kram. el. p.* 340.

Anas longicauda. *Brif. av.* 6. *p.* 369. *n.* 16. *t.* 34. *f.* 1. 2.

Anas Seevogel. *Aldr. Orn.* 3. *p.* 229.

Canard à longue queue. *Buff. Hift. Nat.* 9. *p.* 199. *t.* 13.--*Pl. Enl. N.* 954.

B 2   SEA

PLATE CXXIX.

SEA PHEASANT, or CRAKER. *Ray, av. p.* 147. *A.* 5.
PIN-TAIL. *Arct. Zool.* 2. *p.* 566. *n.* 500.
Brit. Zool. 2. *n.* 282.

———————

A native of Europe, Afia, and America. Sometimes thefe birds appear in fmall flocks on the Englifh coafts during the winter feafon, but depart again to the northward early in the fpring. They are feen at this feafon in Sweden and Denmark, and are fuppofed to breed in the north of thofe kingdoms. They abound throughout the Ruffian territories as far as Kamtfchatka, at which latter place they breed and live till the approach of winter. According to Decouver, the fpecies appears at particular feafons, in troops of feveral hundred together, on the borders of the river Don. It is likewife faid to breed at Hudfon's Bay, in America.

The male of the Pintail Duck is an interefting bird, poffeffing, in addition to a beautifully varied plumage, a peculiar degree of elegance in its manners and general afpect. Its length is twenty-eight inches, and weight twenty-four ounces. The female, as ufual in the duck tribe, fmaller, and more dufky, and is in particular diftinguifhed by having a fpot of ftraw-colour on the wing inftead of violet. The young males remain of a greyifh brown, not very unlike the plumage of the females, till the fpring after they are hatched, when they affume their proper plumage. The flefh is excellent.

PLATE

PLATE CXXX.

CORVUS MONEDULA.

JACKDAW.

PICÆ.

GENERIC CHARACTER.

Bill convex, acutely edged : noftrils covered with fetaceous recumbent feathers : tongue cartilaginous and bifid : feet formed for walking.

SPECIFIC CHARACTER

AND

SYNONYMS.

Blackifh brown : hind head hoary : front, wings, and tail black.

CORVUS MONEDULA : fufco nigricans, occipite incano, fronte alis
     caudaque nigris. *Gmel. Syft. Nat.* 1. *p.* 367.—
     Linn. Syft. 1. *p.* 156. 6.—*Fn. Suec. No.* 89.
Cornix garrula. *Klein. av. p.* 59. 4.
Dohle. *Gunth. Neft. und Eyer. p.* 51. *t.* 11.
Taccola. *Cett. uc. Sard. p.* 72.
Mulacchia hera. *Zinnan. Uov. p.* 71. *t.* 10. *f.* 62.
Jackdaw. *Arɗ. Zool.* 2. *p.* 251. C.
     Br. Zool. 1. *No.* 81. *t.* 34.

The

PLATE CXXX.

The Jackdaw, one of the moft familiar and well-known birds of its tribe in this country, does not appear to be by any means fo abundant throughout the reft of Europe as fome others which are lefs frequent with us. In England they are feen at all feafons; in France and Germany, which countries they inhabit likewife, they are migratory. In Sweden and Denmark they are pretty conftant inhabitants. The fpecies occurs alfo in the weftern part of Siberia, and a fuppofed variety extends as far as Perfia.

In a wild ftate, the haunts of this bird are the moft retired places among rocks, or ruined edifices, in the cavities of which they conftruct their nefts: occafionally they build in trees, but not commonly. Their eggs, from five to fix in number, are of a paler colour, and have a fmaller number of fpots, than the hooded crow.

This fpecies feeds on infects, grain, and feeds. It is of a docile difpofition, and may be eafily rendered tame, but invariably retains its pilfering inclinations even in a ftate of domeftication. The note of this bird is very peculiar :—an ejaculation of the words *Jakdaw*, *Jakdaw*, not indiftinctly articulated, and repeated at regular intervals; this note is often heard while the bird is on the wing.

PLATE

PLATE CXXXI.

PERDIX RUFA.

RED-LEGGED PARTRIDGE.

GALLINÆ.

GENERIC CHARACTER.

Bill convex, and ſtrong : noſtrils with a prominent margin : orbits papillous : legs naked, and moſtly armed with a ſpur.

SPECIFIC CHARACTER
AND
SYNONYMS.

Legs and bill ſanguineous : chin white, ſurrounded with a black band, and ſpotted with white.

PERDIX RUFA. *Lath. Ind. Orn.* 647. *n.* 12.

TETRAO RUFUS: pedibus roſtroque ſanguineis, gula alba cinâa faſcia nigra albo-punâata. *Kram. el. n.* 357. *n.* 5.

TETRAO RUFUS: reâricibus cinereis, ſuperiore medietate hinc inde rufis. *Linn. Fn. Suec.* 171.

Perdix græca. *Briſſ. av.* 1. *p.* 241. *n.* 12. *t.* 23. *f.* 1.

Bartavelle. *Buff. Hiſt. Nat.* 2. *p.* 420.

RED PARTRIDGE. *Albin. av.* 1. *p.* 27.

GREEK PARTRIDGE, or Great Red Partridge. *Will. Orn. p.* 169.

This

PLATE CXXXI.

This is a bird of confiderable beauty, and is found in fuch vaft abundance in the ifland of Guernfey, as to have obtained the appellation of Guernfey Partridge. Birds of the fame fpecies have been alfo fhot in a wild ftate on the coafts of Norfolk, Kent, and Suffolk, which latter circumftances tend, in our minds, more fully to eftablifh its claim to a place in the Britifh Fauna, than its being an indigenous inhabitant of Guernfey ifland. Some attempts have been made to naturalize the fpecies in the fouthern counties of England, which have not, however, been ultimately attended with the fuccefs anticipated. The flefh is in much efteem, and the birds, for this reafon, are often brought over to England from Guernfey, or, in times of peace, from France, to fupply the tables of the affluent.

Throughout the whole of the fouth of Europe, and alfo in Afia and Africa, thefe birds occur in the greateft plenty ; fo much, indeed, that in fome of the Greek iflands, the natives deftroy as many of their eggs as poffible, in order to diminifh their numbers ; a precaution highly neceffary for the prefervation of their corn-harvefts, which often fuftain vaft injury from their depredations, as thefe birds affociate in immenfe flocks, and fubfift principally on grain during that feafon. In many refpects, their manners accord with thofe of the common partridge, though, in the latter particular, they differ materially, the partridge being a folitary bird ; and it is obfervable likewife, that the Red-legged Partridges occafionally perch on trees, which is altogether unufual with the common kind. There appears to be two or more varieties of this fpecies of partridge.

PLATE

PLATE CXXXII.

ANAS CLANGULA.

GOLDEN EYE DUCK.

GENERIC CHARACTER.

Bill convex, obtufe, the edges divided into lamellate teeth : tongue fringed and obtufe : three fore toes connected, the hind one folitary.

SPECIFIC CHARACTER

AND

SYNONYMS.

Varied with black and white : head tumid, and violet; at each corner of the mouth a large white fpot.

ANAS CLANGULA : nigro alboque varia, capite tumido violaceo : finu oris macula alba. *Linn. Fn. Syec.* 192.— *It. oel.* 49.

CLANGULA. *Geffn. av.* 119.—*Aldr. ofn.* 3. *p.* 224.

Garrot. *Buff. Hift. Nat.* 9. *p.* 222. *pl. enl. n.* 802.

Kobelente. *Frifch. av. t.* 183. 184.

Quakerente. *Bloch Befch. der Berl. Naturf. Fr.* 4. *p.* 599. *n.* 9. *t.* 17. *fig.* 1. 2.

Smaller Reddifh Duck. *Will. Orn. p.* 369. *Ray av. p.* 143. *n.* 1.

VOL. VI.   C   GOLDEN

PLATE CXXXII.

GOLDEN EYE.   *Albin.* 1. *t.* 96.

              Arct. Zool. 2. *p.* 557. *n.* 486.

              Brit. Zool. 2. *n.* 276.

This species of Duck rather exceeds the common size, measuring nearly twenty inches in length, and weighing thirty ounces. This relates to the male only, the female being much smaller. The latter differs likewise very considerably in the colour of its plumage, which is principally of an obscure brown, varied with black and ash. The head is dark and reddish, the breast and belly white; the middle quill-feathers are also white, but the rest of the wings, except the coverts and scapulars, in which the grey prevails, is black. The tail is of the last-mentioned colour; the legs dusky.

The Golden Eye is of the migratory kind, and visits us only in the winter, at which season it is seen in small flocks on many of our sea-coasts. In the spring these birds retire northward, remaining, during the breeding season, in Russia, Norway, and Sweden. In America it is found throughout the summer in Hudson's Bay, where it is observed to frequent fresh water lakes. Its principal food consists of shell-fish, frogs, and other reptiles, mice, &c. The nest, which is composed of grass, and lined with feathers, is of a rounded form, the eggs from seven to ten in number, and white. Linnæus tells us, this species sometimes builds in trees, a circumstance rather singular, but nevertheless, we believe, correctly stated, as some other species of the duck tribe have been known to build occasionally in the same or similar situations.

PLATE

133

PLATE CXXXIII.

PELECANUS CARBO.

CORVORANT.

ANSERES.

GENERIC CHARACTER.

Bill ſtraight, hooked at the point, and furniſhed with a nail : noſ-trils an obliterated ſlit : face rather naked : legs equally balancing the body : toes four, palmated.

SPECIFIC CHARACTER
AND
SYNONYMS.

Tail rounded : body black : head ſubcreſted.

PELECANUS CARBO : cauda rotundata, corpore nigro, capite
ſubcriſtato. *Linn. Fn. Suec.* 145.

Phalacrocorax. *Geſn. av.* 683

Corvus aquaticus. *Aldr. Orn. 3. p.* 261. *t.* 263.

Cormoran. *Buff. Hiſt. Nat.* 8. *p.* 310. *t.* 26.

Corvorant. *Arɛt. Zool.* 2. *p.* 581. *n.* 509.

Brit. Zool. 2. *n.* 291.

C 2

The

PLATE CXXXIII.

The Corvorant meafures three feet in length, in breadth four feet, and weighs feven pounds.

Towards the northern parts of Europe thefe birds are more abund-ant than in the fouth. They breed in Kamtfchatka, Greenland, Ice-land, and other countries of the north, and from their habits of life are almoft conftantly found contiguous to the fea. Their breeding-places are the higheft and moft inacceffible cliffs impending over the fea fhore. It does not appear that they evince much ingenuity in the formation of a neft, and they lay only about three or four eggs, which are the fize of thofe of a goofe, and of a pale green colour. The Corvorant is an extremely voracious bird, and preys chiefly on fifh, in purfuit of which it is continually feen fwimming and darting into the water. The flefh is very indifferent, and eaten only by the Greenlanders, who, however, are by no means partial to it, and their eggs are fo difgufting that they are never eaten. The fkins, in common with thofe of moft other birds, furnifh the natives with an article of drefs. In Britain the Corvorant is uncommon on the fouthern coaft, but is ftill more abundant on the fea coafts of the northern counties.

In China the Corvorant is trained up for the purpofe of fifhing, in which employ they are very ufeful to the fifhermen; many of whom keep feveral of them, and derive a good livelihood from their labours. Thefe birds are taught to plunge into the water at their mafter's command, and feize the fifh in his bill, or with the bill and talons together, and bring it to his mafter; or if the fifh be too large, two of the Corvorants affift each other. To enfure obedience, it is however neceffary to faften a ring round their necks, which prevents them from fwallowing their prize; and when the bufinefs of fifhing is over for the day, the mafter takes off the ring, and rewards them

with

PLATE CXXXIII.

with a fhare of the fpoil. According to Willughby, this mode of taking fifh was practifed formerly in England : the Corvorants were inftructed to dive into the water, and after laying the captive fifh at the feet of his mafter, perch upon his arm. The beft writers give little credit to this account, or confider it as a rare inftance, at leaft, of the docile difpofition of the Corvorant. A leather thong was faftened round the throat of the bird, which anfwered the fame pur-pofe as the ring put round their neck by the Chinefe.

PLATE

134

PLATE CXXXIV.

MOTACILLA TROGLODYTES.

WREN.

PASSERES.

GENERIC CHARACTER.

Bill fubulate, and ftraight: the mandibles nearly equal: noftrils oval: tongue lacerated at the end.

SPECIFIC CHARACTER

AND

SYNONYMS.

Grey: eyebrows white: wings waved with black and cinereous.

MOTACILLA TROGLODITES; grifea, alis nigris cinereoque undu-
        latis. *Linn. Fn. Suec.* 261.
               Scop. ann. 1. *n.* 239.
               Nozem. nederl. Vogel. t. 58.

Regulus. *Briff. av.* 3. *p.* 425. *n.* 24.

Reattino. *Olin. ucc. t.* 6.

Roitelet. *Buff. pl. enl. n.* 651. *f.* 2.

Zaunkoenig. *Frifch. av. t.* 24. *f.* 3.

WREN. *Arct. Zool.* 2. *p.* 414. *n.* 322.

      Ray av. p. 80. *n.* 11.

      Brit. Zool. 1. *n.* 154.

This

PLATE CXXXIV.

This pretty little bird is efteemed the fmalleft of the European fpe-cies, the Golden Crefted Wren excepted. It inhabits the temperate parts of northern Europe, and the fouth of Afia.

The Wren remains in Britain throughout the year. The neft is of an oval form, compofed of mofs lined with feathers, and having a fmall entrance in the middle. The female has two broods annually, one in April, the other in June, at each time laying from ten to fix-teen eggs, which are of a white colour, and marked at the end with reddifh. Its neft is commonly found affixed againft outhoufes, or old walls, or, if built in the woods, ftands generally on a low ftump among bufhes near the ground.

The note of this bird is a pleafing warble, heard at all feafons of the year, but only in the day-time.

PLATE

135

PLATE CXXXV.

ANAS FERRUGINEA.

FERRUGINOUS DUCK.

ANSERES.

GENERIC CHARACTER.

Bill convex, obtufe, the edges divided into lamellate teeth : tongue fringed and obtufe ; three fore toes folitary.

SPECIFIC CHARACTER

AND

SYNONYMS.

Chefnut, breaft and belly paler : bill dilated and rounded at the tip ; and with the legs blueifh.

ANAS FERRUGINEA : fpadicea, roftro dilatato et apice rotundato pedibufque cærulefcentibus. *Gmel. Linn. Syft. Nat.* 528. *n.* 99.

ANAS RITULA, *Fn. Suec. n.* 134.

FERRUGINOUS DUCK. *Arɛt. Zool.* 2. *p.* 576. N. *Lath. Syn.* 3. 2. *p.* 526. *n.* 71.

PLATE CXXXV.

This Duck is about the middle fize, meafuring in length fifteen inches, and weighing twenty ounces. The fpecies has been found in the Swedifh rivers, and in Denmark, but very rarely. A folitary fpecimen was killed in Lincolnfhire fome years ago, a circumftance recorded by Mr. Pennant in his Britifh Zoology.

PLATE

136

PLATE CXXXVI.

ALAUDA ARBOREA.

WOOD-LARK.

PASSERES.

GENERIC CHARACTER.

Bill cylindric, fubulate, and ftraight : the mandibles equal, and a little gaping at the bafe : tongue bifid : hind claw ftraight, and longer than the toe.

SPECIFIC CHARACTER

AND

SYNONYMS.

Head furrounded by a white annular fillet.

ALAUDA ARBOREA: capite annulari alba cinƈto. *Linn. Fn. Suec.*
3.—*Scop. Ann.* 1. *n.* 186.

Alauda reƈtricibus fufcis : prima oblique dimidiato-alba, fecunda, tertia quartaque macula alba cuneiformi. *Fn. Suec.* 1. *n.* 192.

Alauda arborea. *Briff. av.* 3. *p.* 340. *n.* 2. *t.* 20. *f.* 1.

Alouette de bois ou Cujelier. *Buff. Hift. Nat.* 5. *p.* 25.—*Pl. enl.* 660. *f.* 2.

WOODLARK. *Arƈt. Zool.* 2. *p.* 395.
Ray av. p. 69. *n.* 2.
Albin. av. 1. *t.* 42.

D 2

The

PLATE CXXXVI.

The Woodlark is believed to be a general inhabitant of Europe and Siberia, extending as far as Kamtfchatka. In its general appearance it affimilates much more with the fkylark, than in its manners of life, and is obferved to be far lefs common than that fpecies. The fkylark delights in the open fields and meadows, the other is a more retired and timid bird, prefers woody fituations, and often perches on trees, which the fkylark never does. The Woodlark whiftles like the blackbird, and, like the fkylark, emits its note in flight; it fings alfo during the night while perched on the boughs of trees.

Thefe birds build their neft on the ground, and lay five eggs of a light colour, blotched with brown: the neft, like that of the fkylark, is compofed of dry grafs, lined with foft hair. It pairs earlier in the feafon than the fkylark, and has two broods in the year.

PLATE

137

PLATE CXXXVII.

PICUS MEDIUS.

MIDDLE SPOTTED WOODPECKER.

Picæ.

GENERIC CHARACTER.

Bill angular, ftraight, cuneated at the tip: noftrils covered with recumbent fetaceous feathers: tongue round, worm-fhaped, very long, and offeous, miffile, daggered, and befet at the point with briftles bent back: tail feathers ten in number, hard, rigid, and pointed: feet climbers.

SPECIFIC CHARACTER
AND
SYNONYMS.

Variegated with black and white: crown crimfon: fpace round the eyes and fides of the neck white.

Picus medius: albo nigroque varius, criffo pileoque rubris. *Linn. Syft.* 1. *p.* 176. 18.—*Fn. Suec. No.* 101.— *Gmel. Syft.* 1. *p.* 436.—*Lath. Ind. Orn.* 229. *n.* 14.

Picus varius minor. *Ray. Syn. p.* 43. 5.

Picus varius. *Briff.* 4. *p.* 38. 14. *t.* 2. *f.* 1.

Der

PLATE CXXXVII.

Der mittlere Buntſpechte. *Wirſing. Vog. t. 37.*
Pic varié à tête rouge. *Buff. Pl. Enl. 611.*
MIDDLE SPOTTED WOODPECKER. *Arct. Zool. 2. p. 278. D.*
Brit. Zool. 1. No. 86. t. 37.

A native of Europe ; in its manners of life reſembling the reſt of its tribe, living chiefly in woods, and ſubſiſting principally on inſects, which it picks out of the trunks of decayed trees. Its length is about eight inches and a half.

Whether the Middle and the Greater Spotted Woodpeckers are of the ſame ſpecies, differing only in the tranſition of the plumage from an incomplete to a more perfect ſtate, or that it is in reality diſtinct, appears to have excited conſiderable doubts in the opinions of ornithologiſts ; ſome contending they are, and others that they are not the ſame. It is poſſible the former ſurmiſe may be correct, though at the ſame time it cannot but be obſerved, that the characters of the two kinds appear ſufficiently conſtant and obvious to authoriſe a different concluſion. The principal diſtinction that prevails in the two birds conſiſts in the Middle Spotted Woodpecker having the whole crown of the head crimſon, while in the Greater Spotted, the crimſon ſpace is confined to a broad band on the hind head. The latter bird is rather larger, and meaſures half an inch more in the length than the other. Except the difference above-mentioned, the plumage in both pretty nearly alike.

PLATE

138

PLATE CXXXVIII.

FRINGILLA SPINUS.

SISKIN.

PASSERES.

GENERIC CHARACTER.

Bill conic, ſtraight, and pointed.

SPECIFIC CHARACTER

AND

SYNONYMS.

Quill feathers yellow in the middle, the firſt four without ſpots : tail feathers yellow at the baſe, and at the tip black.

FRINGILLA SPINUS: remigibus medio luteis: primis quatuor
immaculatis, rectricibus baſi flavis apice nigris.
Linn. Fn. Suec. 237.
Scop. Ann. 1. n. 222.
Acanthus avicula. *Geſn. av. 1.*
Ligurinus. *Briſſ. av. 3. p. 65. n. 4.*
Tarin. *Buff. Hiſt. Nat. 4. p. 221.*
Abadavine, *Albin. av. 3. t. 76.*
Zeifchen. ˋ *Friſch. av. t. 11.*
Georg. it. p. 174.
SISKIN, or Aberdevine. *Brit. Zool. 129. t. 53.*

The

PLATE CXXXVIII.

The Siskin is a pretty little species of the finch tribe, about the size of the common linnet : the crown of the male is black, the back greenish, and the throat brown : in the female the head and neck is greenish ash, with brown spots, and the chin whitish. The plumage of the male is brighter than that of the female, though in other respects their appearance is not materially different.

This bird is common throughout most of the temperate countries of Europe : it occurs also in the western and southern parts of Russia, but does not inhabit Siberia. In the winter season it visits Britain, and departs again in Spring. It feeds chiefly on feeds of various kinds, is of a docile disposition, and breeds freely with the common canary-bird. Its note is indifferent, notwithstanding many rank it among the birds of song.

PLATE

139

PLATE CXXXIX.

PROCELLARIA PELAGICA.

STORMY PETREL.

ANSERES.

GENERIC CHARACTER.

Bill toothlefs, a little compreffed, and hooked at the point: mandibles equal: noftrils cylindrical, tubular, truncated, and placed at the bafe of the bill: feet palmated, three toed forward, and armed with a fpur behind inftead of back toe.

SPECIFIC CHARACTER

AND

SYNONYMS.

Black : rump white.

PROCELLARIA PELAGICA: nigra uropygio albo. *Linn. Fn. Suec.* 143.—*Act. Stockh.* 1745. *p.* 93.—*Gmel. Linn. Syft. Nat. t.* 1. *n.* 2. *p.* 561. *n.* 1.

Procellaria. *Briff. av.* 6. *p.* 140. *n.* 1. *t.* 13. *f.* 1.

Oifeau de tempete. *Buff. Hift. Nat.* 9. *p.* 327. *t.* 23.—*Pl. enl.* *n* 993.

PLATE CXXXIX.

This is our fmalleft kind of Petrel, being in fize not larger than a fwallow, and meafuring in length only fix inches. Thefe birds are feldom feen on land except in the breeding feafon, but are met with in moft latitudes at fea; large flocks of them often fettle about fhips to reft themfelves, efpecially in the Atlantic ocean. When they fly low, and hover clofe round the ftern of the fhip, it is confidered by mariners as the certain prelude of a ftorm.

The Stormy Petrel fwims and dives extremely well, and is obferved to remain much longer under water than almoft any other bird. They are generally on the wing, fkimming the furface of the waves, or dipping into the water. It is aftonifhing to fee with what a perfect degree of fafety this little creature can brave the perils of the tempeft, gliding with the utmoft velocity over the furface of the waves, then plunging into the deep, and rifing again upon the fummits of the billows.

The food of thefe little birds appear to be the fmall fifh and marine worms which they catch in fwimming or diving. In the night-time they are very noify, though feldom heard in the day, unlefs in cloudy weather. They are fuppofed to breed in the northern ifles of Scotland.

P L A T E CXXXIX.

land. Stragglers have been occasionally found inland, but very rarely. We possess one specimen, formerly in the Leverian collection, which is affirmed to have been shot at Walthamstow, in Essex.

PLATE

140

PLATE CXL.

ANAS FULIGULA.

TUFTED DUCK.

Anseres.

GENERIC CHARACTER.

Bill convex, obtufe, the edges divided into lamellate teeth : tongue fringed and obtufe : three fore toes connected ; hind toe folitary.

SPECIFIC CHARACTER

AND

SYNONYMS.

Creft pendent ; body black : abdomen and wing fpot white.

Anas fuligula: crifta dependente, corpore nigro, abdomine fpeculoque alarum albis. *Linn. Fn. Suet.* 132.
—*Scop. ann.* 1, *n.* 78.—*Kram. el. p.* 341. *n.* 12.
Anas fuligula. *Gefn. Av.* 107.
Aldr. Orn. 3. *p.* 221.
Murillon. *Buff. Hift. Nat.* 9. *p.* 227. 231. *t.* 15.—*Pl. Enl. n.*
1001.
Tufted Duck. *Arct. Zool.* 2. *p.* 573. G.
Br. Zool. 2. *n.* 274.

A winter

PLATE CXL.

A winter inhabitant of the Britifh ifles. This bird is fixteen inches in length; the male in general black, finely gloffed with purple and green, with the belly white, and the creft long and pendent. The female is like the male, except in having the colour of the plumage more inclining to brown, and being deftitute of a creft. In the young birds, the head, neck, and breaft are chefnut; the back, wings, and tail black.

The Tufted Duck inhabits moft parts of Europe, and northern Afia, vifiting, like many other of the Duck tribe, the fouthern climates in winter, and retiring northward in the fummer to breed; the flefh is excellent.

PLATE

141

PLATE CXLI.

TRINGA INTERPRES.

TURNSTONE.

GRALLÆ.

GENERIC CHARACTER.

Bill roundifh, as long as the head: noftrils fmall, linear: tongue flender: feet four toed: the hind toe of one joint, and raifed from the ground.

SPECIFIC CHARACTER

AND

SYNONYMS.

Legs red: body black, variegated with white, and ferruginous: breaft and belly white.

TRINGA INTERPRES: pedibus rubris, corpore nigro, albo ferru-
gineoque vario, peЄore abdomineque albo.
Linn. It. Gotl. 217.—*Fn. Suec.* 178.—*Gmel.*
Linn. Syft. Nat. t. 1. *p.* 2. *p.* 671. *n.* 4.

Le Coulon-chaud. *Brif. Orn.* 5. *p.* 132. 1.

Le Tourne-pierre. *Buff. Oif.* 8. *p.* 130. *pl.* 10.

TURNSTONE, or SEA DOTTEREL. *Ray Syn. p.* 112. *A.* 5.
Lath. Gen. Syn. 3. *p.* 188. *n.* 37.

This

PLATE CXLI.

This species of Sandpiper is about eight inches and a half in length. It inhabits the sea-coasts both of Europe and America, and has obtained the name of Turnstone from its peculiar method of turning up the stones on the sea-shore by means of its bill, when in search of the smaller littoral worms and fishes on which it feeds.

In Britain these birds are local, and almost confined to the most remote and unfrequented shores. It occurs at the extremity of the western promontory about Penzance, on the northern shores of Wales, and in the Hebrides. They build only a slight nest, which is deposited on the ground, and lay four eggs of an olive colour, spotted with black. These birds are seen most commonly in flocks of three or four together.

PLATE

142

PLATE CXLII.

FRINGILLA CŒLEBS.

CHAFFINCH.

PASSERES.

GENERIC CHARACTER.

Bill conic, ſtraight, and pointed.

SPECIFIC CHARACTER

AND

SYNONYMS.

Limbs black: quill feathers white on both ſides, the three fir
without ſpots: two of the tail feathers obliquely white.

FRINGILLA CŒLEBS: artubus nigris, remigibus utrinque albis:
tribus primis immaculatis, rectricibus duabus
oblique albis. *Linn. Fn. Suec. 232. t. 2. f.* 199.
—*Gmel. Linn. Syſt. Nat. t.* 1. *p.* 2. *p.* 901.
n. 3.

Fringilla. *Geſn. av.* 387.—*Ald. Orn.* 2. *p.* 815.

Fringilla ſylvia. *Scop. Ann.* 1. *n.* 217.

Fringuello. *Olin. ucc. t.* 31.

Pinſon. *Buff. Hiſt. Nat.* 4. *p.* 109. *t.* 4.

PLATE CXLII.

CHAFFINCH. *Ray. av. p.* 88. *n.* 16.

Will. Orn. p. 253. *t.* 45.

Arct. Zool. 2. *p.* 381. *F.*

Brit. Zool. 1. *n.* 125.

———————

One of the moſt abundant ſpecies of the Finch tribe found in Britain. The male Chaffinch is a bird of very beautiful, and elegantly varied plumage; the female more obſcure, and inclining to yellow; and it is alſo deſtitute of the vinaceous reddiſh hues conſpicuous on the breaſt, and other parts of the male bird.—There are numberleſs varieties of this ſpecies differing in the colours of their plumage, the moſt curious of which are thoſe either entirely white or black, or with the crown and collar white.

The Chaffinch is a native of Europe, and ſome parts of Aſia, and is more or leſs migratory in different countries. With us both ſexes are ſeen at all ſeaſons of the year, from whence it is concluded, that if they do migrate from Britain, it is only in a very partial degree. It is a ſingular circumſtance, that the males do not commonly migrate with the females, whole flocks of the latter being frequently ſeen in flight from one part to another, unaccompanied by any males. In Sweden in particular, it is well known, that they migrate in flocks to Holland every year, and conſtantly leave the males behind.

The neſt of this bird is compoſed of dried vegetables, fibres, and moſs, lined with hair, wool, or feathers, and is
uſually

PLATE CXLII.

ufually found in the midft of thickfet bufhes, at no great height from the ground. The eggs, five or fix in number, are of a pale reddifh grey, marked at the broadeft end with blackifh fpots.

PLATE

143

PLATE CXLIII.

PICUS TRIDACTYLUS.

NORTHERN THREE TOED WOODPECKER.

PICÆ.

GENERIC CHARACTER.

Bill angular, ſtraight, cuneated at the tip : noſtrils covered with recumbent ſetaceous feathers : tongue round, worm-ſhaped, very long, and oſſeous, miſſile, daggered, and beſet at the point with briſtles bent back : tail feathers ten in number, hard, rigid, and pointed : feet climbers.

SPECIFIC CHARACTER
AND
SYNONYMS.

Variegated with black and white : legs three toed.

PICUS TRIDACTYLUS : albo nigroque varius, pedibus tridaĉtylis.
Linn. Fn. Suec. No. 103.—*Gmel. Syſt.* 1. *p.* 439.
—*Borowſk. Nat.* 2. *p.* 138. 8.
THREE-TOED WOODPECKER. *Arĉt. Zool.* 2. *No.* 168.
Lath. Ind. Orn, 1. *p.* 243. *n.* 56.

The

P L A T E CXLIII.

The Northern Three-toed Woodpecker is an inhabitant of the colder climates of Europe, as Sweden, Lapland, and Ruffia, as far as the Don river. Towards the fouth it extends to Auftria and Switzerland, in the laft of which it appears to be moft frequent, the fpecies delighting in the higheft mountainous fituations. The fpecies, though fo widely diffufed, is not common, and in Britain particularly is very rare. A folitary individual of this kind was lately fhot in the north of Scotland, upon the authority of which the fpecies is inferted among the migratory vifitants of the Britifh ifles.

In point of fize, this bird rather exceeds the greater fpotted woodpecker in bulk, and meafures in length nine inches: the female is the fize of the male, and refembles it in every refpeft, except in the colour of the crown, which in the male is yellow, and in the female white. Should the fouthern three-toed woodpecker prove to be a variety of this fpecies, as is generally believed, this is the only three-toed kind of woodpecker at prefent known, the reft of the genus having four toes, two forward, and two behind.

PLATE

144

PLATE CXLIV.

URIA GRYLLE.

BLACK GUILLEMOT.

ANSERES.

GENERIC CHARACTER.

Bill ftraight, and fubulate; the tip of the upper mandible flightly bent, the bafe fub-plumofe: noftrils linear, and at the bafe: tongue nearly fame length as the bill: legs compreffed, tridactyle, and all placed forward.

SPECIFIC CHARACTER

AND

SYNONYMS.

Body deep black: wing-coverts white.

URIA GRYLLE. *Lath. Ind. Orn.* 2. *p.* 797. *n.* 2.

COLYMBUS GRYLLE: corpore atro, tectricibus alarum albis, *Linn. Fn. Suec.* 148.

Brun. No. 113.

Colymbus Grœnlandicus. *Klein. av. p.* 168. 2.

Uria minor nigra, Columba Grœlandica. *Briff.* 6. *p.* 76. 3.

Le Petit Guillemot noir. *Buff.* 9. *p.* 354.

Greenland Dove, or Sea Turtle. *Albin.* 2. *t.* 80.

Black Guillemot. *Arct. Zool.* 2. *No.* 437.

Br. Zool. 2, *No.* 236.

This

PLATE CXLIV.

This is a very scarce species in Britain. It is confined chiefly to the isle of St. Kilda, and Bass island in Scotland; and the Farn isles on the coast of Northumberland, and has been also seen on the rocks in the north of Caernarvonshire, but rarely. To the north of Europe, as far as Greenland, this bird occurs in vast numbers. The principal food is fish, in pursuit of which it dives and swims with singular dexterity. These birds have a most awkward gait in walking; in flight they appear mostly in pairs, the male accompanying the female.

The length of this bird is about fourteen inches: the plumage very deep black, with a white patch more or less obscured with dusky spots, according to the age of the bird, and the legs scarlet. The plumage is also observed to vary much in different seasons, as well as in the younger birds, the black being often intermixed with dusky brown, and whitish. These birds breed in crevices of the rocks in maritime situations, and lay either one or two eggs, the size of those of the common hen, the colour white, with grey patches, and spots of black.

PLATE

145

PLATE CXLV.

STURNUS VULGARIS.

COMMON STARLING.

PASSERES.

GENERIC CHARACTER.

Bill fubulate, angular, depreffed, bluntifh; the upper mandible entire, fomewhat open at the edges : noftrils furrounded with a prominent rim : tongue notched, pointed.

SPECIFIC CHARACTER

AND

SYNONYMS.

Bill yellowifh : body black, with white dots.

STURNUS VULGARIS: roftro flavefcente, corpore nigro punctis albis. *Linn. Fn. Suec.* 213.—*Gmel. Linn. Syft. Nat. t.* 1. *p.* 2. *p.* 801. 1.

Sturnus. *Gefn. av.* 747.

Aldr. Orn. 2. *p.* 631.

Storno. *Ol. ucc. t.* 18.

Etourneau. *Buff. Hift. Nat,* 3. *p.* 176. *t.* 15.

Stare, or Starling. *Ray av. p.* 67. *n.* 1.

Brit. Zool. 1. *n.* 104.

VOL. VI.   G   The

PLATE CXLV.

The common Starling is from eight inches and a half to nine inches in length ; the male rather larger than the female, and brighter in the colours of the plumage : the prevailing colour is blackifh, in fome parts brown, very fplendidly gloffed with purple, green, and gold, and fpotted nearly throughout with milky white.

The Starling is a native of Europe, Afia, and Africa, and is feen in Britain in large flocks during the winter feafon. They build in the hollows of decayed trees, rocks, and ruined edifices, forming a neft of very flight contexture, confifting of leaves and twigs : their eggs are from five to fix in number, and of a greenifh-afh colour. In a wild ftate, they feed on infeΩs, and various kinds of grain : they are of a docile difpofition, and eafily taught to fpeak. During the winter feafon, they are not unfrequently killed in vaft numbers, and expofed for fale in the markets, notwithftanding that the flefh is bitter, and ill flavoured.

PLATE

146

PLATE CXLVI.

CHARADRIUS CALIDRIS.

SANDERLING PLOVER, or CURWILLET.

GRALLÆ.

GENERIC CHARACTER.

Bill roundifh, obtufe, ftraight : noftrils linear : feet formed for running, and three-toed.

SPECIFIC CHARACTER

AND

SYNONYMS.

Bill and legs black : lores and rump greyifh ; body beneath white, without fpots.

CHARADRIUS CALIDRIS : roftro pedibufque nigris, loris uropygioque fubgrifeis, corpore fubtus albo immacu-
lato. *Georg. It. p.* 172.

Calidris grifea minor. *Briff. av.* 5. *p.* 236. *n.* 17. *t.* 20. *f.* 2.

Sanderling, or Curwillet. *Brit. Zool. n.* 212. *t.* 73.—*Ray av. p.* 109.
n. 11.—*Will. Orn. p.* 303.—*Lath. Syn.* 3. 1. *p.*
197. *n.* 4.

G 2

The

PLATE CXLVI.

The name of Sanderling is indifcriminately applied, by different writers, to two very diftinct birds of the Grallæ order, namely, the Common Purre, or Ox-bird (Tringa Cinclus) and the prefent fpecies, Charadrius Calidris; and to obviate mifunderftanding in this refpect, it is conceived the latter may, with fome propriety, be denominated the Sanderling Plover.

About eight inches is the ufual length of this bird: its colours above cinereous, with the head, back of the neck, and fides of the breaft, dafhed with black ftreaks: wings greyifh and brown, with the edges of the feathers pale, and quill feathers dufky: tail brownifh, with pale margins. The fpecies is fubject to occafional variation in the colour of the plumage.

Thefe birds appear in fmall flocks on the fandy fhores of Cornwall during the winter feafon; and are rarely obferved, it is believed, on any other part of the Englifh coaft. On the continent of Europe the fpecies is not common; in North America it is more abundant.

PLATE

147

PLATE CXLVII.

ANAS COLLARIS.

COLLARED DUCK.

ANSERES.

GENERIC CHARACTER.

Bill convex, and obtufe, the edges divided into lamellate teeth: tongue fringed, obtufe : three fore toes connected, the pofterior one folitary.

SPECIFIC CHARACTER.

ANAS COLLARIS. Black-brown: beneath white: head gloffed with green-violet : neck encircled with a fub-ferruginous ring : wing-fpot cinereous.

A fpecimen of this curious Duck occurred to us in the month of January, 1801, among a number of other wild fowl expofed for fale in Leadenhall-market, London ; and fo far as we have been hitherto able to determine, it appears to be not only new as a Britifh bird, but altogether undefcribed. Its fize rather exceeds that of the common widgeon. The colour above is blackifh, as is likewife the head and neck, the former of which is richly gloffed with purple and green,

and

PLATE CXLVII.

and the latter furrounded in the middle with a pretty and very diftinct collar of deep ferruginous : the lower part of the throat, and upper part of the breaft, are black; the belly white, mottled with dufky towards the pofterior end, and in the region of the vent deep fufcous. The bill and legs dufky. This bird is certainly of the male fex, and is fuppofed to have been taken in the fens of Lincolnfhire.

PLATE

148

PLATE CXLVIII.

ALAUDA MINOR.

FIELD-LARK.

PASSERES.

GENERIC CHARACTER.

Bill cylindrical, fubulate, and ftraight: the mandibles equal, and a little gaping at the bafe: tongue bifid: pofterior claw ftraight, and longer than the toe.

SPECIFIC CHARACTER
AND
SYNONYMS.

Reddifh brown, fpotted beneath: chin and belly white: throat and breaft obfcure yellow: exterior edge of the two outer tail fea-thers white.

ALAUDA MINOR: ex rufefcente fufca, rectricibus extimis duabus extrorfum albis. *Gmel. Linn. Syft. Nat. t.* 1. *p.* 2. *p.* 793. *n.* 12.

ALAUDA MINOR: rubro-fufca fubtus maculata, gula abdomineque albis, jugulo pectoreque obfcure flavefcentibus. *Lath. Ind. Orn.* 2. *p.* 494. 8.

LESSER FIELD LARK. *Will. Orn. p.* 207.

FIELD LARK. *Arct. Zool.* 2. *p.* 395. D. *Brit. Zool. No.* 139.

Inferior

PLATE CXLVIII.

Inferior in fize to the fky-lark, and larger than the tit-lark, form-ing, in this refpect, an intermediate fpecies between the two.

The Field-lark, in the general afpect of its plumage, refembles the tit-lark; its colour reddifh brown, with dufky fpots; the chin and belly white; throat and breaft yellowifh, dafhed with dufky. In its haunts and manners of life, it bears more affinity to the wood-lark, preferring woody fituations, and often perching on trees. Its note is diftinct and melodious. The neft is faid to be moft commonly built on the ground, or among the loweft bufhes.

INDEX to VOL. VI.

ARRANGEMENT

ACCORDING TO THE

SYSTEM of LINNÆUS.

ORDER I.

PICÆ.

ORDER II.

ANSERES.

VOL. VI.   H   ORDER

I N D E X.

O R D E R III.

G R A L L Æ.

O R D E R IV.

G A L L I N Æ.

O R D E R V.

P A S S E R E S.

V O L.

INDEX.

VOL. VI.

ALPHABETICAL ARRANGEMENT.

FINIS.

Law and Gilbert, Printers, St. John's, Square, London.

THE

NATURAL HISTORY

OF

BRITISH BIRDS;

OR, A

SELECTION OF THE MOST RARE, BEAUTIFUL, AND INTERESTING

BIRDS

WHICH INHABIT THIS COUNTRY:

THE DESCRIPTIONS FROM THE

SYSTEMA NATURÆ

OF

LINNÆUS;

WITH

GENERAL OBSERVATIONS,

EITHER ORIGINAL, OR COLLECTED FROM THE LATEST
AND MOST ESTEEMED

ENGLISH ORNITHOLOGISTS;

AND EMBELLISHED WITH

FIGURES,

DRAWN, ENGRAVED, AND COLOURED FROM THE ORIGINAL SPECIMENS.

By E. DONOVAN.

VOL. VII.

LONDON:

PRINTED FOR THE AUTHOR; AND FOR F. C. AND J. RIVINGTON,
No. 62, ST. PAUL'S CHURCH-YARD, 1816.

149

PLATE CXLIX.

COLUMBA ALBINOTATA.

SPOTTED NECKED, OR PANCOU TURTLE.

PASSERES.

GENERIC CHARACTER.

Bill ftraight, defcending towards the tip : noftrils oblong, half covered with a foft tumid membrane.

SPECIFIC CHARACTER
AND
SYNONYMS.

Feathers of the fides of the neck black, with a round white fpot
near the tip of each.

COLUMBA TURTUR β: T. pennis ad colli latera omnibus nigris,
     macula prope apicem rotunda alba notatis. *Gmel.*
     Linn. Syft. Nat. 786. 82.
SPOTTED NECKED TURTLE. *Lath. Syn. V.* 4. *p.* 645. 40 *A. var.*
     Ind. Orn. T. 2. *p.* 606.

————

There is nothing perhaps in the prefent ftate of Natural Hiftory
that can more effectually impede its promotion than the alteration of

VOL. VII.   B   names

PLATE CXLIX.

names fufficiently eftablifhed to be underftood by the generality of collectors. Novelty, in this refpect, fhould be always regarded with caution, becaufe at the leaft it is calculated to confufe and miflead, and when it arifes from affectation merely, becomes highly reprehenfible. Neither is it lefs injurious to the true purpofes of Science to weaken the credit of approved opinions without ample reafon : the hand of innovation may deftroy that which it cannot rebuild : vanity may injure that which it cannot repair ! With thefe impreffions conftantly upon our mind, it has ever been our endeavour, on all occafions, to improve upon rather than replant ; and preceding authorities we are inclined to believe, have been in general as feduloufly retained by us when they appeared admiffible, as by moft authors who have purfued the fame paths of enquiry as ourfelves.

But in oppofing innovation we reft perfectly affured that our endeavours have never degenerated into a pertinaceous refiftance againft amendments, founded on the progreffive increafe of knowledge. The true interefts of fcience oftentimes demand both alterations and improvements, and when thefe appear evident, no authority, however great, fhould in our opinion operate as a barrier againft their progrefs. We only wifh to inculcate, as an unerring principle, that in the fubverfion of the authority of thofe who have preceded us, we fhould act with a cautious and unbiaffed mind, and with every inclination to award that portion of credit, even to the opinions we difpute, to which in candour they are entitled.

We have infenfibly fallen into this train of curfory obfervation at the commencement of our new Volume, from reflecting generally upon thofe differences of opinion which are found to prevail in the

minds

PLATE CXLIX.

minds of British Ornithologists at this period, and those especially which it becomes our peculiar province to examine with more than ordinary attention in the course of the succeeding pages. The digression will serve also as a prelude, among the rest, to the little alteration we are induced to propose in designating the character of the Spotted Necked Turtle, the bird at present under consideration.

It is well known to every Ornithologist, for the Leverian Collection, and the works of Dr. Latham, and Professor Gmelin, are known sufficiently to justify the conclusion, that there does exist a kind of Turtle nearly allied to the common sort, and, in most respects, according with it, but which differs particularly in the form and number of the white spots on the black patch of the collar, and is hence esteemed a variety of the common Turtle.

This bird was preserved originally in the Leverian Museum, and was described from thence by Dr. Latham, whose description has furnished subsequent authors with the particulars of this supposed variety, and if we mistake not, with nearly all the information they possess respecting it.

For various reasons we propose to offer, with deference to future observation, the bird in question appears in our mind to be entitled to consideration, rather as a distinct species than as a variety of the common Turtle. We say rather, because it does seem likely to be distinct, and yet we are far from wishing to conceal our suspicions, that it may have no real claim to be esteemed such: if, however, it should prove a different species, the possibility of which may be inferred, there can be no objection, we apprehend, to distinguish it in future by an appropriate specific appellation; we propose the name

B 2

albi-

PLATE CXLIX.

albinotata, as expressive of the characteristic spots of white on the black patch of the neck; at the same time, that we must observe, there are other particulars in which the plumage differs from the common Turtle, and might afford a suitable specific title, should that selected by us be thought liable to objection.

The first account of this remarkable bird that appears we believe on record, is comprehended in the very concise inscription affixed to a certain case in the late Leverian Museum, containing a specimen and duplicates of the same bird. Dr. Latham mentions only one, but there were more. The inscription briefly denominates them, the " Spotted Necked Turtle," and relates that they were shot in Buckinghamshire. These birds were included in the Museum at the time it was arranged in Leicester House, and possessed by Sir Ashton Lever, as we perceive by the reference of Dr. Latham in the third volume of his General Synopsis, the place in which the first account of this bird occurs; the description is in the following words :—In the Leverian Museum, is a bird, shot in Buckinghamshire, which differs from the common one, in having almost the whole side of the neck black, instead of a patch only; and instead of each feather being tipped with white, there is a round spot of white on each, very near the end, giving the sides of the neck a most beautiful appearance."

In the succeeding work of the same author, (Index Ornithologicus) the same opinion is continued, it being still considered as a variety only of the common Turtle. " A priore variat lateribus colli nigris, apicibus pennarum macula alba notatis."

During the interval that elapsed between the publication of these two works, the improved edition of the Linnæan System, by Gmelin, had

PLATE CXLIX.

had appeared. In this work, Gmelin notices this bird as a variety of the common Turtle, with the following diftinctive character :— " Turtur pennis ad colli latera omnibus nigris macula prope apicem rotunda alba notatis." It is worthy of remark, that though Gmelin refers for the Common Turtle to Latham's Synopfis, and notwithftanding that, he mentions the refpective authors, upon the teftimony of whom, every other fuppofed variety of the fpecies is recorded, he is entirely filent, as to the fource, from whence his knowledge of this particular variety is derived. From the literal accordance of his latin character with the defcription given in the General Synopfis, it is neverthelefs eafy to perceive that the authority upon which he refts muft be Dr. Latham's.

Deeming it a fubject of more than ufual intereft, our inveftigation of authors did not terminate with Gmelin, we fought further information refpecting this curious bird, but with no material fuccefs The bird appears to be unknown to continental writers, fo far as we can obferve. In a collation of the genus Touterelle by Viellot, the continuator of the Hiftory of Birds by Audebert, there is a flight account of it which only tends to confirm our fufpicion that the bird is unknown upon the continent, for he mentions it as an accidental variety of the Common Turtle which has been killed in England, and is defcribed by Latham : it refembles, he fays, the common Turtle, except in having the fides of the neck black, with a round white fpot near the extremity of each feather *.

---

* There appears to be fome mifconception, or perhaps it may be an overfight on the part of the French tranflator, in rendering the meaning of our Englifh author into his own language ; for his words are thefe :—" A les côtés du cou noirs, dont chaque plume eft terminée de blanc, avec une tache ronde blanche vers fons extrémité."

The

PLATE CXLIX.

The laft work to which we can refer, is the Ornithological Dictionary of Mr. Montagu, in which the like repetition from the General Synopfis is detailed: the "*Dove-Turtle fpotted-necked*," "appears (fays this writer) to be a mere variety of the common Turtle. The difference confifts in the whole fide of the neck being black, and inftead of thofe feathers being tipped with white, there is a round fpot of white on each, very near the end. Dr. Latham fays this bird was fhot in Buckinghamfhire."

From the preceding obfervations there can be no difficulty in afcertaining the parent fource from whence the different accounts of this bird that have hitherto appeared, originally emanated, namely, the defcription of the bird in the Leverian Mufeum, inferted in the Synopfis of Dr. Latham; this bird remained in that Mufeum till the period of its diffolution, and then paffed into our poffeffion. There were altogether three examples of it, the whole fuite of which we obtained and have now before us.

In the earlieft defcription of this bird, the Author of the General Synopfis affures us, that he has feen this variety well expreffed in two collections of Chinefe drawings, and that in China it is known by the name of Pancou. He had obferved it likewife among a parcel of birds brought either from the South Seas or the Cape of Good Hope. In the "Index Ornithologicus," it is recorded finally as a native of Europe, Africa, and Afia; and we are befides affured that it occurs alfo in Cayenne. All thefe teftimonies prove moft clearly that this kind of Turtle, whether a diftinct fpecies or a variety, is widely diffufed over moft parts of the Globe; and that in every climate it exhibits the fame permanent diftinction from the common Turtle which we obferve in thofe individuals of our own country.

This,

PLATE CXLIX.

This, in addition to other circumftances, muft tend materially, in our opinion, to confirm its identity as a diftinct fpecies. We have lefs diftruft on this fubject than in its being aborigine in this country, a point apparently acceded to, with great implicitnefs, by other writers on Britifh Ornithology that have preceded us.

To whatever caufe the appearance of this bird in a wild ftate in Britain is to be attributed, does not reft with us to determine. Dr. Latham, in the works fo fully quoted in the preceding obfervations, informs us, that many birds of this fort have been obferved in this country: we befides learn, that it was, in particular, not unfrequent in Buckinghamfhire; and hence, among collectors, it was as diftinctly known by the local appellation of the Buckinghamfhire, as the Spotted Necked Turtle. If it be really a native of Europe, the filence of continental authors may truly excite furprife; it is obvious from the above remarks, that the continental authors derive their information from the naturalifts of this country. It is far more congenial with our ideas to believe, that the bird as an European is peculiar to Britain, or rather that it is an extra European fpecies, introduced by fome fortuitous circumftance into the vicinity of Buckinghamfhire, and which having become naturalized in that part, has gradually diffufed itfelf over the neighbouring counties; the latter is very probable.

At the firft glance we might eafily conceive that this bird partook, in a remote degree at leaft, of fome peculiarities of the common Stock or Wood Pigeon, or that it formed an intermediate link between that bird and the common Turtle. It differs from the Turtle in being rather larger; the wings are comparatively longer, and this

difference

PLATE CXLIX.

difference is yet more obvious in the length of the tail than even in the wings. The general colour of the plumage is more vinaceous: the wings incline more to grey, with scarcely any of the ochraceous hue observable in those of the Turtle, and the black or dusky marks in the disk of the feathers are totally dissimilar; these in the Turtle form a distinct subtriangular spot of a very dark hue, approaching to black, while in the other there is no indication whatever of such a spot, except an obscure longitudinal dash down the middle of each of the feathers upon the scapulars and wing coverts. Its size exceeds that of the common Turtle, the length of the latter being about twelve inches, that of the Spotted Necked Turtle about fourteen.

It is assuredly a matter of some surprize to us, that these essential differences, the existence of which, on an accurate comparison of the two birds, is so palpably obvious, should escape remark till the present moment; but it appears in truth that the bird itself is scarcely known, and that these distinctions being unnoticed in Dr. Latham's work, those who have compiled on his authority, were not aware that any such distinctions prevail. When these characters are considered duly, in addition to the remarkable and more conspicuous, but not more permanent, distinction of the numerous white dots upon the collar or black space of the neck *, we are inclined to apprehend, no very trivial reasons are advanced for considering the two

---

* In this bird, the whole side of the neck is black; there is a black space on the neck of the common Turtle, but it only forms a small patch: in the Spotted Necked Turtle the white dots are numerous, every black feather exhibiting one *near* the end. In the common Turtle the end of the feather itself is white, but there is no spot in the disk of the feather, and the shape is different; that of the bird before us being distinctly round; while in the Common Turtle it forms the segment half of a circle, or is rather lunate.

birds

PLATE CXLIX.

birds as fpecifically diftinct. Indeed the only doubt that can poffibly arife, muft refult from the differences that may be obfervable in the Spotted Necked Turtle in its various ftates of plumage; and this can fcarcely be fuppofed to affect it fo materially as to change its plumage to that of the Turtle-dove *! In its prefent afpect, there can exift, we are perfuaded, but one opinion upon the fubject, and that muft be in confirmation of our firft conclufion, that the fpecies is diftinct.

---

* We are fully aware that varieties of the Common Turtle do exift, in which the plumage differs a little from our Englifh variety. The Portugal Turtle-Dove, one of the admitted varieties, is, however, now fuppofed by fome to conftitute a different bird, and this may hereafter be the opinion with refpect to other varieties.

Vol. VII.   C   PLATE

150

PLATE CL.

MOTACILLA SIMPLEX.

GREATER PETTYCHAPS.

PASSERES.

GENERIC CHARACTER.

Bill fubulate, ftraight; the mandibles nearly equal: noftrils oboval: tongue lacerated at the end.

SPECIFIC CHARACTER

AND

SYNONYMS.

Greenifh, fufcous, beneath; and eyebrows whitifh: quill and tail feathers dufky brown.

SYLVIA HORTENSIS β.: viridi-fufca fubtus fuperciliifque albida, remigibus rectricibufque fufco-obfcuris. *Lath. Ind Orn. T. 2. 507. 3.*

SYLVIA SIMPLEX *Lath. Ind. Brit. birds. Suppl. I. p. 287.*

GREATER PETTYCHAPS, *ib.*

PETTYCHAPS. *Lath. Gen. Syn. T. 4. p. 413. 3. Lewin Br. Birds. 3. t. 100. Walcot. Syn. 2. t. 230.*

PETTYCHAPS GREATER. *Mont. Orn. Dict. V. 2.*—PETTYCHAPS, GREATER Sylvia hortenfis, *Mont. Orn. Dict. Suppl.*

C 2

There

PLATE CL.

There is an expreffion of the late Dr. Johnfon that applies, if we miftake not, with peculiar aptitude to the difcuffions which have taken place among Ornithologifts, refpecting the warbler before us: they have literally tended to " elucidate" the fpecies " into obfcurity." We can fcarcely hefitate in admitting this, while we trace the confufion that prevails among authors upon the fubject of the greater and the leffer Pettychaps, the Linnæan Motacilla hortenfis, and the Fauvette of Buffon, all which have been at times confounded together as a fingle fpecies.

Before we attempt to unravel the web of obfcurity, in which thefe birds and their congenors are entangled, it may not be amifs to confider the greater Pettychaps feparately, as the few remarks we have to offer on the other birds involved in the enquiry, may be introduced with moft propriety after that bird is duly noticed: indeed, a correct defcription of the greater Pettychaps, with a few remarks on its haunts and habits of life, muft lead, in no very inconfiderable degree, to point out the precife diftinctions that prevail between this and the analogous kinds of warblers, with which it has been confounded.

The difcovery of the greater Pettychaps in Britain, is attributed to the zealous affiduity of the late Sir Afhton Lever; having been obferved by that indefatigable collector, in the firft inftance, in Lancafhire, and communicated by him from thence to Dr. Latham, for the purpofe of defcribing in his general fynopfis of Ornithology.

This bird, as it appears, was deemed, at that time, of fufficient intereft, to induce Dr. Latham to record its defcription in its proper

place,

PLATE CL.

place, even after that portion of the fourth Volume, in which the warblers are defcribed, was worked off; the defcription being printed upon a fingle leaf, in a fmaller type, and fubfequently affixed in the volume, that it might appear in its proper feries, among the warbler tribe.

The fpecimens from which Dr. Latham's defcription was taken, paffed into our poffeffion, with the reft of his collection of Britifh birds, about twenty years ago; and thofe of Sir Afhton Lever, in the year 1808, the period in which we obtained Sir Afhton's collection of Britifh birds, through the diffolution of that well known eftablifhment, the Leverian Mufeum. We thus poffefs every individual bird defcribed originally under the title of the Greater Pettychaps, and, confequently, thofe upon which all difcuffion refpecting the identity of a fpecies fo much miftaken, muft, in a material degree, depend: it is from thofe our figure and defcription are taken.

This bird is about the fize, or rather fmaller, than the hedge fparrow; the length, between five and fix inches. The upper parts of the plumage greyifh brown, tinged with a greenifh hue: the under parts dufky white, with a little brown, inclining to blackifh; acrofs the breaft and over the thighs, on the latter of which, the colour is darkeft: the quills are brown, the edges of the feathers with a greenifh tint like the upper parts of the plumage, and over the eye, a pale or whitifh ftreak, which paffing from the bafe of the bill, forms a lobate or rounded fpot behind the eye. All the tail feathers are uniformly dull brown, the bill and legs brown. Both fexes are nearly alike, except that the colours of the male are rather darkeft.

Dr.

PLATE CL.

Dr. Latham, its original defcriber, affigns no latin fpecific ap‑ pellation to the Greater Pettychaps, in the firft inftance; as he efteemed it only a variety of the Fauvette of Briffon. In the Index of the Britifh birds, contained in the Supplement, it is, however, diftin‑ guifhed as a new fpecies by the name of *Simplex;* but after that time, the opinion of Dr. Latham was again changed, for in theIndex Ornitho‑ logicus, which fucceeded the former mentioned Supplement, it is determined to be new only as a Britifh bird, and no other than a variety of the Linnæan fpecies, Motacilla Hortenfis.

As the hiftory of this bird became better known, the fpecies was obferved in other parts befides the vicinity of Lancafhire. But it was rather heard than feen: its extreme fhynefs, added to the difficulty of penetrating the deep thickets where it ufually remains concealed, fe‑ curing it from the intrufion of all, except the more inquifitive, or impertinently curious. Its fong is peculiar, and as this cannot fail to excite attention, from the fweetnefs, melody, and brifknefs, as well as compafs of its notes, the fituation of its hiding place is often‑ times betrayed. By this means its vifits to the more fouthern coun‑ ties of Britain have been afcertained in the months of April and May, and as far weftward as Devonfhire in June. The fpecies is certainly local. Mr. Montagu obferved it frequently between the eaftern parts of Somerfetfhire, and no where more abundant than between Spalding and Bofton in the latter mentioned county, where it occurs even in the few hedges about the village of Wainfleet, and in the thickets furrounding the decoys of the fens in that neighbourhood. Commonly, however, their haunts are in more retired fituations. The note, which is by fome authors, compared with the whiftle of a blackbird; in the opinion of others, is little inferior to that of the

nightingale;

nightingale; and indeed refembles it fo nearly, that the bird is faid, on that account, to have obtained a name on the continent, fynony-mous with the englifh epithet of Mock, or Baftard Nightingale.

In reverting to the authors quoted in our lift of fynonyms, it will be perceived, that the lateft writers on this fubject, confider the Pettychaps as a variety of the Linnæan Motacilla hortenfis. That we cannot be miftaken as to the true Pettychaps, is fufficiently de-monftrated, and with the individual fpecimens originally defcribed be-fore us, we muft be allowed to qualify our acquiefcence to the popu-lar opinion, with at leaft fo much hefitation as to leave it a matter of opinion for future confideration, whether they are really the fame or not. Dr. Latham had his doubts on the fubject formerly; thofe on our mind are not obliterated, we are ftill inclined to think the two birds may be fpecifically diftinct.

The difference of thefe birds is obvious in their general appearance: the plumage in *hortenfis* is greyifh, or afh-coloured brown; that of the Greater Pettychaps, brown, tinged with greenifh: in *hortenfis*, the mandible is black, with the bafe of the lower one paler, in the latter the bill is brown. Motacilla hortenfis, according to the plate in Buffon, (*Pl. Enl.* 579.) has a white fpot between the bill and the eye, defcribed fometimes as a ftreak: and there is a faint whitifh mark over the eye of the Pettychaps, which is, however, more inconfpicu-ous. Befides fome difference in colour of the quill feathers there is a material diftinction in the character of the tail; the latter being uniformly dufky brown, while in the tail of hortenfis, the exterior feather on each fide is white on the outer web, and marked on the inner web, near the tip, with a dirty white fpot.

Thefe

PLATE CL.

Thefe diftinctions are effential, but fhould they yet appear infufficient to authorife our diffent from the prevalent opinion, we may add a few words further in confirmation of our fcruples.—We are but partially acquainted with the hiftory of the Greater Pettychaps; it is a bird by no means fufficiently common in England to leave nothing of its manners unknown, or to render the obfervations of continental writers, of trivial import, fince among them it may be more abundant than with us, and may alfo have been regarded with more attention.

The Fauvette of Buffon, the bird confidered to be the Motacilla hortenfis of Linnæus, arrives in France in the month of April, the males preceding the females by a few days : they are faid to frequent fields and gardens, often building on the pea flicks; more commonly, however, they build on thick bufhes in the hedges or among the low thickets; their neft is compofed of dry herbs with a little green mofs outwardly, and fome hair within, as a lining. The neft of the Greater Pettychaps is compofed of dried fibres of plants, flightly conftructed, and lined with a few hairs; and is depofited in a low bufh near the ground. The difference in the formation of the nefts, is lefs obfervable than in the appearance of the eggs, thofe of the greater Pettychaps being dirty white, marked with irregular dufky blotches of various fizes, particular about the middle, and here and there a fcratch of black.—The egg of the Fauvette is of a dirty white, marked all over with fpots of light brown, which are moft numerous at the larger end.

The defcription of the eggs of our Greater Pettychaps, is repeated in the words of the original defcriber, as he received it from Sir Afhton Lever, for though, in our collection of the eggs of Britifh

birds,

birds, we have fome fpecimens which feem to accord with the defcription pretty nearly, we are not fanctioned by any authority in referring
them to the particular fpecies before us ; we muft reft our opinion on the
accuracy of thofe by whom the eggs of the two birds have been feen
and afcertained, and by thofe they are defcribed as being very dif
ferent from each other.—Upon this fubject we muft however obferve,
that the author of the Ornithological Dictionary defcribes the eggs of
our greater Pettychaps, in the fame terms as thofe of the Fauvette are
defcribed above. This we fufpect may arife from a defire in the
author to render the hiftory of the bird complete ; he concludes the
greater Pettychaps muft be the true Hortenfis, and confequently the
Fauvette, and under this perfuafion, may have been induced to
defcribe the egg of the latter as that of the Greater Pettychaps ; this
fuggeftion will fcarcely appear doubtful, when the language of its
author is collated with that of the French naturalift *. It appeared
very material to the difcrimination of the two fpecies to digrefs on
this point, or we fhould not have mentioned it. The diffimilarity
between the eggs of thefe birds fhould be carefully regarded. It cannot be unknown that the diftinction in the eggs are as permanent as the
characters obfervable in the bird, and fhould thefe exhibit the difference ftated, as the authorities mentioned, incline us to believe
there cannot remain a doubt that the two birds are diftinct.

---

* " It lays four eggs, about the fize of a Hedge Sparrow's, weighing about thirty-fix
grains, of a dirty white, blotched all over with light brown, moft numerous at the larger
end, where fpots of afh appear." *Orn. Dict.*—" La femelle y dépofe ordinairement quatre
œufs pefant chacun trente-fix grains, d'un blanc fale, avec de petites taches brunâtres
affez nombreufes et plus rapprochées au gros bout."

PLATE CL.

Authors concede, with apparent probability, that the Fauvette of Buffon and the Motacilla hortenfis of the Linnæan fyflem, are the fame; there muft ever remain fome little obfcurity refpecting the birds intended by Linnæus, the fpecimens he defcribes being long fince deflroyed or loft, and the figures in authors to which he refers being lefs fatisfactory than might be defired: it was once fuggefled to us by a Swedifh naturalift, that it could not be hortenfis, but whether the objection arofe from a due confideration of the two birds, we are not enabled to determine.

With refpect to the Leffer Pettychaps, the Motacilla Hippolais of Linnæus, it is fo definitively diftinct from the Greater Pettychaps, that we fhould have fcarcely deemed it neceffary to enter upon the fubject in this place, if it had not been obferved, that a late ingenious author has defcribed it under the name of Hippolais, and thus confounded the Greater and Leffer Pettychaps together as one fpecies. The firft we have already defcribed at length, the other is a fmall bird of very delicate flructure, and not larger than the little, or Golden-crefled Wren, and is the fmalleft of the feathered tribe that inhabits Britain.—The Leffer Pettychaps will be found delineated in one of the immediately fucceeding Plates.

PLATE

151

PLATE CLI.

MOTACILLA OENANTHE.

WHEAT EAR, OR WHITE RUMP.

PASSERES.

GENERIC CHARACTER.

Bill fubulate, ftraight; the mandibles nearly equal: noftrils oboval: tongue lacerated at the end.

SPECIFIC CHARACTER

AND

SYNONYMS

Back hoary: front, line above the eyes, rump and bafe of the tail white; through the eyes a black band.

MOTACILLA OENANTHE : dorfo cano, fronte alba, oculorum faf-
cia nigra. *Linn. Fn. Suec.* 254. *Gmel. Syft. Nat.*
966. 15.
Fabr. Faun. Groenl. p. 122. 84.
Scop. Ann. I. No. 230.
Brünn. No. 276.
Kramer. El. p. 374—4.
Nozem. nederl. Vogel. t. 81.

D 2   SYLVIA

PLATE CLI.

Sylvia Oenanthe: dorfo cano, fronte linea fupra oculos uro-
 pygio bafique caudæ albis per oculos falcia ni-
 gra. *Lath. Ind. Orn. p. 529. 79.*
Curruca major, pectore fubluteo, *Frifch. av. t. 22.*
Oenanthe f. vitiflora. *Aldr. orn. 2. p. 762. t. 763.*
 Jonft. Av. 123. t. 45. f. 13.
Vitiflora, *Brif. 3. p. 449. 33.*
Culo bianco *Zinnan, p. 41. t. 6. f. 24.—Cetti úc. Sard. p. 223.*
Le Cul-blanc, Vitrec, ou Motteux. *Buff. 5. p. 237.—Pl. Enl. 554.*
 f. 1, 2.

Wheat-ear, Fallow-fmich, white tail, white rump. *Ray av.*
 p. 75. n. 1.—Will. Orn. p. 233. t. 41.—Albin.
 av. 1. t. 55. 3. t. 54.—Edwards av. pref. p. 12.
 Brit. Zool. 1. n. 157.—Orn. Dict.—Bewick v. 1.
 229.
Wheat-ear. *Lath. Gen. Syn. 4. p. 465. 75.—Id. Supp. p. 182.*
Snorter. *Pult. Cat. Dorfet. p. 9.*

The Wheat-ear is a bird of very pleafing afpect: there is a peculiar delicacy in its appearance and manner, which, added to the fingular contraft of colours its plumage exhibits, entitles it to more
than ordinary attention.

This bird is of the migratory kind, refiding in Britain during the
fummer months. The males preceded ufually by the females about
ten days or a fortnight, arrive on our fhores in the month of March,

or

PLATE CLI.

or the beginning of April, and from that time till late in May: about the end of September the Wheat-ears affemble and depart, the laft flight in October. A few birds occafionally remain in England when the feafon is mild during the whole winter.

As a bird of paffage the Wheat-ear is a fpecies very widely dif-fufed over the globe : towards the north it has been traced as far as the remoteft of the Scottifh iflands, Norway and Iceland, and by Fabricius afcertained even as a native of Greenland, Edwards, and after that writer Latham, fpeak of it as an inhabitant of the Eaft Indies, and from the late obfervations of Sonnini, Wheat-ears are by no means uncommon in Egypt. In the Index Ornithologicus it is no-ticed as a native of Africa.

There are feveral varieties of this fpecies, the principal of which may be reduced to three, the Grey Wheat-ear of Pennant's Britifh Zoology (Cul-blanc gris *of Briffon*) the Afk-coloured (Cul-blanc cendré *of Briffon*) and the Dwina Wheat-ear. var. δ.

The firft of thefe differs from the ufual kind in being olive or tawny above, with a mixture of whitifh and fulvous ; the lower part of the neck marked with very fmall grey fpots ; the two middle tail feathers wholly black, the reft, as in the Common Wheat-ear, and fringed with pale rufous ; the bill and legs brown.

In the Afh-coloured Wheat-ear, the plumage as the name implies is of an afhen colour, at leaft, on the upper parts of the body, and irregularly mixed with grey brown ; the rump of the fame colour

PLATE CLI.

lour inftead of white. The forehead white as in the common kind.

That found about the Dwina is white above; the throat, wings, and almoft the whole of the two middle tail feathers black ; and two fpaces of black on the outer feathers of it.

In our Mufeum we poffefs alfo a buff-coloured Wheat-ear, a fuppofed variety of this fpecies ; perhaps allied to, if not the bird intended by Linnæus under the fpecific name of Stapazina : the red or ruffet-coloured Wheat-ear of Edwards.

According to the obfervations of Buffon the Wheat-ear does not attain the adult plumage till the third year. This may induce a perfuafion that fome, if not all the birds admitted heretofore as varieties may prove to be no other than the common kind in an immature ftate of plumage. Before the firft moult the young birds are a mixture of reddifh, or ruffet, with brown on the head, the neck and upper part of the body to the rump faintly ftreaked with blackifh, the rump itfelf white : the under parts reddifh dotted with blackifh or dufky, with the lower part of the belly white. The afh-coloured variety is conceived to be the young of the male bird. The plumage of the female is marked like that of the male, but the colours are more obfcure. Preparatory to the commencement of the breeding feafon, the Wheat-ear feeks fome convenient depofitary in the ground for the conftruction of a neft, felecting for this purpofe a hollow under the fhelter of a ftone, or clod of earth, or not unfrequently the deferted burrow of a rabbit. The neft is formed of grafs, or mofs, mixed

with

PLATE CLI.

with the wool of sheep and other animals which it easily collects in the places of its haunts, and lined with a few hairs and feathers. The eggs are of a light blue colour, with a circle of deeper blue at the large end. Stragglers have been known to breed here, but the circumstance is rather unusual.

The chief food of the Wheat-ear consists of insects, and worms, upon which latter they are observed to thrive well and become very fat.

The vicinity of Eastbourn in Sussex is a celebrated resort of this little bird, which, generally speaking, though it may occasionally abound elsewhere, is a local species : they seem dispersed, seldom appearing in great numbers together in any one spot. Mr. Pennant attributes their appearance in such abundance, about this particular situation, to the presence of a certain fly which feeds on the wild thyme, that abounds in the neighbouring hills ; these flies, in his opinion, constituting the favourite food of the Wheat-ear. Besides the vicinity of Eastbourn we have observed these birds very common in the open grounds to the westward of the coast of Sussex, and no where in greater plenty than about the salt marshes, towards the sea from Chichester to the borders of Hampshire. They fly low, and settle frequently to pick up the worms, and insects that are found on the ground. The peasantry, the shepherds especially, snare them in traps contrived simply in the manner boys take sparrows with traps of brick, except that those for the Wheat-ear are made with stone or clods of earth instead of brick. There is besides another mode of capture which consists merely in placing two clods of earth edgeways, so as

to

PLATE CLI.

to form a kind of tent or awning, with a ſtick at the opening to which a running nooſe of hair is faſtened : in the night time, and not unfrequently when diſturbed or frightened, the Wheat-ears enter theſe traps for ſafety and are taken.

The excellence of the Wheat-ear as an article of food has ob-tained it the emphatic name of the Engliſh Ortolan, for which rea-ſon they are ſought after with avidity. Mr. Pennant obſerves, that the numbers enſnared in his time, in the diſtrict of Eaſtbourn alone, amounted to 1840 dozen, which uſually ſold for ſixpence a dozen. Mr. Montagu informs us, it is a common cuſtom in thoſe parts where the Wheat-ear is taken, to viſit the traps ſet by the the ſhepherds, take out the bird, and leave a penny in each as a re-ward to the owners ; remarking further, that theſe birds uſually ſell for a ſhilling a dozen.

Nor was the moderate price abovementioned unuſual in the parts adjacent. In Portland, where theſe birds are called the Snorter, and are entrapped in great numbers, Dr. Pultney aſſures us, the price is one ſhilling per dozen. More than thirty dozen, adds this accu-rate writer, are ſaid to have been taken in a day, by one perſon. In 1796, an inſtance, he ſays, occurred, of even fifty dozen being caught in one day. It is further ſtated, that a perſon in the Wey-mouth market, had paid thirty pounds to one man for Wheat-ears, in the year 1794, at one ſhilling a dozen ; and that the ſame vender had been ſupplied with fifty dozen more than could be diſpoſed of.

Dr. Latham obſerves, that quantities of theſe birds are eaten on

the

PLATE CLI.

the fpot about Eaftbourn, by the neighbouring inhab'tants; others ate picked and fent up to the London poulterers, and many are potted, being as much efteemed in England as the Ortolan on the continent.

It is true, the Wheat-ear abounds in thofe parts as in former times; but the paffing ftranger, who, from thefe details, might be induced to expect in his vifits to any of the towns or villages in the vicinity, a difh of thefe Englifh ortalans, at a moderate charge, will be ferioufly difappointed. The influx of vifitors from the metropolis into thofe parts during the bathing feafon, (the time in which thofe birds are common,) has effected a change fo material in this refpect within the laft few years, that we may almoft with as much confidence confult the regulations of the " 8th Hairy" for the prices of the prefent London markets, as confult the authors of ten or twenty years ago for the prefent price of Wheat-ears in the neighbourhood of Eaftbourn. Five, ten, or fifteen fhillings a dozen, is fometimes paid for thefe birds, and thought not very immoderate;—it certainly is not, compared with that of the potted Wheat-ear, the price of which, in the fummer of 1813, as we found charged by a purveyor in thefe delicacies refident in the town of Brighton, was three half-crowns a pot, each pot containing two birds!—Such are the charges for frefh and potted Wheat-ears, at the watering places of Suffex, at leaft to the occafional vifitors.

152

PLATE CLII.

MOTACILLA ATRICAPILLA.

BLACK-CAP WARBLER.

PASSERES.

GENERIC CHARACTER.

Bill fubulate, ftraight, the mandibles nearly equal: noftrils oboval, tongue lacerated at the end.

SPECIFIC CHARACTER

AND

SYNONYMS.

Teftaceous, beneath cinereous: cap dufky-black.

Motacilla Atricapilla: teftacea, fubtus cinerea, pileo ob-
        fcuro. *Linn. Fn. Suec.* 256.—*Scop. Ann.* 1. *n.*
        229.—*Brünn. Orn. n.* 278, 279.—*Kram. el. p.*
        377. *n.* 15. *Gmel. Syft. Nat.* 970. 1.
Curruca Atricapilla. *Brif. av.* 3. *p.* 380. *n.* 6.
        Klein av. p. 79. 14.
Atricapilla, f. Ficedula *Gefn. av.* 348.
        Aldr. orn. 2. *p.* 756. *t.* 757.
        Ray. Syn. p. 79. *A.* 8.—*Will. orn. p.* 162. *t.* 41.
        E 2   Capinera,

PLATE CLII.

Capinera. *Olin. Ucc. t.* 9.

      Zinnan. Uov. p. 56. 58. *f.* 45.

      Cetti uc Sard. p. 216.

La Fauvette à tête noire, *Buff.* 5. *p.* 125. *t.* 8. *f.* 1.—*Pl. Enl. t.* 580. *f.* 1—2.

Meiſſen Moenche, *Gunth. Neſt. u. Ey. t.* 68.

Kloſterwenzel, *Friſch av. t.* 23. *f.* 1.

BLACK-CAP. *Penn. Brit. Zool.* 1. 148.—*Arct. Zool.* 2. *p.* 418. *Ray. av.* 79. 8.

BLACK-CAP (Sylvia atricapilla). *Lath. Syn.* 4. *p.* 415. 5.—*Ind. Orn.* 508. 6.

As the turtle is the emblem of conjugal fidelity, the Black-cap is that of parental tendernefs; no bird providing for its young with greater induſtry, or more anxious folicitude. In this refpect the male is not lefs attentive than the female; while one quits the neſt-lings in fearch of food, the other remains to watch the neſt and pro-tect them. When the young are able to fly and leave the neſt they aſſociate together in families, accompany each other in their daily excurfions, and at night rooſt together on the fame branch, the male on one fide, the female on the other, and the infant brood in the middle between them; the whole huddled together as clofe as poſſible for the fake of warmth.

The vocal powers of the Black-cap are not furpaffed by many of the feathered tribe: it is deficient in that wild variety which conſti-

                            tutes

PLATE CLII.

tutes a chief excellence in the fong of the nightingale, but it never-thelefs refembles it, and, with that exception, its noteis fo little inferior, that it has obtained the appellation of the Mock Night-ingale.—An inftance is recorded by Buffon, of a Black-cap, tutored by the fong of the nightingale, whofe note at length fo far excelled in melody and compafs as to filence the fong of its inftructor.

The Black-cap is a fmall bird, its length fcarcely exceeding five inches; the general colour above is afhen, with a tinge of greenifh; the fides of the head and body beneath greyifh, becoming white towards the vent: the top of the head black. This is the defcription of the male bird, from which the female differs in having the head ferruginous.

When this bird firft arrives, which is early in the fpring, before the infect race appears in fufficient number to afford it fuftenance, the berries of various plants furnifh it fupport: thefe it afterwards rejects as infects become more common, except the fruit of the ivy, which is faid to conftitute a favourite food at all times.

The neft, which is compofed of dried ftalks, with an intermixture of wool and green mofs, the fibres of roots and horfe-hair as a lining, is placed generally in a bufh very near the ground: the eggs are five in number, of a pale reddifh, mottled with a deeper colour, and a few dark fpots.

The

PLATE CLII.

The Black-cap is a fummer refident with us, coming to England in fpring and returning in autumn. It is not undeferving of remark, that though this bird is not unfrequent in the north as well as fouth of Europe, it is by no means common in this country.

PLATE

153

PLATE CLIII.

EMBERIZA SCHOENICULUS.

REED SPARROW.

PASSERES.

GENERIC CHARACTER.

Bill conic; mandibles receding from each other from the bafe downwards, the lower with the fides narrowed in, the upper part with a callofity or hard knob within.

SPECIFIC CHARACTER

AND

SYNONYMS.

Head black ; body grey and black ; outmoft tail-feathers with a white cuneate fpot.

EMBERIZA SCHŒNICLUS : capite nigro, corpore grifeo nigroque, rectriubus extimis alba cuneiformi. *Linn. Fn. Suec.* 231.—*Syft.* 1. 311. 17.—*Gmel. Syft. Nat.* 1. *p.* 881.
Brünn. Orn. n. 2. 251

Müller.

PLATE CLIII.

Müller. n. 254.

Nozem. nederl. Vogel. t. 44.

Georg. it. p. 174.

Emberiza capite nigro, maxillis rufis, torque albo, corpore rufo nigricante. *Linn. Fn. Suec.* 1. n. 211. *Kram. El. p.* 37. *n. 5.*

Hortulanus arundinaceus. *Briff. av. 3. p.* 274. *n. 5.*

Paffer arundinaceus. *Gefn. av.* 652.

Paffer torquatus f. arundinceus. *Raii, Syn. p.* 93, *A. 3.—Will. p.* 196.

Emerling *Gunth. Neft. u. Ey. t.* 17.

Ortolan de rofeaux. *Buff. hift. nat.* 4. *p.* 315.—*Pl. Enl.* 247. 2. *(male)* 477. 2. *(female.)*

REED BUNTING: *Brit. Zool. n.* 120.—*Arct. Zool.* 2. *p.* 368. *E. —Lath. Gen. Syn. 3. p.* 173. 9.—*Suppl. p.* 157.

Greater Reed Sparrow. *Ray. av. p.* 93. 3.—*Will. Orn. p.* 269.— *Albin av.* 2. *t.* 51.

The Reed Bunting is a local bird: an inhabitant chiefly of marfhes that abound in reeds: thefe, with the willow and other trees that grow in watery places, are its favourite haunts, from the neighbourhood of which it is obferved to wander over the adjacent cultivated places during the day time, and returning again to its neftling places in the marfhes before night, at leaft during the feafon of incubation. Thefe birds fly low, feldom perching, except in hedges, or low bufhes: their ufual food confifts of grain of all kinds, corn efpecially, and infects generally. The fruit of the reed in particular af-

fords

PLATE CLIII.

fords it ftore of palatable food during part of the year without the trouble of fearch beyond its ordinary places of refort. The Reed Bunting never affociates in large flocks, feldom more than fix or eight together; they frequent the fame fpots as the Sedge Warbler, and it is imagined by fome late writers that the fine melodious note attributed to the Reed Bunting belongs to the former bird. As the Reed Bunting, though a timorous bird, is lefs fhy than the Reed Warbler, we can eafily conceive a miftake of this kind might have arifen, when both fhould happen to be in the fame bed of rufhes, the one concealed, the other confpicuous to the view of the liftener; the fong of the Warbler concealed might, under fuch peculiar cir-cumftances, ftrike upon the ear, and be miftaken for that of the Reed Bunting. The note of the Reed Bunting, according to Mr. Montagu, confifts only of two notes, the firft three or four times repeated, the laft fingle and more fharp.

Thefe birds form their neft of the dry ftalks of grafs, fome-times with an intermixture of mofs, and line it with various fubftances of a foft and light texture, fuch as the down of the reed, horfe hair, and, in fome inftances, it has been known to employ for this purpofe the hair of the calf. The neft is ufually placed among the rufhes, or in a low bufh, and not always impending over the water. The eggs are of a dirty blueifh white, marked with veins of purplifh brown, which are moft confpicuous at the larger end.

This bird inhabits northern Europe, as far as Sweden, Ruffia and Siberia: towards the fouth it extends no further than Italy,

VOL. VII.   F   where

PLATE CLIII.

where it is efteemed very rare. It is uncertain whether they migrate or not from the country, many are known to remain with us throughout the year.

PLATE

154

PLATE CLIV.

CERTHIA FAMILIARIS.

COMMON CREEPER

OR

TREE CREEPER.

PICÆ.

GENERIC CHARACTER.

Bill arched, flender, fomewhat triangular, pointed : tongue vari‑
ous, feet formed for climbing.

SPECIFIC CHARACTER

AND

SYNONYMS.

Grey, beneath white : quill feathers brown, ten of them with a
white fpot.

CERTHIA FAMILIARIS : grifea, fubtus alba, remigibus fufcis, de‑
cem macula alba. *Linn. Syft. Nat.* 1. *p.* 184. 1.
—*Fn. Suec. n.* 106.—*Gmel. Linn. Syft.* 1. *p.* 469.
Lath. Ind. Orn. T. 1. 280.

F 2

CERTHIA

PLATE CLIV.

CERTHIA MINOR. *Frifch. t.* 39.

CERTHIA. *Raii Syn. p.* 47. *a.* 5.—*Will. Orn. p.* 100. *t.* 23.

Falcinellus arboreus noftras, *Klein. Av. p.* 106. 1.

Ifpida cauda rigida, *Kramer El. p.* 337. 2.

Picchio, Piccolo, *Zinnan. Uov. p.* 75. *t.* 11. *f.* 66.

Common Creeper, *Br. Zool.* 1. *No.* 92. *t.* 39.—*Lath. Gen. Syn.* 2. *p.* 701.—*Id. Sup. p.* 126.

Of the many fpecies contained in the Genus Certhia, this is per-haps the only one that can be introduced with implicit confidence among the feries of Britifh Birds.   There is another, Certhia Mura-ria, or Wall-Creeper, an European bird of very uncommon fcarcity, and no lefs remarkable for its richly varied plumage of black and crim-fon, which has obtained a place in the Britifh Fauna, on the authority of a zealous colleĉtor, and from refpeĉt to that authority it cannot be amifs, in the prefent inftance, to ftate the circumftance.   It muft be neverthelefs acknowledged, that there is no very fatisfaĉtory evidence of its being Britifh, and, therefore, at leaft, for the prefent, the Certhia familiar is isto be confidered as the only well authenticated fpecies of its genus found in Britain*.

---

* Dr. Latham alludes to the prevalence of this opinion in the following obfervation on Certhia muraria. "Buffon does not rank it as a bird of France; and I will venture to fay that it was never found in England whoever may affert to the contrary." *Synop.* V. 2. *p.* 1.731. As there is no precife reference to the information on which this general opi-nion refted at that time, we cannot pretend to fay whether the late opinion of its being Britifh might be derived from the fame fource of authority or not. Yet under all its cir-cumftances one faĉt is certain that Certhia Muraria has been confidered as a Britifh bird, however queftionable may be the authority on which it refts.

The

PLATE CLIV.

The habits of the Common Tree-Creeper refemble thofe of its congeners : it fubfifts on infects which it collects on trees, and in ueft of which this little bird is obferved running up and down the trunks of trees, fearching among the mofs or pecking into the cre vices. The moment it perceives itfelf watched, it turns to the oppofite fide of the tree, and refts quiet and concealed : and fhould the obferver follow it to that fide, it again fhifts its pofition fo as conftantly to conceal itfelf on the fide oppofite the fpectator. They breed in the hollows of trees, forming their neft in a conveni-ent recefs which they line with dried vegetables, fragments of light rotten wood and a few feathers. Their eggs, from fix to eight in number are white powdered or fpeckled with ferruginous.

This bird is an inhabitant of Afia and America, as well as Europe, and is faid to be no where more abundant than in this country. With us it is a refident throughout the year. Its note is weak and mono-tonous.

155

PLATE CLV.

FRINGILLA DOMESTICA.

HOUSE SPARROW.

PASSERES.

GENERIC CHARACTER.

Bill conic, ftraight and pointed.

SPECIFIC CHARACTER

AND

SYNONYMS.

Quill and tail feathers brown ; body grey and black ; wings with a fingle white band.

FRINGILLA DOMESTICA : remigibus rectricibufque fufcis, corpore grifeo nigroque, fafcia alarum alba folitaria. *Linn. Fn. Suec.* 242.—*Gmel. Syft. I. p.* 925.—*Lath. Ind. Orn. T.* 1. *p.* 1. 432. 1.
Kramer El. p. 369. 10.
Frifch. t. 8.
Georgi. p. 174.
Fn. Arag. p. 87.
Borowfk. Nat. 3. *p.* 144. 11.

Paffer

PLATE CLV.

Paſſer domeſticus. *Raii Syn. p.* 86. *A.*

Will. Orn. p. 182.

Briſs. Orn. 3. *p.* 72. 1.

Schæff. El. t. 53.

Roman. Orn. 1. *p.* 99. *t.* 16. *f.* 1.

Paſſer domeſtica. *Zinnan. Uoo. p.* 79. *t.* 11. *f.* 70.

Olin. uc. t. p. 42.

Cetti uc. Sard. p. 204.

Rauch-Sperling. *Gunth. Neſt. u. Ey. t.* 57.

Le Moineau. *Buff.* 3. *p.* 474. *t.* 29. *f.* 1.

Houſe Sparrow. *Will. Orn. p.* 249. *t.* 44.

Albin. 1. *t.* 62.

Br. Zool. 2. *p.* 382.

The Common Sparrow is worthy of particular obſervation for the ſingular diverſity, we might almoſt ſay the beauty of its plumage; the male bird eſpecially. In the immediate vicinity of great towns, and cities, theſe familiar inmates of our houſe tops acquire an habitual ſootineſs from their conſtant reſidence about the ſmoky flues of chimnies, the places to which they ſeem to reſort for the ſake of warmth; but in the country, where they are more hardy and cleanly, the ſparrows exhibit a brightneſs of colouring that renders their appearance very different from thoſe which reſide in populous places.

In its ordinary ſtate of plumage no bird can be more generally known than the common Sparrow : there are however variations in which its appearance is more ambiguous, particularly thoſe of the pied

kind.

PLATE CLV.

kind. Sparrows wholly white occur not very unfrequently : thofe alfo which are varied with white and dark brown or chefnut, or rufous, occur occafionally, and fometimes the Sparrow is found entirely black. Among other extraordinary varieties of this bird, in our Mufeum, is one in which all the fore toes are double, fo that each foot is furnifhed with fix toes in front, the back toe which is fingle as ufual on one fide, is treble on the other, except which there is no material appearance of diftortion, every toe is diftinctly formed, and armed with its proper claw. We are the more explicit in detailing the circumftances of this fingular bird as it prefents one of the moft remarkable varieties of the Sparrow we are acquainted with.

The Sparrow is proverbially a bold and familiar bird, and feems to evince fo far an attachment for man that it is obferved only in fuch places as are inhabited by the human race ; and it is affuredly the leaft of all other birds, the fport of that cruelty and caprice this "terreftrial lord" is prone to exercife over the minor race of creatures. Sparrows are faid to be injurious to gardens, but wherever Sparrows have been deftroyed in any confiderable numbers, the fruits of our gardens, as the corn of our fields have fuffered infinitely greater mifchief from the hofts of infect depredators which thefe birds would have timely deftroyed. Sparrows therefore are ufeful, and are the natural and welcome refident of the cottage, and to a certain extent may be ufeful alfo in cities. Their general food confifts of infects, grain, and fruits of all kinds, befides which they will eat other animal food. The neft is often conftructed under the eaves, and other projections of houfes, in broken walls, dry gutters, and even the tops of chimnies : The eggs

VOL. VII.   G   of

PLATE CLV.

of the fparrow are well known; they are of a pale or afhen colour with numerous fpots and dots of brown.

This bird is an inhabitant of Europe, Afia and Africa, fubject however to fome trifling variation of plumage in different climates; the male bird conftantly darker, and the colours more diftinct than in the female.

PLATE

PLATE CLVI.

MOTACILLA HIPPOLAIS.

LESSER PETTYCHAPS.

PASSERES.

GENERIC CHARACTER.

Bill fubulate, ftraight: the mandibles nearly equal: noftrils ob-oval: tongue lacerated at the end.

SPECIFIC CHARACTER

AND

SYNONYMS.

Greenifh-afh, beneath yellowifh: abdomen filvery: limbs fuf-cous: eyebrows whitifh.

MOTACILLA HIPPOLAIS: virefcente-cinerea fubtus flavefcens abdomine argenteo, artubus fufcis, fuperciliis albidis. *Linn. Syft. Nat. I. p.* 330. 7.
Gmel. Syft. I. p. 954.

SYLVA HIPPOLAIS. *Lath. Gen. Syn.* 4. *p.* 413. 3 *.—Ind. Orn.* 507. 4.

Ficedula Septima Aldrovandi (Pettychaps) *Raii Syn. p.* 79. *A.* 7.—*Will. p.* 158.—*Ind. Angl. p.* 216.

G 2

LESSER

PLATE CLVI.

LESSER PETTYCHAPS. *Br. Zool. I.* 149.—*Arɛt. Zool.* 11. *p.* 418.
—*Lath. Gen. Syn.* 4. 414.—*Lev. Mus.*

There can be no doubt this bird has been fometimes confounded with the Willow Wren, and that even by Englifh authors of refpeɛtability. The Willow Wren has already obtained a place in this work, and the fpecimen delineated and defcribed was the individual bird of Dr. Latham's colleɛtion, now in our poffeffion. The example of the Leffer Pettychaps at prefent before us is no lefs well authenticated, as being the original bird found by the Duchefs of Portland, at Bulftrode, and communicated by her grace to Sir Afhton Lever. No one can, therefore, queftion, that upon the identity of two fpecies fo nearly analogous and yet diftinɛt, we poffefs the beft authority that can be obtained, and are thus enabled to fpeak with confidence on a fubjeɛt which might otherwife remain involved in doubt.

The Leffer Pettychaps is a very diminutive bird, the fmalleft almoft, without exception, found in England. It is fuppofed to be rare; but is, perhaps, more abundant in this country, in the milder parts efpecially, than may be imagined; though, from the fhynefs of its manners, it is not often feen. As a migratory fpecies it is one of our early fpring vifitors; indeed fo early, that it is imagined, and with fome probability, ftragglers may remain with us throughout the winter when the weather is mild. Inftances of this nature occur at leaft in the warmer parts of Britain. Its note, which is a mere *twit*, *twit*, or, as fome liken it, to the words chip chop, is

heard

PLATE CLVI.

heard fometimes as early as March. It feeds on infects, which it takes occafionally on the wing ; and it is obferved to be moft frequent about plantations of fir-trees.

Dr. Latham defcribes the neft as being of an arched form, compofed of dry bents, mixed with a little mofs, and thickly lined with feathers: it is placed on the ground under a tuft of grafs, or at the bottom of a bufh. The eggs, five in number, white and fprinkled all over with red fpots, efpecially at the largeft end.

PLATE

PLATE CLVII.

PHALAROPUS HYPERBOREUS.

RED PHALAROPE.

GRALLÆ.

GENERIC CHARACTER.

Bill roundifh, ftraight, and ufually fomewhat infle&ted at the top :
noftrils minute: feet lobate, or furnifhed with lateral membranes,
generally fcalloped.

SPECIFIC CHARACTER
AND
SYNONYMS.

Cinereous beneath, rump and band on the wing white: breaft
cinereous: neck at the fides ferruginous.

PHALAROPUS HYPERBOREUS. *Lath. Ind. Orn. T. 2. 775. 1.—*
Gen. Syn. 5. p. 270. 1.

TRINGA HYPERBOREA. Cinereus, fubtus uropygio fafciaque
alarum albis, pectore cinereo, colli lateribus ferrugineis. *Linn. Syft. Nat. 1. p. 249. 9.—Fn.*
Suec. Nº 179. Gmel Syft. 1. p. 675. Faun. Groenl.
Nº 75 (mas).

PHALAROPUS

PLATE CLVII.

PHALAROPUS CINEREUS. *Brif.* 6. *p.* 15. 2.
Cock Coot-footed Tringa. *Edw. t.* 143.
Red Phalarope. *Br. Zool* 2. *n.* 219. 17.—*Lath. Gen. Syn.* 5. 270. 1.
TRINGA FULICARIA. *Linn. Syft.* 1. *p.* 676. 6. *(fem.)*

There are certain birds which it is well known to the experienced Ornithologift, exhibit at the two remote periods of the year the winter and the fummer, a moft ftriking difference in their appearance, and this indeed fo very confiderable in certain inftances as to mif-guide the beft informed, if they have not the good fortune to afcertain the fame bird under both the circumftances of the winter and fummer plumage.

Befides thefe very prominent varieties in which the diftinctions are ftrongly marked, there are intermediate tranfitions which denote the progreffive advancement of one ftate of plumage to the other, and thefe are fometimes no lefs ambiguous, or lefs calculated to miflead. Thefe remarks will apply generally to all birds which inhabit equally the cold and warmer climates; to the whole of the fandpiper tribe in a particular manner, and as may be conceived to the varieties of that fection which are denominated Phalaropes, the natural order to which the bird before us appertains.

This bird appears to conftitute one of thofe ambiguous varieties, but which is probably ambiguous in its firft appearance only, for with a little caution it may eafily be reduced we think to its legitimate

fpecies ?

PLATE CLVII.

ſpecies : there can be no doubt in our mind that it is no other than a very pretty and intereſting variety of the bird already deſcribed by writers under the title of the Red Phalarope; a variety, the effect of ſeaſon merely, and not entitled as ſome conſider it to the appellation of the " New Red Phalarope." The Red Phalarope in its ordinary ſtate of plumage, is by no means common in Britain; it is better known to the more northern naturaliſts, as it is an inhabitant of the regions more approximating to the polar circle, and is deſcribed by Linnæus under the title of *Tringa hyperborea*, Later authors have determined the Linnæan Tringa hyperborea, and Tringa fulicaria to be the two ſexes of the ſame ſpecies, which opinion has been in its turn amended by ſubſequent reſearch, the reſult of which has proved the exiſtence of two ſexes in both fulicaria and hyperborea. Dr. Latham was led to believe them to be the two ſexes of one ſpecies.

The very ſingular ſtructure of the feet in the Phalaropes render them particularly curious, beſides which their plumage generally is very pleaſingly diverſified ; there is an unuſual delicacy in that of the Grey Phalarope, and the Red Phalarope in point of gaiety compenſates for the minor defect of being leſs chaſte in its colours and variegations than its very analogous ſpecies.

There is ſome account extant, that the Red and Grey Phalaropes have been ſhot in Yorkſhire, and it was concluded on this or ſimilar authority, that they were the two ſexes of the ſame ſpecies, an opinion, as before obſerved, which has been ſince removed by the moſt cogent of all evidence, the demonſtration of anatomical inveſtigation.

Vol. VII.   H   Beſides

PLATE CLVII.

Befides thefe two birds, Dr. Latham defcribes a third kind from the Bankfian collection, found far northward between Afia and Africa, the particulars of which appear in the fifth volume of the Synopfis, and a figure of it in the frontifpiece of that volume. This he defcribes as a variety of the Red Phalarope (fem) in the Synopfis, and the fame opinion is retained in his fubfequent work (Ind. Orn.) it is this variety which accords more nearly with the bird before us than any that has been defcribed to that period.

We have been long in poffeffion of this bird in our own collection of Britifh birds, and have given it fufficient publicity; neverthelefs it was confidered among collectors as a new bird. In the year 1807, a communication appeared in the eighth volume of the Tranfactions of the Linnæan Society, from the pen of Mr. Simmonds, F.L.S. and in this paper, befides other ufeful information, will be found a general defcription of the fame variety.

Mr. Simmonds conceiving it might be fpecifically different from the variety of Dr. Latham's laft defcribed, propofed to give it the name Phalaropus Williamfii, in compliment to his friend Mr. J. Williams of Dartford: to this there could be no objection, excepting only that it does not appear to be diftinct, and would therefore only create confufion fhould it be defcribed under any other than its original denomination. Thefe birds were found at the edge of two or three frefh water lakes in Sanda and North Ronalfha, the two moft northern of the Orkney iflands: in the ftomachs of feveral were found the remains of Monoculi and Onifci. We poffefs the egg which Mr. Simmonds feems to be unacquainted with, as he expreffes

his

PLATE CLVII.

his regret that the fearch after the nefts was not attended with the defired fuccefs, and refpecting the eggs he is entirely filent. The egg is of an olivaceous colour, mottled with dots and fplafhes of blackifh. Mr. Simmonds finally remarks, that as none of the inhabitants had obferved them before, they had no provincial name: the bird has been fince difcovered in the breeding feafon in fome plenty among the iflands, from which we may conclude, that although they had efcaped the pofitive obfervation of the natives, we are not to regard their appearance at that time as a recent or accidental vifitation.

Upon a comparifon of the Red with the Grey Phalarope it will be found, that independently of other differences, the form of the bill affords a diftinction between the two birds, by which they may be readily determined; in the Grey Phalarope the bill is ftouter, rather more compreffed, and broader: in the red kind it is weaker, the tip pointed and a little bent. It has been urged againft this diftinction, which appears to us fatisfactory that by improper treatment in the drying and preparation of the bird in the hands of an injudicious preferver, thefe characters may be fo far difguifed by partial contractions as to be no longer worthy of dependence; an argument in our opinion of little weight, fince it will apply equally to the diftinctions drawn from the characteriftic form of the bill in every other bird as well as this. The variety we have figured is of the fame fize as the red Phalarope of other authors, and meafures in length feven inches.

H 2

158

PLATE CLVIII.

HIRUNDO RUSTICA.

SWALLOW.

GENERIC CHARACTER.

Bill fmall, weak, curved, and awl fhaped, with the bafe depreffed ; gape larger than the head : tongue fhort, broad, cleft : wings long : tail in general forked.

SPECIFIC CHARACTER

AND

SYNONYMS.

Blueifh black ; beneath white, front and chin chefnut : tail feathers, except the two middle ones, with a white fpot.

Hirundo Rustica: nigro-cœrulefcens fubtus albida, fronte gulaque caftaneis, rectricibus lateralibus macula alba notatis. *Lath. Gen. Syft.* 2. 572. 1.

Hirundo Rustica: rectricibus, exceptis duabus intermediis, macula alba notatis. *Linn. Fn Suec.* 270.— *Gmel. Syft. Nat.* 1. *p.* 1015.—*Scop. Ann.* 1. *N°* 249.—*Brun. N°* 289. *Kram. El. p.* 380. 1.— *Georgi. p.* 175.—*Frifch. t.* 18.—*Klein. Av. p.* 82. 2.—*Schæff. El. Orn. t.* 40.

Hirundo

PLATE CLVIII.

Hirundo Domestica. *Raii Syn. p.* 71. *A.* 1.—*Will. Orn.*
p. 155. *t.* 39.—*Brifs.* ii. *p.* 486. 1.—*Id. 8vo.* 1.
p. 294.

Hirondelle de Cheminée. *Buff.* 6. *p.* 591. *t.* 25. *f.* 6.—*Pl. Enl.* 543.
f. 1.

Schwalbe, *Gunth. p.* 62. *t.* 15.

Rondine Minore, Zinnan, Uov. *p.* 48. *t.* 7. *f.* 35.

Chimney Swallow, or Common Swallow. *Phil. Tranf.*
51. *p.* 459.—*Id.* 53. *p.* 101.—*Id.* 65. *p.* 528.
343. *Br. Zool.* 1. *N°* 168. *t.* 58.—*Arct. Zool.*
11. *N°* 330.—*Alb.* 1. *t.* 45.

The habits and manners of the Swallow are fo well known, as
render much general information unneceffary. They occur in almoft
every part of the globe at different periods of the year, and are alfo
found in great plenty in America. They vifit England in immenfe
flights about the latter end of March, and as invariably collect in
amazing numbers in September, taking at that time their departure
for warmer climates to avoid the rigour of our approaching winter.

The Swallow ufually builds its neft, which is compofed of mud,
mixed with ftraw or hair, and lined with feathers, in the walls of
chimnies, or under the projecting ridges of the roofs of houfes,
churches, and fometimes trees: we poffefs a neft with the eggs,
built in the hollow of a conch-fhell as it laid in the garden of Sir
Afhton Lever, at Arlington, in Lancafhire.

The

PLATE CLVIII.

The Swallows of our cities, blackened and difcoloured with the fwarthy vapours of our chimnies exhibit an appearance only of footy blacknefs, varied with fpots of a dingy white. When in a high flate of plumage the Swallow is really a very beautiful bird, the black being of a jetty blacknefs, finely gloffed with fhining blue, partaking in different lights of purple and azure, and the white of unfullied purity; the front and chin a rich brown inclining to chefnut.

Varieties of the Swallow wholly white, occur occafionally, and are not more uncommon than the white varieties of the common birds in general. The Swallow is fix inches in length: it has two broods in a year; the eggs from four to fix in number, of a white colour, and fpeckled with reddifh. Its chief fubfiftence confifts of infects.

PLATE

159

PLATE CLIX.

TRINGA ALPINA.

ALPINE SANDPIPER,

OR,

DUNLIN.

GENERIC CHARACTER.

Bill roundifh, and as long as the head : noftrils fmall and linear : tongue flender: feet four toed, the pofterior toe of one joint, and raifed from the ground.

SPECIFIC CHARACTER

AND

SYNONYMS.

Brown teftaceous : breaft blackifh : tail feathers whitifh afh : legs brownifh.

TRINGA ALPINA: teftaceo-fufca, pe&tore nigricante, re&ricibus cinereo-albidis, pedibus fufcefccntibus. *Linn. Fn. Suec.* 180.
Linn. Syft. 1. *f.* 249. 11.
Gmel. Syft. 1. *p.* 676.

VOL. VII.　　　　　　I　　　　　　　*Fabr.*

PLATE CLIX.

Fabr. Fn. Groenl. n. 77.

Frifchav. t. 241.

Cinclus Torquatus. *Brifs.* 5. *p.* 216. *n.* 11. *t.* 19. *f.* 2.

Gallinago Anglicana, Le becaffine d'Angleterre. *Brif.* 5. *p.* 3095.

Le Cincle. *Buff.* 7. *p.* 553.—*Pl. Enl.* 852.

La Brunette. *Buff.* 7. *p.* 493.

Dunlin. *Ray Syn. p.* 109 *A.* 11.—*Will. Orn.* 305. *Brit. Zool.* No 205. *Lath. Gen. Syn.* 5. *p.* 185. 33.—*Id. Suppl. p.* 249.—Tringa Alpina, *Ind. Orn.* 736. 37.

A fcarce Britifh bird, found as the name implies, chiefly in Alpine fituations. It occurs in Afia and America as well as Europe. Length between nine and ten inches.

PLATE

160

PLATE CLX.

SCOLOPAX RUFA.

RUFOUS GODWIT.

GRALLÆ.

GENERIC CHARACTER.

Bill roundifh, obtufe, longer than the head: noftrils linear: face covered: feet four toed: hind toe confifting of many joints.

SPECIFIC CHARACTER

AND

SYNONYMS.

Ferruginous rufous: head, and back of the neck dafhed with black: wings cinereous, the feathers white at the edges: tail barred alternately with white and dufky.

SCOLOPAX HUDSONICA VAR? *Lath. Gen. Syn. Suppl. p.* 246— *Ind. Orn.* 720. 20.

———————————

There is much reafon for prefuming that this bird, though not till very lately, introduced to the acquaintance of the more expe-rienced Ornithologift, is by no means to be confidered as a recent

I 2   acquifition

PLATE CLX.

acquifition to the Britifh Fauna. The bird has very probably long exifted in fome collections of the country, which from their privacy have remained unknown, and where it has been perhaps confidered as the Red Godwit, or a mere variety of that rare and interefting fpecies.

This conjecture is rendered more than probable from various local circumftances within our knowledge which are not material to repeat. It was not, however, till within the laft four or five years that this bird has appeared to be correctly known as a Britifh fpecies; when befides the difcovery of a few birds in different parts of the country which fell by accident into the hands of curious fporting gentlemen, fome few were captured by the fowlers, and brought for fale to the London market. Thefe, very fortunately for the gra-tification of the Englifh collectors, were purchafed by a dealer in London *, and we believe that moft of the London collections at leaft, were fupplied with examples of this curious bird from that accidental fource and circumftance.

The impreffion that ftruck our mind upon the firft view of this bird was the very ftrong affinity it bore to the Scolopax Hudfonica of Dr. Latham, which we had feen fome years before, and fubfe-quent reflection does not feem to leffen the fimilitude in our ideas. We lament fincerely on this account that the means of comparifon are too remote either to eftablifh or to remove conjecture, for the individual fpecimen defcribed by Dr. Latham, under the name of

---

* Mr. Corbet.

Scolopax

PLATE CLX.

Scolopax Hudfonica exifts no longer in the country; we faw it for the laft time in the poffeffion of an intelligent German naturalift *, the day previous to its departure for the Imperial cabinet at Vienna about twelve years ago. We have no notes upon the fubject, but fo far as memory can be relied upon, the fize and general afpect could not be materially different, with the exception only of the tail, which was black and white, but not difpofed in alternate bands as in the prefent bird; the bafe was white and the pofterior half, or rather more, black without bands. But for this latter circumftance we fhould have little hefitation in believing them the fame fpecies; nor are we yet entirely fatisfied that the diftinctions which do exift fhould be attributed to any other caufe than a difference in the ftate of plumage arifing from the effect of climate.

We have feen this bird placed in the arrangement of Britifh birds, under the name of Scolopax Noveboracenfis, and generally fpeaking, it has obtained the name of Red-breafted Snipe, and Red Godwit: with refpect to the latter it is exceptionable, becaufe it may poffibly lead to fome confufion between this kind and the bird already known under the fame Englifh appellation. As to its analogy with the fpecies defcribed under the name of Noveboracenfis we muft confefs we cannot perceive it; nor fhall we venture to place it in our lift of fynonyms, with the mark of doubt, allowing even for the imperfect accounts we happen to poffefs of the latter bird.

The length of this bird is between fifteen and fixteen inches.

---

* Mr. Leopold Fichtel.

PLATE

161

PLATE CLXI.

EMBERIZA CIRLUS.

CIRL BUNTING.

PASSERES.

GENERIC CHARACTER.

Bill conic: mandibles receding from each other from the bafe downwards, the lower with the fides narrowed in, the upper with a hard knob within.

SPECIFIC CHARACTER

AND

SYNONYMS.

Above brown, varied: beneath yellowifh, breaft fpotted: eyebrows pale yellow: two outmoft tail feathers with a white cuneated fpot.

EMBERIZA CIRLUS: fufca, peĉtore maculato, fuperciliis luteis, reĉtricibus duabus extimis macula alba cuneata. *Gmel. Linn. Syft. Nat. T.* 1. *p.* 2. *n.* 879.

EMBERIZA CIRLUS: fupra varia, fubtus lutea, peĉtore maculato, fuperciliis luteis, reĉtricibus duabus, extimis macula alba cuneata. *Lath. Ind. Orn. T.* 1. *p.* 1. *p.* 401.

EMBERIZA SEPIARIA. *Brifs.* 3. 263. 2.

Le

PLATE CLXI.

Le Bruant Laye, *Buff.* 4. *p.* 347.—*Pl. Enl.* 653. *f.* 1, 2.
CIRL BUNTING. *Lath. Syn.* 3. *p.* 190. 2 B.

The Cirl Bunting is a bird not very uncommon in various countries of the fouth of Europe, but which till a very late period, remained unknown as an inhabitant of this country. With us it feems to be very local, and confined exclufively to the moft fouthern diftriEts. From its clofe affinity to the Yellow Bunting, or as it is more ufually called the Yellow Hammer, it may be eafily miftaken by the fuperficial obferver for that fpecies. It is, neverthelefs, diftinEt.

The natural food of this bird confifts of worms and infeEts as well as the feeds of various plants. Length fix inches and a half.

PLATE

162

PLATE CLXII.

TRINGILLA LINARIA β.

TWITE.

P A S S E R E S.

GENERIC CHARACTER.

Bill conic, ftraight and pointed.

SPECIFIC CHARACTER

AND

SYNONYMS.

Above varied, beneath reddifh, abdomen whitifh: eye-brows and band on the wings pale rufous: crown and rump red.

FRINGILLA LINARIA: fupra varia, fubtus rufefcens, abdomine
albido, fuperciliis fafciaque alarum rufefcentibus,
vertice uropygioque rubris. *Lath. Ind. Orn.*
459. 83.

TWITE. *Albin.* 3. *t.* 74.
Lev. Muf.

The Twite was formerly confidered as a variety of the mountain Linnet *, and at a later period has been placed with doubt as a variety

---

* The Mountain Linnet itfelf has borne fucceffively the fpecific names piplans and montium.

PLATE CXLII.

of the leffer Linnet. We are aware that much confufion has, and ftill continues to prevail among the Linnet tribe, we can however venture to fpeak precifely as to the bird before us being the Twite of the Leverian Mufeum, and therefore the individual bird defcribed by Dr. Latham and feveral other writers under the appellation of the Twite: it paffed immediately from that collection into our poffeffion.

The length of this bird is four inches and a half. The red colour of the rump is generally confidered as a decifive character of the Twite.

PLATE

163

PLATE CLXIII.

ANAS HISTRIONICA.

HARLEQUIN DUCK.

ANSERES.

GENERIC CHARACTER.

Bill convex, obtufe, the edges divided into lamellate teeth : tongue fringed and obtufe : three fore toes connected, the hind one folitary.

SPECIFIC CHARACTER

AND

SYNONYMS.

Varied fufcous blue and white; ears, double line on the temples, collar, and pectoral band white (male). Grey, ears white : primary quill feathers blackifh.

MAS.

ANAS HISTRIONICA : fufco albo cæruleoque varia, auribus, temporibus linea gemina, collari fafciaque pectora li albis. *Fabr. Fn. Groenl. n.* 46.

ANAS HISTRIONICA. *Linn. Syft.* 1. *p.* 204. 35.—*Gmel. Syft.* 1.

K 2

p. 534.—

PLATE CLXIII.

p. 534.—*Brun. N° 84, 85.—Phil. Tranf.* 62.
p. 417.—*Frifch. t.* 157.—*Faun. Amer. p.* 16.
Anas torquata ex infula terræ novæ. *Brifs.* 6. *p.* 362. 14.
Anas Brimond, *Olaff. Ill* 2. *t.* 34.
Le Canard à Collier de terre neuve, *Buff. ix. p.* 250.—*Pl. Enl.* 798.
STONE DUCK. *Hift. Kamtfch. p.* 160.—DUSKY AND SPOTTED
    DUCK. *Edw. av. t.* 99.—HARLEQUIN DUCK.
    Arct. Zool. 2. *N°* 490.—*Lath. Syn.* 6. *p.* 485.
    88.—*Ind. Orn.* 849. 45.

FEM.

ANAS MINUTA: fufca (vel grifea) auribus albis, remigibus pri-
    moribus nigricantibus. *Fabr. Faun. Groenl. n.*
    46.—*Brun. Orn. n.* 86.—*Linn. Syft.* 1. *p.* 204.
    36.—*Gmel. Syft.* 1. *p.* 534.
Querquedula freti Hudfonis. *Brifs.* 6. *p.* 469. 41.—*Id. 8vo.* 11.
    p. 483.
Le Canard brun, et le Canard brun et blanc. *Buff. ix. p.* 253.—
    Pl. Enl. 1007.
La Sarcelle brune et blanc. *Buff. ix. p.* 287.—*Pl. Enl.* 799.
Little brown and white Duck. *Edw. t.* 157.
    Cat. Car. 1. *t.* 98.
HARLEQUIN DUCK. (female) *Lath. Syn.* 6. *p.* 485. 38.

———————————

It is fome years fince the Harlequin Duck was firft introduced
among collectors as a bird appertaining to the Britifh Fauna; and
fifteen years have at leaft elapfed fince we became poffeffed of the

3   fpecimen

PLATE CLXIII.

fpecimen which is figured in the annexed plate : during the whole of that period we underſtand that it has been only found occaſionally perhaps to the amount of three or four ſpecimens at the utmoſt, and hence we may conclude, that it ranks among the rarer ſpecies of the Duck tribe found in this country.

As is frequently the caſe with birds the two ſexes of which differ very materially in plumage, the male and female of this bird has been conſidered as diſtinct ſpecies. The figure of the male which is ſhewn in our plate may ſuperſede the neceſſity of any particular deſcription of that ſex, except with reſpect to ſize, which correſponds with that of the common wigeon, and is about ſeventeen, inches in length. The female is leſs by three inches ; very little exceeding in length thirteen inches: the general colour grey ; forehead and between the eye white: lower part of the breaſt and belly barred with pale rufous and white, and the lower part, together with the thighs rufous and brown : quills, tail and legs duſky.

The ſpecies inhabits all the northern parts of Europe and America, retiring a little to the ſouthward as the winter become intenſely ſevere. Its chief food conſiſts of the vaſt variety of teſtaceous animals, with which the waters of all the northern lakes and rivers abound, and eſpecially the innumerable hoſt of gnats and other aquatic infects which are there produced in numbers beyond all conception during the ſhort period of the polar ſummer.

PLATE

164

PLATE CLXIV.

PARUS PALUSTRIS.

MARSH TITMOUSE.

PASSERES.

GENERIC CHARACTER.

Bill very entire, narrow and fomewhat compreffed; ftrong, hard, pointed and covered at the bafe with briftles: tongue truncated, briftly at the end: toes divided to the origin, the pofterior one large and ftrong.

SPECIFIC CHARACTER

AND

SYNONYMS.

Head black: back cinereous: temples white.

PARUS PALUSTRIS: capite nigro, dorfo cinereo, temporibus albis.

> *Linn. Syft.* 1. *p.* 341. 8.—*Faun. Suec.* N° 269.
>
> *Gmel. Syft. Nat.* 1. *p.* 1009.—*Lath. Gen. Syn.* 4.
>
> *p.* 541. 8.—*Ind. Orn.* 565. 9.

MARSH TITMOUSE, or BLACK CAP. *Albin.* 3. *t.* 53. *f.* 1.

> *Will.* 241. *t.* 43.

Allied

PLATE CLXIV.

Allied to the Colemoufe, and has been confidered as no other probably than a variety of that bird. This like the reft of its tribe, is a moft prolific bird laying a vaft number of eggs, and hatching frequently. Its fize is that of the Colemoufe : the length four inches, Authors defcribe it as an inhabitant of northern Europe.

Dr. Latham defcribes it as being fond of bees.

PLATE

165

PLATE CLXV.

STRIX PULCHELLA.

SMALLER PENCILLED,

OR,

SIBERIAN EARED OWL.

ACCIPITRES.

GENERIC CHARACTER.

Bill hooked: no cere: noftrils oblong, covered with briftly recumbent feathers : head, auricles and eyes large: tongue bifid.

SPECIFIC CHARACTER

AND

SYNONYMS.

Head eared: body above and wings grey, powdered and waved with rufty and black, and varied with white fpots : beneath whitifh.

STRIX PULCHELLA. *Pallas. It.* 1. *p.* 456. 8.—*Lepech. It.* 11. *t.* 4.—*Nov. Com. Act. Petrop. Vol.* 15. *p.* 490. *t.* 26. *f.* 1.

STRIX PULCHELLA: minima, capite aurito, corpore pulveratim cinereo-undulato fubtus albido, alis fafciato-pul-

VOL. VII.   L   veratis,

PLATE CLXV.

veratis, litura ad nares alba. *Lath. Ind. Orn.* 1.
57. 19.

SIBERIAN EARED OWL. *Lath. Gen. Syn.* 1. *p.* 130. 16. *t.* 5. *f.* 1.

A fpecimen of this elegant bird has been in our poffeffion for a
confiderable period, and as our Mufeum of natural hiftory in which
it is contained was allowed to remain open to general infpection for
the fpace of fome years, we conceive the bird muft have obtained
every requifite publicity to entitle us to the credit, if there be any
due, of having firft introduced it to the notice of our country as a
Britifh bird.

Some of our Ornithologifts have placed this bird in their arrange-
ments as a new Britifh fpecies under the fpecific name of Scops,
believing it to be the fame kind with *le petit duc* of Buffon, a bird
found in France, and the fouth of Europe, and diftinguifhed by
Gmelin under the name of Scops. We are ever unwilling to differ
from our cotemporary collectors, and would not on any very trivial
occafion be inclined to exprefs our diffent to what appears an eftablifhed notion, but in the prefent inftance there really appears to us
the exiftence of an error material to be corrected. The bird before
us is affuredly not the Gmelinian Scops: we have ever confidered it
as the Strix Pulchella defcribed by Dr. Pallas, and have little hefitation in conceiving on a more attentive inveftigation the opinion of
thofe who think the contrary will gradually yield to ours.

With

PLATE CLXV.

With refpect to the general hiftory of this curious bird our knowledge is very circumfcribed: to the beft of our information it was fhot in Yorkfhire a few years ago. It is unqueftionably one of the moft elegant of the Owl tribe: its length is fcarcely more than fix or feven inches.

PLATE

166

PLATE CLXVI.

ANAS MARILA.

SCAUP DUCK.

ANSERES.

GENERIC CHARACTER.

Bill convex, obtufe, the edges divided into lamellate teeth; tongue fringed, obtufe: three fore toes connected, the pofterior one folitary.

SPECIFIC CHARACTER

AND

SYNONYMS.

Black: fhoulders waved cinereous: belly and fpot on the wings white, *male.* Ferruginous brown, fpot on the wing abdomen, head and ring at the bafe of the bill white, *female.*

ANAS MARILA: nigra, humeris cinereo-undulatis, abdomine fpe- culoque alari albis. *Maf. Act. Angl.* 62. *p.* 413. *Gmel.* 509. 8.—Fufco ferruginea, fpeculo alarum abdomine capitifque annulo ad roftri bafin albis *(Fem.) Lath. Ind. Orn.* 853. 54.

SCAUP

PLATE CLXVI.

SCAUP DUCK. *Will. Orn. p. 365. Ray. av. p. 142. A. 6. Lath. Syn. 3. 2. p. 500. n. 49.*

Length about twenty inches. The Scaup Duck is a general inhabitant of the colder regions, of Europe, the north of Afia and America: feeds chiefly on aquatic infects and the teftaceous animals of the frefh waters, and migrates to warmer climates in winter.

PLATE

167

PLATE CLXVII.

PEDICEPS HEBRIDICUS.

BLACK-CHIN GREBE.

ANSERES.

GENERIC CHARACTER.

Bill ftraight, acute, noftrils linear, lores naked: tongue fomewhat bifid: tail obfolete: legs compreffed with a double feries of dentations behind: toes furnifhed on each fide with a broad plain membrane.

SPECIFIC CHARACTER

AND

SYNONYMS.

Head fmooth, body blackifh: belly cinereous intermixed with filvery: chin of the male black, throat ferruginous.

COLYMBUS HEBRIDICUS: capite lævi, mento nigro, gutture ferrugineo, abdomine cinereo et argenteo. *Gmel. Linn. Syft. Nat. T.* 1. *p.* 2. 594. 28.

PODICEPS HEBRIDICUS: nigricans, gula nigra, jugulo ferrugineo,

6

PLATE CLXVII.

gineo, abdomine cinereo argenteo vario. *Lath.*
Gen. Syn. 5. 292.

BLACK-CHIN GREBE. *Penn. Br. Zool. 2. 227. f. 79.*

With the fingle exception of the Red-neck Grebe figured in the
6th volume of this work, the bird before us may be efteemed the
rareft of the tribe found in Britain ; and none among the number
which altogether includes feven kinds, can be confidered common,
exclufive of the Dobchick, or little Grebe, which occurs in moft
fenny places: The moft frequent of the larger kinds is the Crefted
Grebe, and that is confined chiefly to the fens of Lincolnfhire, and
two or three other parts of Britain.

The black-chin Grebe, called alfo the Hebridal Grebe is almoft
entirely confined to the waters of the ifland of Tirée, one of the
Hebrides, and thence it has received the appellation of Hebridal
Grebe. It is rather larger than the Little Grebe, and is in length
about eleven inches.

PLATE

168

PLATE CLXVIII.

FRINGILLA LINOTA.

COMMON GREY LINNET.

PASSERES.

GENERIC CHARACTER.

Bill conic, ftraight, and pointed.

SPECIFIC CHARACTER

AND

SYNONYMS.

Chefnut brown, beneath whitifh; wings with a longitudinal white band: tail feathers each fide edged with white.

FRINGILLA LINOTA: fufca caftanea, fubtus albida, fafcia alarum longitudinali alba, reftricibus nigris, marginibus undique albis. *Gmel. Linn. Syft. Nat.* 1. *p.* 916. *Lath. Ind. Orn.* 457.

COMMON LINNET.

This ranks among the fong birds of Britain; it is very common in England and throughout Europe. It feeds on feeds, principally

VOL. VII.   M   thofe

PLATE CLXVIII.

thofe of the hemp, which it is obferved to peel before it eats. The rofy colour which appears confpicuous on the breaft of this bird denotes the male in full plumage, the female and young birds being deftitute of this diftinction. The length of this bird is five inches.

The eggs amounting in each neft to five in number are of a whitifh colour fpotted with chefnut.

PLATE

169

PLATE CLXIX.

SCOLOPAX PYGMÆA.

PIGMY CURLEW.

GRALLÆ.

Bill roundifh, obtufe, longer than the head: noftrils linear: face covered: feet four toed, hind toe confifting of many joints.

SPECIFIC CHARACTER

AND

SYNONYMS.

Arched bill and legs black: body varied with ferruginous brown and white, beneath white.

SCOLOPAX PYGMÆA: roftro arcuato pedibufque nigris, corpore ex ferrugineo, fufco et albo vario, fubtus albo. *Gmel. Syft. Nat.* 655. 20.

NUMENIUS PYGMEUS: fufco ferrugineo alboque variegatus, corpore fubtus uropygioque albo, remigibus rectricibufque exterioribus albo marginatis. *Lath. Ind. Orn.* 713. 11.

M 2

This

PLATE CLXIX.

This may be confidered as a curious variety of that rare **Britifh** bird called the Pigmy Curlew, or the Pigmy Sandpiper in the plumage it affumes in autumn. It was obligingly communicated by Mr. Weighton: the bird was fhot by Mr. Lenard, at Holyavon, on the 26th of Auguft, 1812.

PLATE

170

PLATE CLXX.

MUSCICAPA GRISOLA.

SPOTTED FLY CATCHER.

PASSERES.

GENERIC CHARACTER.

Bill nearly triangular, notched each fide, bent in at the tip, and befet with briftles at the root: toes moftly divided to their origin.

SPECIFIC CHARACTER

AND

SYNONYMS.

Brownifh, beneath whitifh: neck longitudinally fpotted: vent pale rufous.

MUSCICAPA GRISOLA: fubfufca, fubtus albicans, colli longitu-
dinaliter maculato, criffo rufefcente. *Gmel. Syft. Nat.* 949. *fp.* 20.
Gobe Mouche. *Buff. Oif.* 4. *p.* 517. *t.* 25. *f.* 2.
SPOTTED FLY CATCHER. *Lath. Syn.* 2. 1. *p.* 323. *n.* 1.

The

PLATE CLXX.

The fpotted Fly-catcher is a bird of the migratory kind in England, arriving in the fpring, and taking its departure in autumn.

The length is between five and fix inches: it frequents gardens, and is obferved befides other fruits, to be particularly fond of cherries. The eggs from four to five in number, are of a white colour fpotted with reddifh.

PLATE

171

PLATE CLXXI.

VELVET DUCK.

ANSERES.

GENERIC CHARACTER.

Bill convex, obtufe, the edges divided into lamellate teeth : tongue fringed, obtufe : three fore toes connected, the pofterior one folitary.

SPECIFIC CHARACTER

AND

SYNONYMS.

Black : lower eye-lid and fpot on the wing white.

ANAS FUSCA : nigricans, palpebra inferiore, fpeculoque alarum albis.   *Linn. Fn. Suec.* 109. *Gmel. Syft. Nat.* 507. 6.

GREAT BLACK DUCK.   *Ray. Orn. p.* 141. *A.* 4. *Will. Orn. p. 363. t.* 70.

VELVET DUCK.   *Lath. Gen. Syn.* 3. 2. *p.* 482. *n.* 37.

Allied to the Scoter Duck but is more rare. Length twenty inches.

PLATE

6

172

PLATE CLXXII.

TRINGA HYPOLEUCOS.

COMMON SANDPIPER.

GRALLÆ.

GENERIC CHARACTER.

Bill roundifh and as long as the head: noftrils fmall and linear: tongue flender: feet four toed, the pofterior toe of one joint, raifed from the ground.

SPECIFIC CHARACTER

AND

SYNONYMS.

Bill fmooth: legs livid; body cinereous with black ftripes, beneath white.

TRINGA HYPOLEUCUS: roftro lævi, pedibus, lividis, corpore
      cinereo lituris nigris, fubtus albo. *Linn. Fn.*
      Suec. 181.—*Gmel. Syft. Nat.* 678. 14.

COMMON SANDPIPER. *Lath. Syn.* 5. 178.

Length feven inches and a half: lays its eggs in fand banks; thefe are four or five in number, and of a dirty yellowifh colour with pale fpots. Inhabits Europe and America.

VOL. VII.   N   INDEX

INDEX to VOL. VII.

ARRANGEMENT

ACCORDING TO THE

SYSTEM of LINNÆUS.

ORDER I.

ACCIPITRES.

ORDER II.

PICÆ.

ORDER III.

ANSERES.

ORDER

INDEX.

ORDER IV:

GRALLÆ.

ORDER V.

PASSERES.

VOL.

·I N D E X.

V O L. VII.

ARRANGEMENT

ACCORDING TO

LATHAM's SYNOPSIS of BIRDS.

ORDER

INDEX.

ORDER IV. Columbine.

DIVISION II. Water Birds.
ORDER VII. With Cloven Feet.

ORDER IX. Web-footed.

INDEX:

INDEX.

VOL. VII.

ALPHABETICAL ARRANGEMENT.

Printed by R. & R. Gilbert, St. John's Square, London.

THE

NATURAL HISTORY

O F

BRITISH BIRDS;

OR, A

SELECTION of the MOST RARE, BEAUTIFUL, and INTERESTING

B I R D S

WHICH INHABIT THIS COUNTRY:

THE DESCRIPTIONS FROM THE

S Y S T E M A   N A T U R Æ

O F

L I N N Æ U S:

WITH

GENERAL OBSERVATIONS,

EITHER ORIGINAL, OR COLLECTED FROM THE LATEST
AND MOST ESTEEMED

ENGLISH ORNITHOLOGISTS;

AND EMBELLIHED WITH

F I G U R E S,

DRAWN, ENGRAVED, AND COLOURED FROM THE ORIGINAL SPECIMENS.

By E. DONOVAN.

VOL. VIII.

LONDON:

PRINTED FOR THE AUTHOR; AND FOR F. C. AND J. RIVINGTON,
No. 62, ST. PAUL'S CHURCH-YARD. 1817.

173

PLATE CLXXIII.

TRINGA ISLANDICA.

RED SANDPIPER.

GRALLÆ.

GENERIC CHARACTER.

Bill roundifh, as long as the head: noftrils fmall, linear: tongue flender: feet four-toed: the hind-toe of one joint, and raifed from the ground.

SPECIFIC CHARACTER
AND
SYNONYMS.

Bill and legs fufcous: back variegated with fufcous, rufous, and black: body beneath ferruginous: fecondary quill-feathers edged with white.

TRINGA ISLANDICA: roftro pedibufque fufcis, dorfo fufco rufo nigroque variegato: corpore fubtus ferrugineo remigibus fecundariis margine albis.

TRINGA ISLANDICA: roftro pedibusque fufcis, corpore fubtus ferrugineo remigibus fecundariis margine albis. *Gmel.* 682. 24.

VOL. VIII.   **B**   TRINGA

PLATE CLXXIII.

TRINGA FERRUGINEA. *Brünn. Orn. n.* 180.
RED SANDPIPER. *Brit. Zool.* 2. *n.* 202. *b.* 72.
ABERDEEN SANDPIPER. *Brit. Zool.* 2. *n.* 203.—*Lath. Gen.*
Syn. 3. *I. p.* 186. *n.* 84.

———————

The fpecimen of the Red Sandpiper, originally depofited in the Leverian Mufeum, and thence defcribed by Englifh authors, has been in our poffeffion from the period in which that Mufeum was difperfed; and as our own Mufeum was allowed to remain open to public view for fome years fince that time, the appearance of this bird may be perhaps familiar to many of our readers, for it can fcarcely be imagined that any vifitor would have neglected to beftow fome attention towards this elegant and truly interefting fpecies.

As a Britifh Bird, the Red Sandpiper was always, we believe, confidered rare. It is indeed recorded to have appeared occafionally, and even flocks of them have been obferved; but from peculiar cir-cumftances few of thefe had been procured, and it remained a rarity till the fpring of 1812 afforded this curious bird in fome plenty. A flock of them, as we underftand, was feen at that time in the fens, and feveral of thefe being captured in the fowling-nets, they paffed very fortunately from the wild-fowl dealers of the London market into the hands of Mr. Corbet, an ingenious preferver of birds, by whom they were prepared for the cabinet, and who has fince furnifhed feveral of the principal collections with the fpecimens they poffefs.

One

PLATE CLXXIII.

One of thefe laft-mentioned birds is now in our own poffeffion : it differs in no material particular from the original bird, which is at prefent before us, and with which we have compared it with due attention ; the only diffimilarity confifts in the general hue of the upper parts of the plumage, in the more recent fubject being fomewhat darker, and the lower more deeply rufo-ferruginous.

Notwithftanding the apparent fcarcity of this kind of Sandpiper, it fhould be remarked however, that there is fome reafonable grounds for believing, that it is only rare to us in this more lively ftate of plumage; that in its ordinary drefs we fhould recognize it to be no other than the Tringa Canutus, the bird familiarly known to our poulterers by the name of the Knot-bird. This idea was firft fuggefted by the appearance of one of the Knot-birds, which we met with fome years ago among a parcel of the common kind in Leadenhall market. The whole of the lower furface had affumed a richly varied intermixture of brown with the white feathers, and dufky crefcent-like marks of the neck, breaft, and fides of the abdomen ; and befides this fpecimen, there was another, the breaft of which had began to affume the fame ruddy afpect. The correfpondence in this refpect was ftriking ; but there was nothing in the upper furface of the plumage to diftinguifh it from the common Knot; unlike the true Icelandic Sandpiper, it was entirely deftitute of that elegant intermixture of black and oblong fpots of ruft colour, which appears confpicuous in the example we have reprefented, and which Linnæus confiders characteriftic of the Iceland fpecies.

Thefe laft-mentioned birds were placed in the fpring of 1812, with other varieties of the Knot-bird, in our Mufeum, in a fituation immediately contiguous to the original fpecimen of the Red Sandpiper, in

B 2

order

- 231 -

PLATE CLXXIII.

order that every obferver might be enabled to form his own judgment as to the relative connection of the two birds, and it is, we may be allowed to conceive, from the publicity thus afforded them, that an opinion has arifen of the Common Knot and the Iceland Sandpiper, being only the winter and fummer drefs of the fame bird.

In a future plate we fhall fubmit a portrait of the rufous bellied Knot, above defcribed, to the attention of our readers; together with fuch information as we may conceive likely to elucidate the obfcurity that feems to prevail refpecting them.

Length of the original fpecimen of the Red Sandpiper eleven inches; of the one recently captured about ten inches.

PLATE

174

PLATE CLXXIV.

LARUS MARINUS γ.

WAGEL GULL.

ANSERES.

GENERIC CHARACTER.

Bill ftraight, fharp edged, a little falcated at the tip, and toothlefs : lower mandible gibbous below the point : noftrils linear, broader on the fore part, and placed in the middle of the bill.

SPECIFIC CHARACTER

AND

SYNONYMS.

Varied brown, afh and white with brown fpots : band of black on the tail, tip white.

LARUS MARINUS γ. *Lath. Ind. Orn.* 814. 6.

LARUS NAEVIUS : albus, dorfo cinereo, rectricibus apice nigris. *Gmel. Linn. Syft.* 598. 5.

LARUS VARIUS. *Brünn. Orn. n.* 150.

LARUS MACULATUS. *Brünn.* N° 146. *young bird?*

Le Goiland varié, ou le Grifard. *Brifs. Orn.* 6. *p.* 167. 5. *pl.* 15.— *Buff. Oif.* 8. *p.* 413. *pl.* 33.—*Pl. Enl.* 266.

Wagel, Burgo-mafter of Groeland, Great Grey Gull, *Raii Syn. p.* 130. *A.* 13.—*Will. Orn. p.* 349. *pl.* 66.

Wagel. *Br. Zool.* 2. *n.* 247. *A.*—*Arct. Zool.* N° 453.

To

PLATE CLXXIV.

To reconcile the various opinions that have prevailed among naturalists refpecting the identity of fome fpecies of the Gull tribe would be a tafk of more than ordinary difficulty: the tranfitions of their plumage, refulting from the effects of climate, the feafons of the year, the difference of fex, and the various periods of their growth, preclude the poffibility of fpeaking with any pofitive degree of precifion in fome inftances, and in no one more it would appear upon the teftimony of the beft writers, than in the identity of the fpecies before us.

Among the firft of thefe opinions we may mention that of Linnæus, who confidered it as the Herring Gull in the firft year's plumage. Fabricius, the author of *Fauna Groenlandica*, on the contrary conceived it to be the young of the Black-backed Gull, and Pennant concludes that it is not an immature bird, but the female of the Black-backed Gull.

It would have tended much to confolidate the two firft of thefe opinions had the conclufions offered by Dr. Latham in his general Synopfis, been eftablifhed by later obfervation; namely, that the Herring Gull and the Black-backed Gull are of the fame fpecies; but the contrary of this opinion has been fo clearly afcertained, that in a fubfequent production, (*Index Orn.*) Dr. Latham himfelf abandons that idea, and admits the Herring Gull and Black-backed Gull to be fpecifically diftinct.—This is our own opinion: we have the two fexes of both fpecies in very perfect order of maturity in our own Mufeum, and from a due comparifon of thefe, we cannot entertain the leaft diftruft of the accuracy of this laft conclufion. In ftating this, the reader will perceive that the idea of the Wagel being the female of the Black-backed Gull, as Mr. Pennant imagines, is not admiffible, the female of that fpecies being altogether diffimilar.

But

PLATE CLXXIV.

But although Dr. Latham divides the Black-backed Gull from the Herring Gull in his Index Ornithologicus, he does not feparate the Wagel from the firft-mentioned fpecies, but places it as a variety, or rather as the young of that bird Larus Marinus γ. This arrangement is, we believe, founded on actual obfervation of the fpecies in the various ftages of its growth, or we might be inclined, from the general appearance of the bird, to follow, in preference, the example of Gmelin, who eftablifhes the Wagel as a diftinct fpecies, fubject to fome variations, under the fpecific name of Naevius. The Wagel is a large bird meafuring, in length, two or three feet, and exceeding, when full grown, the ordinary fize of the Black-backed Gull; it is alfo confiderably larger than the Herring Gull which, in point of fize, is inferior to the Black-backed Gull *.

---

* Mr. Montagu, in his Ornithological Dictionary, remarks, that " the appellation (WAGEL GULL) has been affigned to feveral fpecies of the genus in the mottled infant plumage; and as there is no fuch bird claiming fpecific diftinction, it fhould be erafed as fuch from the pages of Ornithology." *Suppl.* This affurance deferves confideration; for if it be correct it muft prove the fallacy of all difcuffion upon the fubject, as the Wagel will be identified in the young of feveral fpecies of the Gull tribe, which in their perfect ftate are acknowledged to be fpecifically diftinct. But we cannot yield entirely to an obfervation which, in our opinion, teftifies a precipitancy of conclufion beyond even what the author himfelf intended, for he confiders in another place the Wagel as the young of the Black-backed Gull alone. That the young of all the Gull tribe have the plumage mottled in the early periods of their growth, is a fact fufficiently known, but every individual poffeffes yet fome ftriking peculiarity of the adult bird, by which the judgment of the fkilful Ornithologift may be directed, and we think fuccefsfully, to the determination of the fpecies; and in proof of this, we cannot doubt that the true Wagel may always be diftinguifhed from the reft of the Gull tribe in their mottled ftate of plumage.

With refpect to the Wagel when full grown, its ordinary fize is known to exceed that of the Black-backed Gull, a circumftance that does not arife from the greater laxity of the feathers as might be imagined in admitting it to be the younger bird; but from the bird itfelf being actually larger than the Black-backed Gull at the full maturity. This would tend, at leaft in fome degree, to prove that it might not be the " infant offspring" of that bird, although there may be certain inftances in fome fpecies of the young bird being larger than the adult.

Thefe

PLATE CLXXIV.

Thefe birds occur on the fea fhore and in the vicinities of great rivers in various parts of Britain, but it is faid not in any confiderable plenty. In fevere winters they have been known to vifit the banks of the Thames in company with others of the Gull tribe, and one fhot near Richmond, about twenty years ago, by a well-informed fportfman was tranfmitted to us as a rarity. Within a few years paft they appear to have become more common about the banks of the Thames, and in that of 1812 in particular, when they occurred in fome abundance. It may not be amifs to add, that in the fummer of the year 1801, we were favoured with a recent fpecimen of the Wagel (fhot in Cornwall) from the late Mr. Hutchins, a very intelligent Ornithologift, the plumage of which did not differ from that of the Wagels fhot in winter, as before mentioned.

PLATE

175

PLATE CLXXV.

HIRUNDO URBICA.

HOUSE MARTIN.

PASSERES.

GENERIC CHARACTER.

Bill fmall, weak, curved, fubulate, and depreffed at the bafe : gape larger than the head : tongue fhort, broad cleft : wings long : tail generally forked.

SPECIFIC CHARACTER

AND

SYNONYMS.

Blue-black, beneath white : tail without fpots.

HIRUNDO URBICA: nigro-coerulefcens fubtus alba, retricibus im-maculatis. *Lath. Ind. Orn.* 573.

HIRUNDO URBICA : retricibus immaculatis, dorfo nigro-coeru-lefcente, tota fubtus alba. *Linn. Fn. Suec.* 271. —*Linn. Syft. I. p.* 344. 3.—*Gmel. Syft. I. p.* 1017.

Hirundo minor f. ruftica. *Briff. av.* **2.** *p.* 490. *n.* **2.**

Hirundo ruftica f. agreftis Plinii. *Raj. av. p.* 71. *n.* **2.**—*Will. Orn. p.* 155. *t.* 30.

VOL. VIII.   C   Hiron-

PLATE CLXXV.

Hirondelle à cul blanc. *Buf. 6. p.* 614. *t.* 25. *f.* 2.

Le Petit Martinet. *Pl. Enl.* 542. *f.* 2.

MARTIN, MARTLET, or MARTINET. *Will. (Angl.) p.* 213.
t. 39.—*Alb.* 2. *t.* 56. a.

Lath. Gen. Syn. 4. p. 564. 3.—*Id. Suppl.*
p. 192.

The Houfe Martin differs from the Sand Martin, which we have already given in this Work, in fome very ftriking particulars. It is a larger bird, and unlike that fpecies which forms its refidence in the banks of fand, in fituations folitary and remote from the habitations of man; this little intruder, emboldened by familiarity, conftructs its nefts, and breeds under the cornices and eaves of our houfes, againft our windows, or amongft the thatch, wherever a convenient refting place is afforded; but not in chimnies like fwallows. Their neft is compofed of mud and a few twigs or fticks, and lined with feathers; and the eggs, which in the firft laying are five in number, in the fecond four, and in the third three, are of a white colour. There are two or more broods in the year.

This fpecies is more abundant than the common Swallow, and though many build and breed here, the Houfe Martin performs an annual migration like that of the common Swallow, and nearly about the fame time. They arrive in England about twenty days before the Swallow, and depart again in Autumn.

It is a fmall bird, being only five inches and a half in length; the colour above blackifh, very richly gloffed with blue; the wings
brown,

PLATE CLXXV.

brown, and tail uniformly brown gloffed with blue; the plumage beneath white, and the legs covered with a whitifh down. A fuppofed variety that inhabits North America, has the quill and tail feathers tipped with white.

The Houfe Martin, like the reft of its tribe, fubfifts on infeƈts, which it chiefly takes on the wing.

C 2

PLATE

176

PLATE CLXXVI.

LARUS RIDIBUNDUS γ.

BROWN-HEADED GULL.

ANSERES.

GENERIC CHARACTER.

Bill ftraight, fharp-edged, a little falcated at the tip and toothlefs, lower mandible gibbous below the point: noftrils linear, broader on the fore part, and placed in the middle of the bill.

SPECIFIC CHARACTER

AND

SYNONYMS.

White, head blackifh : bill and legs red.

LARUS RIDIBUNDUS. *Linn. Syft. I. p.* 225. 9.

LARUS RIDIBUNDUS: albidus, capite nigricante, roftro pedibufque rubris. *Oedm. Nov. Act. Stockh.* 1783. 2. *n. I. p.* 119. *n.* 9.

LARUS RIDIBUNDUS γ : albidus, capite fufco maculis albis, dorfo cano, rectricibus decem intermediis fafcia nigra. *Lath. Ind. Orn.* 812.

BROWN-HEADED GULL. *Lath. Syn.* 6. *p.* 383. 11.—*Albin.* 2. *pl.* 86.

BROWN-HEADED GULL. *Lev. Muf.*

The

PLATE CLXXVI.

The length of this bird is fixteen inches; the plumage white, with the head and neck moufe colour: tail white, and compofed of twelve feathers; the exterior one of which on each fide is white and immaculate, the interior black at the tip, and the bill and legs red.

This is generally believed by Ornithologifts to be the Black-headed Gull in its fummer ftate of plumage, the dark brown colour of the head and neck affuming a more intenfe or blacker hue towards the winter feafon.

The Brown-headed Gulls are found, we underftand, in plenty along the banks of the Thames and Medway towards the fea; we have ourfelves feen them among the flocks of gulls that haunt the fhallow waters off the north of the ifle of Sheppy, in Kent; and we think alfo about the ifle of Grain, in the fame vicinity. That which we have reprefented is a Leverian fpecimen; and it may not be improper to mention this, as the different fuppofed varieties of Larus Ridibundus feem yet involved in fome obfcurity; at leaft there is a confiderable diverfity of opinion among writers refpecting them.

PLATE

177

PLATE CLXXVII.

TRINGA NIGRICANS.

PURPLE SANDPIPER.

GRALLÆ.

GENERIC CHARACTER.

Bill roundifh, as long as the head: noftrils fmall and linear: tongue flender: feet four-toed: the hind-toe of one joint raifed from the ground.

SPECIFIC CHARACTER

AND

SYNONYMS.

Cinereous brown: back black with fubviolaceous glofs: chin and middle of the abdomen white.

TRINGA NIGRICANS: cinereo-fufco, dorfo nigro fubviolaceo-nitens, mento abdomineque albo.

TRINGA NIGRICANS: Blackifh afh: chin and middle of the belly white: bafe of the bill and legs red. *Linn. Tranf. IV.* 40.

The

PLATE CLXXVII.

The defcription of the Purple Sandpiper, inferted in the fourth volume of the Tranfactions of the Linnæan Society, was taken from a fpecimen killed at Langharne, on the coaft of Caermarthenfhire, in company with the Purre : it was killed in the month of January, and two others of the fame kind having been fhot there the fame winter, it obtained the trivial name of the Welfh Sandpiper, and from its colour, the fpecific name of nigricans.

A figure of the fame bird had previoufly appeared in the work of Walcot, *Syn.* 11. *t.* 155, under the name of the Purple Sandpiper; and as the bird has been fince found on the coafts of Kent and Suffex, the appellation of Purple Sandpiper feems more appropriate than the very local name of Welfh Sandpiper.

Nor is the merit of its original introduction into the Britifh Fauna even due to Walcot, fince the bird muft have been known as one of the rareft of our Sandpipers prior to the time in which his Synopfis was compofed, for long before that period there was a fpecimen of it in the late Leverian Mufeum, and which, if we miftake not, was the firft example of this interefting fpecies introduced to public notice.

Having mentioned this, it is equally due, in juftice, to thofe by whom the fpecies was defcribed to ftate, that whilft it remained among the Leverian birds it bore the name of the " Knot," and it is not impoffible, under that defignation, it might have efcaped remark, or been confidered as a variety of the common Knot. That the fpecies had a place in the Leverian Mufeum can admit of no diftruft, for we are in poffeffion of the original bird with the label annexed.

In fubmitting the preceding obfervations to our readers we are

only

PLATE CLXXVII.

only anxious to award the credit of the difcovery of this elegant Britifh Sandpiper to its true fource : we believe it was Mr. Bolton who difcovered it ; it was affuredly Sir Afhton Lever who afforded it publicity. This flight digreffion, we truft, will be the more readily excufed, when it is further added, that the Purple Sandpiper paffes ufually among collectors, even at this period, as an addition to the Britifh Tringæ of recent date. Our remarks, however, teftify the contrary, and very obvioufly demonftrate that the Purple Sandpiper of the Leverian Mufeum, has fuffered the fate of many other of its valuable articles, in having been for years before the public, and yet appearing to have been as little known, or at leaft as little noticed as if it never had been placed there.

Befides the Leverian fpecimen above recorded, we have another example of the Purple Sandpiper, captured on the Kentifh coaft within the laft four or five years, and which merely differs from that fpecimen, in being fomewhat fmaller or lefs robuft, and in having the general tint of the plumage more inclining to dufky. The length of the former is nine inches ; of the latter, eight inches and three quarters.

Like the reft of the birds that haunt the fea fhores, the Purple Sandpiper feeds on the fmall marine worms, and infects found on the beach as the tide recedes from the fhore.

178

PLATE CLXXVIII.

ANAS FERINA.

GREATER RED-HEADED WIGEON.

POKER, OR POCHARD, *fem.*

GENERIC CHARACTER.

Bill convex, obtufe, the edges divided into lamellate teeth : tongue fringed, obtufe : three fore-toes connected, the hind one folitary.

SPECIFIC CHARACTER

AND

SYNONYMS.

Cinereous waved : head brown : pectoral band, vent, and rump black : wings cinereous.

FEMALE more dufky : head pale reddifh brown : pectoral band obfcure cinereous and brown : wing coverts cinereous.

ANAS FERINA : cinereo-undulata, capite brunneo, fafcia pectorali
criffo uropygicoque nigris. *Linn. Fn. Suec.* 127.
—*Brünn. Orn. n.* 80.—*Gmel.* 530. 31.
Anas erythrocephala. *S. G. Gmelin it. I. p.* 70
Anas fera fufca. *Raii. Syn. p.* 143. *A.* 10.
Penelope. *Brif.* 6. *p.* 384. 19. *t.* 35. *f.* 1.

D 2

Mil-

PLATE CLXXVIII.

Millouin. *Buff.* ix. *p.* 216. *Pl. Enl.* 803.

Rothhals. *Bloch. Bef. der Berl. Nat.* 4. *p.* 603. *t.* 17. *f.* 5, 6.

Poker, Pochard, Red-headed Wigeon. *Br. Zool.* 11. *N°* 284. *Will. (Angl.) p.* 367. *t.* 72.

Thefe birds are common in the fens during the winter feafon, and are often caught in great plenty, and fent for fale, with other wild fowl, to the London markets, where it is known by the name of the Dunbird. This is the fize of the common Wigeon, but is held in rather more efteem for the table. The fpecies is of the migratory kind, and is found to inhabit Afia and America, as well as Europe.

With us it appears to be only an occafional vifitor during the winter feafon. Its ordinary food confifts of the teftaceous tribes of animals, and various fifhes of the fmaller kind, which inhabit fenny fituations. The flight of thefe birds is diftinguifhed by its fwiftnefs.

The length of the male bird is about nineteen inches, that of the female is fomewhat lefs, and the general tints of the plumage more obfcurely cinereous and dufky than in the male bird.

PLATE

179

PLATE CLXXIX.

ARDEA CICONIA.

WHITE STORK.

Grallæ.

GENERIC CHARACTER.

Bill ftraight, pointed, long, fomewhat compreffed, with a furrow from the noftrils towards the tip: noftrils linear: tongue fharp: feet four-toed, and cleft: toes connected at the bafe.

SPECIFIC CHARACTER

AND

SYNONYMS.

White: orbits and quill-feathers black: bill, legs, and fkin, red.

ARDEA CICONIA: alba, orbitis remigibufque nigris, roftro, pedibus cuteque fanguineis. *Liun. Fn. Suec.* 162.—
Scop. ann. 1. *n.* 123.—*Brünn. orn. n.* 154.—
Gmel. 622. *n.* 7.—*Lath. Gen. Syn.* 3. 1. *p.* 47.
n. 9.

CICONIA *Bell. av.* 45. *a.*—*Aldr. orn.* 3. *p.* 291.—*Jonſt. av.* 147.
t. 50.

CICONIA ALBA: *Brif. av.* 5, *p.* 365. *t.* 32.

CICOGNE

PLATE CLXXIX.

Cicogne Blanche. *Buff. Hift. Oif.* 7. *p.* 253. *n.* 12.—*Pl. enl. n.* 866.

White Stork. *Will. (Angl.) p.* 286. *t.* 52.
Albin. 2. *t.* 64.
Lath. Syn. 5. *p.* 47.—*Suppl. p.* 234.

The Stork is a general inhabitant of various parts of Europe, Afia, and Africa, preferring however the more mild and temperate regions, and migrating regularly in flocks under the guidance of a leader, at certain feafons, from one country to another, in order to avoid the extreme of heat and cold. For this reafon, the Stork is never met with between the tropics, and is rarely known to penetrate northward beyond the country of the Swedes, or the fouthern parts of Ruffia. In Holland, it is met with only during the fummer, arriving in the fpring, and departing about Michaelmas. In England the Stork is very rare.

Like the reft of the Grallæ race, the Stork feeds on fifhes, frogs, and other reptiles, and being a large and powerful bird is eminently ferviceable in thofe countries infefted with ferpents, and other noxious creatures, which it attacks and deftroys with impunity. All writers concur in affuring us, that in many of the eaftern countries, the Stork is held in great refpeɛt: in Egypt, the fens of which they vifit in autumn, they are highly venerated; and in Turkey they are never molefted, but allowed to build upon the tops of their houfes, and even fometimes on places of religious worfhip. In Perfia alfo, they are treated with equal lenity; and no European traveller need be informed, that in certain parts of Spain, of France, and Holland, they experience every poffible indulgence, and are allowed to build their nefts

upon

PLATE CLXXIX.

upon the tops of the houfes in return for the benefits they beftow on the inhabitants, in deftroying the reptiles that harbour about their habitations.

The flefh of the Stork is reputed infipid, a circumftance that may have contributed among other caufes towards the prefervation of this ufeful race of birds; for it cannot be denied, that where the gratification of the appetite is in view, we are too apt to difregard every other ufeful property in the lower race of animals. The Stork, however, paffes unmolefted, as there is no temptation to deftroy the bird for the value of its flefh as an article of food.

The neft is compofed of fticks, in which the female lays from two to four eggs, that correfpond in fize with thofe of the goofe, but are rather more elongated, and the colour a fordid yellowifh white. The young, which are hatched in the fpace of a month, are at firft brown, and are watched by the male and female alternately with the greateft affiduity.

Though very rare in Britain, the Stork obtained a place in the Britifh Fauna at an early period; a fpecimen fhot in Norfolk being recorded by Willughby. Albin is lefs explicit than might be wifhed; but he feems to intimate, that although this bird was rare in England, it had occurred occafionally. "They are feldom," obferves this writer, "found in England in the fummer time, without being driven over by a ftorm, or brought over by fome curious perfon. I faw two of thefe birds at his Grace the Duke of Chandos, at Edger, in Middlefex." *Alb. vol.* 2. *p.* 59*.

---

* The author of the General Synopfis remarks, "that two inftances only are on record (of its being met with in England), Willughby mentions one being fhot at Norfolk,

The

PLATE CLXXIX.

The Stork has been known of late years to vifit the fouthern coafts of Britain.   One, Dr. Latham was informed by Mr. Boys, had been picked up dead, but frefh, on the fhore of Sandwich Bay ; another, as ftated by Mr. Macreth, was fhot in the winter of 1785, at South-fleet, in Kent; and Mr. Montagu mentions another fhot in Hamp-fhire, in the autumn of 1808, by the game-keeper of Major Guiton.

---

and Albin a fecond in Middlefex." This latter obfervation we are afraid has arifen from overfight (a circumftance by no means ufual in the writings of Dr. Latham), for the words of Albin, as above tranfcribed, and to which by the fynonyms we are referred, can never be fuppofed to imply, that the Stork had been fhot in Mid-dlefex, but only that he faw Storks at the Duke of Chandos, in Middlefex; and which, from the general tendency of the text, we fhould apprehend, had been intro-duced as curiofities by the Duke from Holland.

PLATE

180

PLATE CLXXX.

ANAS FERINA.

GREATER RED-HEADED WIGEON.

POKER, OR POCHARD, *male.*

GENERIC CHARACTER.

Bill convex, obtufe, the edges divided into lamellate teeth : tongue fringed, obtufe : the three fore-toes connected, the hind-one folitary.

SPECIFIC CHARACTER

AND

SYNONYMS.

Cinereous, waved : head brown : pectoral band vent and rump black : wings cinereous.

ANAS FERINA : cinereo undulata, capite brunneo, fafcia pe ctoral criffo uropygioque nigris. *Linn. Fn. Suec.* 127.

POCHARD, or GREAT RED-HEADED WIGEON. Anas fera fufca. *Albin. V.* 2. 87.—*pl.* 98.

POCHARD. *Lath. Gen. Syn. V.* 6. *p.* 523. 68.

The plumage of the male of the Greater Red-headed Wigeon is diftinguifhed by being of a brighter hue, and in having the charac-

VOL. VIII.   E   teriftic

PLATE CLXXX.

teriftic colours defined with more diftinctnefs than in the female: there is otherwife a general refemblance as confpicuous at leaft as ufually prevails between the two fexes of the Duck tribe; and in point of fize, the female is very little inferior to the male bird.

As the two fexes affociate and are found in the fame places, their habits and manners may be prefumed the fame: they are not fuppofed to breed in Britain, but to vifit us in autumn, and retire again in the fpring: they ufually appear in flocks of thirty or forty together, and are obferved to fly with greater velocity; but not like fome of the Wild Fowl tribe, in a regularly marfhalled manner, under the direction of a leader; they fly rather indifcriminately. During part of the year they inhabit North America.

PLATE

181

PLATE CLXXXI.

SCOLOPAX GALLINAGO.

COMMON SNIPE.

GRALLÆ.

GENERIC CHARACTER.

Bill roundiſh, obtuſe, longer than the head : noſtrils linear : face covered : feet four-toed, hind toe conſiſting of many joints.

SPECIFIC CHARACTER
AND
SYNONYMS.

Bill ſtraight, tuberculated: legs fuſcous : front with four brown lines,

SCOLOPAX GALLINAGO: roſtro recto tuberculato, pedibus fuſcis, frontis lineis fuſcis quaternis. *Linn. Fn. Suec.* 173.—*Gmel. Linn. Syſt.* 662. 7.

Gallinago. *Briſs. Orn.* 5. *p.* 298. *n.* 2. *t.* 26. *f. I.*

Gallinago Minor. *Bell. av.* 54.

Ald. Orn. 3. *p.* 484. 476. *t.* 477.

Bécaffine. *Buff. Oiſ.* 7. *p.* 483. *t.* 26.

Schnepf. *Friſch. av. t.* 229.

E 2

Snipe,

PLATE CLXXXI.

Snipe, or Snite.   *Raj. av. p.* 105. *n.* 2.
                    Will. Orn. p. 290, *t.* 3.
                    Albin. av. I. t. 71.
COMMON SNIPE,   *Brit. Zool.* 2. *p.* 187. *t.* 68.
                    Arct. Zool. p. 463. *n.* 366.
                    Lath. Syn. 111. *I. p.* 134. *n.* 6.

The common Snipe is abundantly diffufed throughout every part of the known globe. In England, the fpecies is common in the fens and fwampy places, during the winter; but after the early part of fpring they difappear; fome, it is believed, migrating, and others retiring to the higher or more mountainous fituations, where it is found during the whole fummer feafon. It is ftated, on the authority of Dr. Heyfham, that the Snipe, as well as the Jack Snipe, are found throughout the year in the hilly parts of Cumberland; and, according to Barrington, they never quit the fens of Lincolnfhire, Wolmar Foreft, and Bodmin Downs. Its food confifts of woims, infects, and fmall fnails.

The neft of the common Snipe is frequently found in the fens and marfhes: it is compofed of dried plants, intermixed with a few feathers; and the eggs, which are from four to five in number, are of an oblong form, and fordid olive colour, marked with dufky fpots.

The length of this bird is about twelve inches; its weight four ounces.

PLATE

182

PLATE CLXXXII.

PELECANUS GRACULUS.

SHAG.

ANSERES.

GENERIC CHARACTER.

Bill ftraight, hooked at the point, and furnifhed with a nail: noftrils an obliterated flit: face rather naked: legs equally balancing the body: all the four toes palmated.

SPECIFIC CHARACTER

AND

SYNONYMS.

Tail rounded: body black, beneath fufcous: tail feathers twelve.

PELECANUS GRACULUS: cauda rotundata, corpore nigro fubtus
fufco, rectricibus duodecim. *Lynn. Fn. Sv.* 146.
Brünn. n. 121.—*Gmel.* 574. 4.

PHALACROCORAX minor. *Briff. av.* 6. *p.* 516. *n.* 2.

Corvus Aquaticus. *Albin. av.* 2. *p.* 74. *t.* 81.

Petit Cormoran, ou Nigaud. *Buff. Hift. Oif. p.* 319.

Shagge, or Crane. *Raj. av. p.* 123. *A.* 4.

Will. Orn. p. 330. *t.* 63.

Shag.

PLATE CLXXXII.

Shag. *Arct. Zool.* 2. *p.* 581. *n.* 508.

   Lath. Gen. Syn. 111. 2. *p.* 598. *n.* 14.

The Shag is an inhabitant of the northern parts of Europe: it occurs on the fea coafts of the Britifh ifles, and alfo thofe of Holland, Sweden, Norway, and Iceland. There is one variety of it found in Cayenne and the Caribbee iflands, and another at the Cape of Good Hope: but the variety found in Britain is that peculiar to the northern regions.

This bird is fmaller than the Corvorant, being only thirty inches in length: like that bird, its haunts are the rocky parts of our fea coafts, where it breeds in the cliffs, and alfo like the Corvorant in trees. Ray informs us, the eggs are long and white. When on land, the Shag is reputed a very ftupid bird; but in the water, it is remarkably active, and is not fhot in that element without great difficulty, as it fwims with only the head and neck above water, and dives with uncommon celerity the inftant it perceives the flafh of the fowling piece.

The general tints of the plumage of this bird are dufky, glcffed with afhen green or olive, and the darker parts with a vivid blue, appearing altogether refplendent, as the light falls upon it in various directions.

PLATE

183

PLATE CLXXXIII.

PAVO CRISTATUS β.

PIED PEAHEN.

GALLINÆ.

GENERIC CHARACTER.

Bill convex, robuſt: head covered with revolute feathers: noſtrils large: feathers of the rump, long, broad, expanſile, and covered with ocellar ſpots.

SPECIFIC CHARACTER

AND

SYNONYMS.

Head with a compreſſed creſt: ſpurs ſolitary.

PAVO CRISTATUS: capite criſta compreſſa, calcaribus ſolitariis β. *Gmel. Linn. Syſt. Nat.* 729. 1.

PAVO VARIUS. *Briſſ. Orn. I. p.* 288. *A.*

Bunter Pfau. *Frich. av. t.* 119.

Paon Panaché. *Buff. Hiſt. Oiſ.* 2. *p.* 327.

The repreſentation of the male bird of this Pied variety of the Peacock in the Fifth Volume of the preſent Work, appears to have

afforded

PLATE CLXXXIII.

afforded fo much fatisfaction to many of our readers, that we are prevailed upon to add in this place a figure of the Hen and Young.

The female of this bird, we muft acknowledge, yields in point of gracefulnefs and commanding gaiety of the plumage to the male; and ftill more confpicuoufly in the deficiency of that fine flowing train of feathers, which conftitute the tail coverts: but it yet pof-feffes a very pleafing air, and its introduction in the prefent Work will affift to complete the hiftory of that much admired variety of the Crefted Peacock.

The Pied Peacock, an intermediate breed or mixture between the common and the white Peacock, is not inferior in point of fize to the common kind, and as in that bird the female is rather fmaller; the creft is alfo fhorter, and the legs rarely armed with fpurs.

PLATE

PLATE CLXXXIV.

TRINGA MACULARIA.

SPOTTED SANDPIPER.

GRALLÆ.

GENERIC CHARACTER.

Bill roundifh, as long as the head: noftrils fmall, linear: tongue flender: feet four-toed; the hind-toe of one joint, and raifed from the ground.

SPECIFIC CHARACTER

AND

SYNONYMS.

Bill at the bafe and the legs flefh coloured: every part of the body fpotted; eye-brows and double band on the wings white.

TRINGA MACULARIA: roftro bafi pedibufque incarnatis, corpore undique maculato, fuperciliis fafciaque gemina alarum albis. *Gmel. Syft. Nat.* 672. 7.

TRINGA MACULARIA: *Linn. Syft. I. p.* 249. 7.

TURDUS AQUATICUS: *Brif. av.* 5. *p.* 255. *n.* 20.

Grive d'Eau. *Buff. Hift. Nat. des Oif.* 8. *p.* 180.

VOL. VIII.   F   SPOTTED

PLATE CLXXXIV.

SPOTTED SANDPIPER. *Brit. Zool. 2. n.* 196.
Lath. Gen. Syn. 5. p. 179. 24.
—*Ind. Orn.* 734. 29.

———

We efteem this one of the rareft of the Britifh Sandpipers: the length of our bird is eight inches and a half.

The fpecies is defcribed as a native of America as well as Europe; and the female is faid to be diftinguifhed by having the under parts of the plumage immaculate.

PLATE

185

PLATE CLXXXV.

TURDUS VISCIVORUS.

MISSEL THRUSH.

PASSERES.

GENERIC CHARACTER.

Bill ftraightifh: the upper mandible a little bending and notched near the point: noftrils naked or half covered with a fmall membrane: mouth ciliated with a few briftles at the corners: tongue jagged.

SPECIFIC CHARACTER
AND
SYNONYMS.

Back brown; neck fpotted with white; bill yellowifh.

TURDUS VISCIVORUS: dorfo fufco, collo maculis albis, roftro
flavefcentibus. *Scop. Ann. I. p.* 132. *n.* 193.
Brünn. Orn. p. 65. *n.* 231.
Kram. el. p. 361. *n.* 6.
Gmel. Linn. Syft. 806. 1.
Turdus Vifcivorus. *Gefn. av.* 759. *t.* 760.
Turdus Vifcivorus major. *Ray. av. p.* 64. *n.* 1.
Turdus Major. *Brifs. av.* 1. *p.* 200. *n.* 1.

F 2

Tordo,

PLATE CLXXXV.

Tordo.   *Olin. ucc. t.* 25.

Draine.   *Buff. Hift. Oif.* 3. *p.* 295. *t.* 19. *f.* 1. *Pl. Enl. n.* 489.

Miffel Bird, or Shrite.   *Will. Orn. p.* 187.

Albin. av. 1. *t.* 33.

Miffel Thrufh.   *Brit. Zool.* 1. *n.* 105.

Arct. Zool. 2. *p.* 341. *B.*

Lath. Gen Syn. 11. 1. *p.* 16. *n.* 1.

The Miffel Thrufh is the largeft of the Turdus tribe; the length eleven inches. The plumage of the female is inferior to that of the male in brightnefs.

This bird inhabits the temperate parts of Europe, where it lives in woods, and feeds on berries of various bufhes, fuch as the holly, ivy, and hawthorn, and, as the name implies, on thofe of the miffel-toe; and befides thefe, it fubfifts on caterpillars and infects.

The neft, which is ufually placed in a low tree or bufh, is compofed of mofs, lichens and leaves; the lining of withered grafs. The eggs, four or five in number, of a dirty flefh colour, marked with reddifh fpots. The note of this bird, though pleafing, is not equal to that of the Thrufh. The flefh is held in fome efteem for the table.

PLATE

186

PLATE CLXXXVI.

SCOLOPAX GALLINULA.

JACK SNIPE.

GRALLÆ.

GENERIC CHARACTER.

Bill roundifh, obtufe, longer than the head: noftrils linear: face covered: feet four-toed: hind-toe confifting of many joints.

SPECIFIC CHARACTER

AND

SYNONYMS.

Bill ftraight, tuberculated: legs greenifh: lores brown: rump varied with violet.

SCOLOPAX GALLINULA: roftro re&to tuberculato, pedibus viref-
        centibus, loris fufcis, uropygio violaceo-vario.
        Scop. Ann. I. n. 139.
        —*Gmel. Syft. Nat.* 662. 8.
Gallinago minor. *Briff. av. 5. p.* 303. *n. 3. t. 26. f. 2.*
Gallinago minima. *Bell. av.* 217.
Cinclus quartus. *Aldr. Orn. 3. p.* 493.

Petite

PLATE CLXXXVI.

Petite bécaffine. *Buff. Hift. Nat. Oif.* 7. *p.* 490.—*Pl. enl. n.* 884.

Halbfchnepfe. *Frifch. av. t.* 231.

Jacksnipe, Gid, Judcock. *Will. (Angl.) p.* 291.—*Albin.* 111.
t. 86.—*Brit. Zool.* 11. 189. *t.* 68.
Lath. Gen. Syn. 5. *p.* 136. 8.

———————

The Jackfnipe has the fame habits and manners of life as the common Snipe, but appears to be lefs abundantly difperfed over the globe than that bird. Dr. Latham remarks that he cannot trace it more South than Aleppo (where, according to Dr. Ruffel, it is not uncommon), and to the North as far as lat. 80. 27.

This is only half the fize of the common Snipe, weighing no more than two ounces; its length eight inches and a half.

PLATE

187

PLATE CLXXXVII*.

PHALOROPUS LOBATUS.

LOBATE, or GREY PHALAROPE.

GRALLÆ.

GENERIC CHARACTER.

Bill roundiſh, ſtraight, and uſually ſomewhat inflected at the top : noſtrils minute : feet lobate, or furniſhed with lateral membranes, generally ſcalloped.

SPECIFIC CHARACTER

AND

SYNONYMS.

Bill ſubulate, the tip inflected : legs pinnate : breaſt undulated white.

PHALAROPUS LOBATUS. *Lath. Ind. Orn.* 776. 2.

TRINGA LOBATA : roſtro ſubulato : apice inflexo, pedibus pin-
                 natis: pectore albo undulato. *Linn. Fn. Suec.*
                 179.—*Brünn. Orn. n.* 171.
                 —*Fab. Fn. Groenl. n.* 75.

PHALAROPUS. *Briſſ. av. 5. p.* 18. *n.* 1.

Phalarope a feſtons dentelés. *Buff. Oiſ. p.* 226.

                                                    Grey

---

* ERRATUM.

Through an error in numbering the Plate of the Grey Phalarope, which belongs to this place in the preſent Volume (Plate CLXXXVII.) it has been improperly introduced Into the former, or Seventh Volume.

                                                    This

PLATE CLXXXVII.

Grey Coot-footed Tringa. *Edw. glean. t.* 308.—*Act. Angl.* 50. *t.* 6.

GREY PHALAROPE. *Brit. Zool. n.* 218. *t.* 76.

Lath. Syn. 3. 1. *p.* 272. *n.* 2.

This elegant fpecies is very rare in England. Its fize is that of the common Purre : the bill black : the legs cinereous.

It is found fparingly in Europe, Afia, and America.

This miftake was not difcovered till a confiderable number of the Seventh Volume had been bound up and fold, and at a time therefore when any alteration in placing the Plates would only have created perplexity.

The true Plate CLVII. is included in the prefent Volume, infcribed with its proper number.

The Plate numbered CLVII., in the former Volume, fhould be CLXXXVII.

By altering the erroneous number CLVII. laft mentioned, into CLXXXVII., and then referring the two Plates to their proper fituations in Vol. VII. and VIII. as indicated by their feveral numbers, the overfight will be corrected.

When the two Plates are thus tranfpofed the letter-prefs will be found to correfpond refpectively with the Plates referred to them ; for it is nothing more than the mifnumbering of one Plate in Vol. VII. that has occafioned confufion.

PLATE

188

PLATE CLXXXVIII.

TURDUS ILIACUS.

RED WING.

PASSERES.

GENERIC CHARACTER.

Bill ftraightifh; the upper mandible a little bending, and notched near the point: noftrils naked, or half covered with a fmall membrane: mouth ciliated with a few briftles at the corners: tongue jagge!.

SPECIFIC CHARACTER

AND

SYNONYMS.

Wings beneath ferruginous: eye-brows whitifh.

TURDUS ILIACUS: alis fubtus ferrugineis, fuperciliis albicantibus. *Linn. Fn. Suec.* 218. *Scop. ann.* 1. *n.* 196.— *Kram. el. p.* 361. *n.* 9.—*Gmel. Syft. Nat.* 808. 3.

Turdus Ilias. *Gefn. av.* 760. *t.* 761.

Aldr. orn. 2. *p.* 597.

Turdus Iliacus. *Nozem. nederl. Vogel. t.* 12.

Klein. av. 66.

Brif. av. 2. *p.* 208. *n.* 3. *t.* 20. *f.* 1.

Mauvis. *Buff. Hift. Nat. Oif.* 3. *p.* 309.—*Pl. enl. n.* 51.

VOL. VIII.　　　　　　　G　　　　　　　Klera

PLATE CLXXXVIII.

Klsra Kladra, Tall-Traſt. *Fn. Suec. ſp.* 218.

Weindroffel. *Friſch. av. t.* 28.

Redwing, Swinepipe, or Wind-Thruſh. *Will. Orn.* 189.

Ray. av. 64. *n.* 4.

Albin. av. 1. *t.* 35.

Lath. Gen. Syn. 2. 1. *p.* 22. *n.* 7.

———————

The Redwing is rather ſmaller than the Fieldfare, to which, in general appearance, as well as in its manners of life, it bears conſiderable reſemblance.

The uſual length of the Redwing is about nine inches, that of the Fieldfare ten inches. They appear in flocks ſo nearly about the ſame time, that they may be conſidered as accompanying each other in their periodical migrations; the flocks of the Redwing arriving uſually at the place of deſtination a few days only before the Fieldfare; and it is alſo known that in the northern parts of Europe, where theſe birds breed, both kinds occur together.

One of the moſt remarkable characters of the Redwing, and by which, independently of other peculiarities, it may be diſtinguiſhed from the Fieldfare, is the colour of the inner ſurface of the wing, that part of the Fieldfare being wholly white, while in the Redwing it is marked with a rufous orange ſpot. The name of Redwing, given to this bird by Ray and other early writers, appears hence to be happily choſen, as it at once implies the diſtinction that prevails between this bird and its very analogous ſpecies.

Linnæus informs us in the *Fauna Suecica*, that in Sweden this

ſpecies

PLATE CLXXXVIII.

fpecies builds its neft in fome low fhrub or hedge; and that the female lays fix eggs, which are of a blue-green colour, and fpotted with black *: it is obferved alfo, that the Redwing perches on high trees in the maple forefts in Sweden, and has a fine note in fpring.

---

* We believe it may be afferted with entire confidence, that the collection of the eggs of Britifh Birds in our poffeffion, is beyond comparifon, the moft perfect that ever has been brought together; and as the introduction of the eggs are confeffedly requifite to complete the Natural Hiftory of our birds, the reader will be induced, we truft, to approve our intention of introducing the figures of thefe in the moft compendious form the Intereft of the fubject will admit, in a diftinctly feparate Supplement to the prefent Work. A genuine work on this interefting fubject is unqueftionably much wanted.

G 2   PLATE

189

PLATE CLXXXIX.

TRINGA PUGNAX, *fem.*

REEVE, or FEMALE RUFF.

GRALLÆ.

GENERIC CHARACTER.

Bill roundifh, as long as the head: noftrils fmall, linear: tongue flender: feet four-toed; the hind toe of one joint, and raifed from the ground.

SPECIFIC CHARACTER

AND

SYNONYMS.

Bill and legs red: three lateral tail feathers without fpots: face with flefh-coloured granulations.

Fem. Pale brown; back fpotted with black: tail brown, the middle feathers fpotted with black: breaft and belly pale.

TRINGA PUGNAX. *Linn. Syft.* 1. *p.* 247. 1.—*Fn. Sv. N°* 175.—
     Gmel. Syft. 1. *p.* 669.—*Lath. Ind. Orn.* 725. 1.
Avis Pugnax. *Aldr. Orn.* 3. *p.* 523. *t.* 526.
Combattant, ou Paon de Mer. *Buff.* 7. *p.* 521. *t.* 29, 30.—*Pl. Enl.*
     305. 306.

          Pavoncella,

PLATE CLXXXIX.

Pavoncella. *Cet. uc. Sard. p.* 253.
Ruff and Reeve. *Will. (Angl.) p.* 302. *t.* 56.
  Albin. 1. *t.* 72. 73.
  Lath. Syn. 5. *p.* 159. 1.

In an early part of the prefent work we prefented our readers with a figure of the Male or Ruff*; the fubjeÊ now chofen is a variety of the Female or Reeve. The length of the male is about twelve inches, that of the female is rather lefs.

It is remarked of this fpecies that the ruff or male birds are fo variable in colour, that it is fcarcely poffible to find two birds that perfectly correfpond in this particular. The Reeves are alfo variable in a lefs confiderable degree.

The males when in full plumage are diftinguifhed by the remarkable tufts of feathers which form a kind of collar or ruff round the neck, from whence it derives the name; but in the moulting feafon this ruff falls off, and the apppearance of the male at that time fo nearly refem‚ bles the female that they cannot eafily be diftinguifhed.

---

* Plate XIX, Vol. I.

PLATE

PLATE CXC.

ANAS PENELOPE.

WIGEON.

ANSERES.

GENERIC CHARACTER.

Bill convex, obtufe, the edges divided into lamellate teeth : tongue fringed and obtufe : three fore-toes connected, the hind one folitary.

SPECIFIC CHARACTER
AND
SYNONYMS.

Tail rather pointed : vent black : head brown : front white : back cinereous waved.

ANAS PENELOPE : cauda acutiufcula, criffo nigro, capite brunneo, fronte alba, dorfo cinereo undulato. *Linn. Fn. Suec.* 124.—*Brünn. n.* 72.—*Kram. el. p.* 342. *n.* 16.

PENELOPE. *Gefn. av.* 108.

Aldr. orn. 3. *p.* 217. *t.* 219. 220.

Jonft. av. 39.

Anas fiftularis. *Gefn. av.* 121.

Canard

3

PLATE CXC.

Canard fiffleur. *Buff. Oif.* 9. *p.* 169. *t.* 10. 11.—*Pl. Enl. n.* 825.

Pfeifente. *Bloch. Befch. der Berl. Naturf.* 4. *p.* 601. *t.* 18. *f.* 5.

Blaffente. *Frifch. av. t.* 164.

WIGEON, WHEWER, or WHIM. *Ray. Orn. p.* 146. *A. S.*

Will. Orn. p. 375. *t.* 12.

Albin. av. 2. *t.* 99.

Lath. Gen. Syn. 3. 2. *p.* 518. *n.* 63.

This fpecies of wild fowl occurs abundantly in various countries of Europe, Afia, and Africa, in all which it is reputed of the migratory kind: it migrates as far as Aleppo and Egypt during the winter, and at different feafons appears in various parts of Europe as far to the northward as Sweden: it is believed to breed more to the northward than the Britifh ifles.

During the winter months the Wigeon is taken in plenty in our marfhes, with others of the Duck tribe, by means of the nets and decoys, and as the flefh is much admired, the London markets are pretty regularly fupplied with the Wigeon throughout the winter, and early part of the fpring.

The length of this bird is twenty inches: the female is rather fmaller, of a brown colour, and undulated: the neck and breaft paler: fcapulars dark brown, and the wings and belly as in the male bird. The young males are obferved to refemble the female till the following fpring after being hatched, when they acquire the full plumage, which however they retain only till the end of the fummer, when they again refemble the female till the enfuing moulting.

PLATE

191

PLATE CXCI.

TURDUS PILARIS.

FIELDFARE.

PASSERES.

GENERIC CHARACTER.

Bill ftraightifh; the upper mandible a little bending, and notched near the point: noftrils naked, or half covered with a fmall membrane: mouth ciliated with a few briftles at the corners: tongue jagged.

SPECIFIC CHARACTER

AND

SYNONYMS.

Tail feathers black, the outermoft at the inner edge tipped with white: head and rump hoary.

TURDUS PILARIS: rectricibus nigris: extimis margine interiore apice albicantibus, capite uropygioque cano. *Linn. Fu. Suec.* 215.—*Scop. Ann.* 1. *p.* 133. *n.* 194.— *Brünn. Orn. p.* 65. *n.* 232.

TURDUS PILARIS. *Gefn. av.* 753.—*Aldr. Orn.* 2. *p.* 595. *t.* 596.

Turdus Pilaris five Turdela. *Briff. av.* 2. *p.* 214. *n.* 5.

VOL. VIII.   H   Litorne

PLATE CXCI.

Litorne ou Tourdelle. *Buff. Hift. Nat. des Oif.* 3. *p.* 301. *t.* 19.
f. 2.—*Pl. Enl. n.* 490.

FIELDFARE, or FELDEFARE. *Brit. Zool.* 1. *n.* 106.

Will. Orn. p. 188. *t.* 37.

Albin. av. 1. *t.* 36.

Lath. Syn. 2. 1. *p.* 24. *n.* 11.

Length ten inches; the bill yellowish, with the tip black, the legs blackish. The male and female much refemble each other.

The Fieldfare, like the Redwing, is of the migatory kind: retiring to the northern regions, as far as Iceland, during the fummer or breeding feafon, and returning fouthward in autumn: it arrives in Britain about the month of October, or later, when the northern winter is more mild than ufual: it alfo vifits France and Italy, but not till about the month of December. In Sweden it is obferved to build in high trees, and to prefer thofe fituations where the juniper grows. It fubfifts chiefly on the berries of various plants, among which, thofe of the hawthorn and the holly appear to be favourite food.

The Fieldfare and the Redwing are frequently taken in the nets during winter: the flefh, though fometimes rather bitter, being held in fome efteem. The capture of thefe birds (Turdi of the Romans) was an object of attention among the ancients as we learn from their hiftorians: they were accuftomed to take thefe birds alive in vaft numbers at particular feafons of the year, and fatten them in aviaries by thoufands, as an article of delicacy and luxury for the table *.

---

* Varro, Plutarch, Martial, &c.

Several

3

PLATE CXCI.

Several varieties of this bird are defcribed by writers, all which differ from the common kind only in having the plumage varied with white, efpecially about the head and neck: the Pied Fieldfare of Albin, which is of this kind, has the head and neck entirely white.

192

PLATE CXCII.

ANAS CRECCA. *Fem.*

COMMON TEAL. *Female.*

ANSERES.

GENERIC CHARACTER.

Bill convex and obtufe, the edges divided into lamellate teeth : tongue fringed and obtufe : three fore toes connected, the hind one folitary.

SPECIFIC CHARACTER

AND

SYNONYMS.

Wing fpot green : line above and below the eye white, *male.*—Head and neck whitifh and brown varied ; wing fpot green : vent white.

ANAS CRECCA : fpeculo alarum viridi, linea alba fupra infraque oculos. *Linns.*—Caput et collum albida fufcoque varia ; alæ ut in mare : criffum totum album. *fem. Linn. Syſt.* 1. *p.* 204. 33.—*Lath. Ind. Orn.* 872. 100.

Querquedula fecunda Aldr. *Raii. Syn. p.* 148.9.

COMMON TEAL. *fem. Lath. Gen. Syn.* 6. 551. 88.

The

PLATE CXCII.

The Common Teal is a native of Europe and Afia, inhabiting the north as far as Iceland, and the fouth to the Cafpian fea, beyond which it extends on the eaft as far as China, India, and fome other adjacent countries.

During the winter feafon there are few fpecies of wild fowl more abundant in this country than the Common Teal, which is caught in vaft numbers for the fupply of the London markets. But it was long before it was clearly afcertained to breed in this country. It is now known to breed in the north of England, and alfo in Scotland.

In France, where the Common Teal inhabits throughout the year, it frequents watery places, making its neft, in April, among the rufhes that grow about the edges of the ponds, and feeds on the feeds of the rufhes, and on the grafs and weeds the ponds afford, as well as on the fmaller kinds of fifhes. The neft is faid to be of a large fize, and fo placed in the water as to rife and fall with it : the eggs, which in point of fize, refemble thofe of a pigeon, are of a whitifh colour, with fmall hazel fpots.

The fubject reprefented in the prefent plate is the female, which differs materially in fome particulars of the plumage from the male.

Length about fourteen inches.

PLATE

193

PLATE CXCIII.

ANAS CRECCA, *Mas.*

COMMON TEAL, *Male.*

ANSERES.

GENERIC CHARACTER.

Bill convex and obtufe, the edges divided into lamellate teeth:
tongue fringed and obtufe : three fore-toes connected, the hinder one
folitary.

SPECIFIC CHARACTER

AND

SYNONYMS.

Wing fpot green : line above and below the eye white.

ANAS CRECCA : fpeculo alarum viridi, linea alba fupra infraque
oculos. *Linn. Fn. Suec.* 129.—*Linn. Syft.* 1. *p.*
204. 33.—*Gmel. Syft. Nat.* 532. 33.—*Lath.*
Ind. Orn. 872. 100.
Petite Sarcelle, *Buff.* ix. *p.* 265. *t.* 17. 18.—*Pl. enl.* 947.
COMMON TEAL. *Br. Zool.* 11. N° 290.—*Arct. Zool.* 11. *p.* 577.
P. —*Lath. Gen. Syn.* 6. *p.* 551. 83.—*Id. Supp.*
p. 276.

In

PLATE CXCIII.

In the preceding Plate we have given a figure of the female Teal, the prefent is that of the male bird, which confeffedly excels the female in point of elegance; there is a general refemblance, and the fize is nearly the fame, but the fuperior brightnefs of its colours, added to the very beautiful appearance of the head of the male bird, affords a very ftriking contraft to the appearance of the female.

There is a fuppofed variety of this bird with the legs cinereous inftead of reddifh brown.

PLATE

194

PLATE CXCIV.

ANAS ACUTA.

PINTAIL. *Fem.*

ANSERES.

GENERIC CHARACTER.

Bill convex and obtufe, the edges divided into lamellate teeth: tongue fringed and obtufe: three fore-toes connected, the hind one folitary.

SPECIFIC CHARACTER
AND
SYNONYMS.

Tail pointed, long, beneath black: hind-head each fide with a white line; back waved cinereous.

Female rufous brown and dufky varied: wing-fpot pale brown, margined above and beneath with a white line: two middle tail feathers not elongated.

ANAS ACUTA: cauda acuminato elongata fubtus nigra, occipite utrinque linea alba, dorfo cinereo undulato. *Linn. Fn.
Suec.* 126.—*Scop. Ann.* 1. *n.* 73.—*Georg. it. p.*
166.—*Lath. Gen. Syn.* 3. 2. *p.* 526. *n.* 72.

VOL. VIII.   I

PLATE CXCIV.

The male of this very elegant fpecies of the Anas tribe has appeared already in the prefent work, being included in the fixth volume, at Plate CXXIX. The female is little inferior in the beauty of her plumage to the male bird, from which, however, on comparifon, it will be found to differ in feveral material particulars. The female is rather fmaller, the head and neck pale brown, and marked with numerous fhort dafhes of fufcous; the back and wing coverts dark brown, with a delicate reddifh buff-coloured margin, and many of the feathers marked acrofs the middle with a curved band of the fame colour: wings pale fufcous; wing-fpot pale brown, bounded above and beneath with a white band or line: the tail cuneate, having the middle feather rather longer and pointed, but not elongated like the two middle tail feathers of the male bird.

PLATE

PLATE CXCV.

HIRUNDO APUS.

SWIFT.

GENERIC CHARACTER.

Bill fmall, weak, curved, fubulate, and depreffed at the bafe: gape larger than the head: tongue fhort, broad, and cleft: wings long: tail in general forked: toes all placed forward.

SPECIFIC CHARACTER

AND

SYNONYMS.

Blackifh, throat white.

HIRUNDO APUS: nigricans, gula alba, digitis omnibus quatuor
anticis, *Linn. Fn. Suec.* 272.—*Scop. Ann.* 1. *p.*
166. *n.* 251.—*Brünn. Orn. p.* 74. *n.* 292.—*Lath.*
Ind. Orn. 582. 32.

Le Grand Martinet. *Buff. Pl. Enl.* 542. *f.* 1.

Martinet Noir. *Buff. Hift. Nat. Oif.* 6. *p.* 643.

Rondone *Zinnan. Uov. p.* 47. *t.* 7. *f.* 33.

Cett. Uc. Sard. p. 231.

Mauerfchwalbe. *Frifch. av.* 17.

I 2 SWIFT.

PLATE CXCV.

Swift. *Brit. Zool.* 1. N° 171. *t.* 57.—*Will. (Angl.) p.* 214. *t.* 39.—*Albin.* 2. *p.* 55.

Lath. Gen. Syn. 4. *p.* 584. 34.

The Swift appears to be an inhabitant of almost every part of the known globe, as we learn from the teftimony of many travellers who have feen it in the different countries they have refpectively vifited: in the Britifh ifles it remains only for a very fhort period, feldom arriving before the early part of May, and retiring again about the beginning of the month of August.

The fpecies is about eight inches in length, the colour footy black, except the chin, which is white; the female differs from the male only in being rather fmaller, the plumage brownifh, and the white on the throat more obfcure than in the male bird.

From the diminutive fize of the feet it walks with much difficulty, and cannot rife from the ground. It is, therefore, never feen in any low fituation, unlefs it fhould chance to fall there by accident, but refts upon the tops and fides of houfes, church fteeples, the lofty craggs of hills or other confiderable elevations, from whence it can conveniently take its flight. In the mornings and evenings it is obferved on the wing, flying and darting in a fwift and peculiar manner in fearch of infects, its cuftomary food; and which it eafily catches in the enormous gape of its mouth, while on the wing. The Swifts fly only in the morning and evening: during the day time, and in the night they lurk in their hiding places to avoid the extremities of either heat or cold.

The Swift is faid to have only one brood in a year: the eggs, ufually five in number, are of a lengthened form, and white colour.

PLATE

196

PLATE CXCVI.

SCOLOPAX ARCUATA.

CURLEW.

GRALLÆ.

GENERIC CHARACTER.

Bill roundiſh, obtuſe, longer than the head: noſtrils linear: face covered: feet four-toed, hind-toe conſiſting of many joints.

SPECIFIC CHARACTER

AND

SYNONYMS.

Bill arched, legs blueiſh: wings blackiſh, with ſnowy ſpots.

SCOLOPAX ARCUATA: roſtro arcuato, pedibus cæruleſcentibus, alis nigris maculis niveis. *Linn. Fn. Suec.* 168.— *It. ſcan.* 333.—*Gmel.* 655. 3.

Numenius. *Briſſ. av.* 5. *p.* 311. *n.* 1.

Nozem. nederl. Vogel. t. 57.

Numenius f. Arquata. *Geſn. av.* 222.—*Aldr. Orn. l.* 20. *c.* 21.— *Marf. dan. t.* 17.

NUMENIUS ARQUATA, cineraſcente nigroque varius, pedibus cæruleſcentibus, alis nigris maculis niveis. *Lath. Ind. Orn.* 710. 1.

Le

PLATE CXCVI.

Le Courlis. *Buff.* 8. *p.* 19.—*Pl. Enl.* 818.
COMMON CURLEW. *Br. Zool.* 2. 176. *t.* 63.
Will. (Angl.) p. 294. *t.* 54.
Albin. 1. *t.* 79.
Lath. Syn. 5. *p.* 119. 1.—*Id. Suppl. p.* 242.

The Curlew inhabits the marſhy parts of Europe and Aſia ; and a variety of the ſpecies is is alſo found in Africa and America. During the ſummer ſeaſon it migrates in flocks to the Northward, and returns again to the ſouthward as the winter approaches.

Theſe birds never leave Great Britain entirely, for although they migrate in great numbers they are found in plenty at all ſeaſons of the year, both on our ſea coaſts and among the fens in the inland parts of the kingdom. With us they breed in April; the eggs are four in number, and of a pale olive colour, with browniſh ſpots.

The remarkable incurvate ſtructure and diſproportionate length of the bill ſeem to indicate the habits of life theſe birds are deſtined by nature to purſue : thoſe which haunt our ſea coaſts are ſeen running in flocks along the ſhore in queſt of worms and cruſtaceous animals, which they draw out of the ſands and from the crevices of the rocks, by means of their elongated bill, with perfect eaſe ; and in like manner they readily obtain the worms, the larva of inſects, and other creatures which inhabit the fens and marſhes to which they retire when they retreat inland from the ſea coaſt. The fleſh of the Curlew is admired generally : thoſe which feed upon the ſea coaſts acquire a certain degree of fiſhy ranknefs in flavour, from the nature of their food, which is

not

PLATE CXCVI.

not obferved in thofe which inhabit the frefh water marfhes. The two fexes much refemble each other.

The Curlew varies in fize: it is ufually from twenty to about twenty-four inches in length, and has a bill of an incurvate form, nearly feven inches long; fometimes they occur much larger. We have the head of one, the bill of which is ten inches long, but the Curlew of this fize is not common, and it was for this reafon the head was pre-ferved by a curious fportfman as a very fingular novelty.

PLATE

INDEX to VOL. VIII.

ARRANGEMENT

ACCORDING TO THE

SYSTEM of LINNÆUS.

ORDER III.

ANSERES.

ORDER IV.

GRALLÆ.

VOL. VIII.   K   ORDER

I N D E X.

ORDER V.

GALLINÆ.

ORDER VI.

PASSERES.

VOL.

INDEX.

VOL. VIII.

ARRANGEMENT

ACCORDING TO

LATHAM's SYNOPSIS of BIRDS.

DIVISION I. LAND BIRDS.

ORDER III. PASSERINE.

ORDER IV. GALLINÆ.

DIVISION II. WATER BIRDS.

ORDER VII. WITH CLOVEN FEET.

K 2   ORDER

INDEX.

ORDER IX. Web-footed.

VOL.

INDEX.

VOL. VIII.

ALPHABETICAL ARRANGEMENT.

Printed by R. & R. Gilbert, St. John's Square, London.

THE

NATURAL HISTORY

OF

BRITISH BIRDS;

OR, A

SELECTION of the MOST RARE, BEAUTIFUL, and INTERESTING

BIRDS

WHICH INHABIT THIS COUNTRY:

THE DESCRIPTIONS FROM THE

SYSTEMA NATURÆ

OF

LINNÆUS:

WITH

GENERAL OBSERVATIONS,

EITHER ORIGINAL, OR COLLECTED FROM THE LATEST
AND MOST ESTEEMED

ENGLISH ORNITHOLOGISTS;

AND EMBELLISHED WITH

FIGURES,

DRAWN, ENGRAVED, AND COLOURED FROM THE ORIGINAL SPECIMENS.

By E. DONOVAN.

VOL. IX.

LONDON:

PRINTED FOR THE AUTHOR; AND FOR F. C. AND J. RIVINGTON,
No. 62, ST. PAUL'S CHURCH-YARD. 1818.

197

PLATE CXCVII.

ANAS SPONSA.

SUMMER DUCK.

ANSERES.

GENERIC CHARACTER.

Bill convex, obtufe, the edges divided into lamellate teeth : tongue fringed, and obtufe : three fore-toes connected, the hind-one folitary.

SPECIFIC CHARACTER

AND

SYNONYMS.

Creſt pendent, double, varied with green, blue, and white.

ANAS SPONSA: criſta dependente duplici viridi cœruleo abloque varia. *Gmel. Syſt. Nat.* i. *p.* 539.—*Luth. Ind. Orn. n.* 871. 97.

ANAS SPONSA. *Linn. Syſt. I. p.* 207. 43.

ANAS AESTIVA. *Briſſ.* 6. *p.* 351. 11. *t.* 32. *f.* 2. *Id. 8vo.* 11. *p.* 455.

Beau Canard huppé. *Buf.* IX. *p.* 245.—*Pl. Enl.* 980, 981.

AMERICAN WOOD DUCK. *Brown, Jam. p.* 481.

SUMMER DUCK. *Cat. Car. I. t.* 97.

Edw. t. 101.

Lath. Gen. Syn. 6. *p.* 546. 85.

VOL. IX.   B   About

PLATE CXCVII.

About fifteen years ago, or rather more, our much esteemed friend the Rev. Thomas Rackett, F.L.S. informed us, that the Summer Duck had been then very recently found in a wild ftate in Dorfet-fhire. Upon this unqueftionable authority, we firft introduced the fpecies into our Lift of Britifh Birds. We have fince that period become poffeffed of a whole family of thefe interefting Birds which had bred in England.

The Summer Duck is fo named by the inhabitants of North America, becaufe it reaches that part of the globe in the fpring, and refides there during the whole of the fummer; it is alfo called the Tree Duck from its ufually building its neft in the hollow trunks of trees, or between their furcated branches, a peculiarity in the habits of the Duck tribe truly fingular, but which is nevertheless not wholly confined to this fpecies. In Mexico, and the iflands contiguous, thefe birds are conftant inhabitants throughout the whole year. The flefh is efteemed delicious.

Dr. Latham informs us, the natives of Louifiana ornament their pipes or calumets of peace with the fkins of thefe birds; thofe of the male birds doubtlefsly, as thefe are extremely beautiful; the colours about the head and fine pendent creft in particular are uncommonly fine.

The total length of the male bird is about nineteen inches: the female is fomewhat fmaller.

PLATE

198

PLATE CXCVIII.

SCOLOPAX GLOTTIS.

GREEN SHANKS.

GRALLÆ.

GENERIC CHARACTER.

Bill roundifh, obtufe, longer than the head: noftrils linear: face covered: feet four-toed, hind toe confifting of many joints.

SPECIFIC CHARACTER
AND
SYNONYMS.

Bill ftraight, lower mandible at the bafe red: body beneath fnowy; legs greenifh.

SCOLOPAX GLOTTIS: roftro re&to bafi inferiore rubro, corpore fubtus niveo, pedibus virescentibus. *Linn. Fn. Suec.* 171.—*Gmel.* 664. 10.

Pluvialis Major. *Aldr. Orn. 3. p.* 535.

     Raj. av. p. 106.

Limofa. *Gefn. av.* 519.

Limofa grifea major. *Briff. av. 5. p.* 272. *n.* 3. *t.* 24. *f.* 2.

Glottis. *Gefn. av.* 520.

Greater Plover of Aldrovand. *Will. Orn. p.* 106. *t.* 55.

Green-

PLATE CXCVIII.

Green-legged Horfeman. *Albin. av. 2. t. 69.*

Greenfhank. *Brit. Zool. 2. n. 183.*

Lath. Gen. Syn. 3. 1. p. 147, n. 18.

A general inhabitant of Afia, America, and Europe : it appears in Britain during the winter feafon only, as it paffes the fummer in Sweden, Norway, and other northern countries : it ufually appears in fmall flocks upon the fea-coafts, and fometimes occurs in the marfhes near the fea.

It is a flender delicate bird about the fize of the Red fhank : length fourteen inches.

PLATE

199

PLATE CXCIX.

TURDUS MUSICUS.

SONG THRUSH.

Passeres.

GENERIC CHARACTER.

Bill ftraightifh : the upper mandible a little bending and notched near the point : noftrils naked, or half covered with a fmall membrane : mouth ciliated with a few briftles at the corners : tongue jagged.

SPECIFIC CHARACTER
AND
SYNONYMS.

Quill-feathers ferruginous at the inner bafe.

TURDUS MUSICUS : remigibus bafi interiore ferrugineis. *Linn. Fn.*
　　　　　　Sv. 217.—*Gmel.* 809. 4.—*Brünn. Orn. n.* 236.
　　　　　　—*Kram. el. p.* 361. *n.* 8.
Turdus Minor. *Briff. av.* 2. *p.* 203. *n.* 2.
Tordo Meffano. *Ol. ucc. t.* 25.
Grive. *Buff. Hift. Nat. Oif.* 3. *p.* 280.
　　　　　　—— *Pl. Enl. n.* 406.
Mavis Throftle, or Song Thrufh.—*Raj. av. p.* 64. *n.* 2.
　　　　　　Will. Orn. p. 188.—*Lath. Syn.* 2. 1. *p.* 18. *n.* 2.
　　　　　　　　　　　　　　　　　　The

PLATE CXCIX.

The Throftle or Song Thrufh is efteemed for the compafs, variety, and fweetnefs of its notes, the chief of our birds of fong. It is of a folitary difpofition, never affociating like the Fieldfare, or Red-wing, in flocks. In northern counties it is feen only during the fummer, as it refides throughout the winter feafon in the more temperate parts of Europe.

This fpecies is inferior in fize to the Miffel Thrufh, being only about nine inches in length, while that of the Miffel Thrufh is eleven inches: it is alfo diftinguifhed among other characters by the form of the fpots on the breaft and belly, thofe in the Miffel Thrufh being roundifh, or oblong; in the Throftle fagittate, or fhaped like the head of an arrow with the point directed upwards: and the whole placed longitudinally. With thefe exceptions, there is a ftriking fimilarity between thefe two birds.

The Song Thrufh breeds early in the year about the beginning of April, and a fecond brood fometimes in May. The neft is ufually compofed of earth and mofs intermixed; the eggs of a blueifh green, marked with a few black fpots.

PLATE

200

PLATE CC.

ARDEA STELLARIS.

COMMON BITTERN.

GRALLÆ.

GENERIC CHARACTER.

Bill ftraight, pointed, long, somewhat compreffed, with a furrow from the noftrils towards the tip: noftrils linear: tongue fharp: feet four-toed, cleft; toes conne&ted at the bafe.

SPECIFIC CHARACTER.

Head fmoothifh; body above teftaceous, with tranfverfe fpots beneath paler, with oblong brown fpots.

ARDEA STELLARIS: capite læviufculo, fupra teftacea maculis tranf-
verfis fubtus pallidior, maculis oblongis fufcis.
Fn. Suec. 164.—*Gmel.* 635. 21.—*Scop. Ann.* 1.
p. 125.—*Brünn. Orn. n.* 155.—*Kram. el. p.* 148.
Georg. it. p. 171.
BOTAURUS. *Briff. av. 5. p.* 444. *n.* 24. *t.* 37. *f.* 1.
BUTOR. *Buff. Hift. Nat. des Oif.* 7. *p.* 411. *t.* 21.—*Pl. Enl.*
n. 786.

BITTOUR,

PLATE CC.

BITTOUR, or BITTERN. *Ray. av. p.* 100. *n.* 11.

Will. Orn. p. 282. *t.* 50. 52.

Lath. Gen. Syn. 3. 1. *p.* 56. *n.* 17.

The common Bittern is an inhabitant of Europe, Afia, and America.

Its ufual haunts are among reeds in marfhy places : its cuftomary food confifts of fifhes and reptiles, which it fwallows whole.

About fun-fet the Bittern rifes in the air to a vaft height in a fpiral direction, making a prodigious and very fingular noife ; befides which it emits on fome occafions another found, which has been compared with that of the bellowing of a bull. When attacked by dogs or other animals, it fights with uncommon vigour. Bittern-hawking was in ancient times a very favourite fport.

The Bittern is about three feet in length. The plumage elegantly varied with ferruginous fpots, difpofed with great regularity in a tranf-verfe manner upon a yellowifh buff-coloured ground : the female is rather fmaller, and darker than the male : the bill brown, beneath greenifh : legs and lores green.

PLATE

201

PLATE CCI.

ORIOLUS GALBULA, *fem.*

FEMALE GOLDEN ORIOLE.

Picæ.

GENERIC CHARACTER.

Bill conic, convex, very fharp and ftraight ; upper mandible a little longer, flightly notched : tongue bifid, fharp pointed : feet formed for walking.

SPECIFIC CHARACTER
AND
SYNONYMS.

Pale yellow; lores and limbs black; outer tail feathers on the hind-part yellow.

ORIOLUS GALBULA: luteus, loris artubufque nigris, rectricibus exterioribus poftice flavis.—*Linn. Syft. Nat.* 1. *p.* 160, 1.—*Gmel. Syft.* 1. *p.* 382.

CORACIAS ORIOLUS. *Linn. Fu. Suec.* N° 95.

GALBULA. *Ray Syn. p.* 68. 5.

Yellow Bird from Bengal. *Alb.* 111. *t.* 19.

GOLDEN THRUSH. *Edwards t.* 185.

VOL. IX.   C   GOLDEN

PLATE CCI.

GOLDEN ORIOLE. *Brit. Zool. App. p. 4. t. 4.*
Lath. Syn. 2. p. 449. 43.
In Orn. p. 186. 45.

The Male of the Golden Oriole appeared in one of the earliest Plates of the prefent Work * : the figure now fubmitted to the reader is that of the female, which however rare the male bird, is ftill more fo than that fex.   The difference in the appearance of the plumage is alfo fo confiderable, that we flatter ourfelves it will be confidered as an acceptable introduction into the prefent volume.

Since the figure and defcription of the male bird appeared, we have had occafion to record the additional difcovery of a pair of thefe rare birds in Britain ; a circumftance of very unufual occurrence : they were found in Cornwall.   This difcovery we announced in the year 1808 †.

---

* Vol. I.

† In the Monthly Magazine for December, 1809, two of the Golden Oriole are mentioned as having been fhot in Cornwall. It does not appear very improbable thofe laft mentioned were the birds we had previoufly recorded.

PLATE

202

PLATE CCII.

PODICEPS CRISTATUS, *var.*

CRESTED GREBE.

ANSERES.

GENERIC CHARACTER.

Bill ftrong, flender, fharp pointed: noftrils linear: lore bare of feathers : tongue fub-bifid: body depreffed : no tail : legs four-toed, fhank compreffed, and befet with a double feries of ferrations behind: toes lobate.

SPECIFIC CHARACTER
AND
SYNONYMS.

Head rufous : collar black: fecondary quill feathers-white.

PODICEPS CRISTATUS: capite rufo, collari nigro remigibus fecun-dariis albis. *Lath. Ind. Orn.* 780. 1.

COLYMBUS CRISTATUS: capite rufo, collari nigro, remigibus fecundariis albis. *Gmel. Linn. Syft.* 589. 7. *Linn. Syft.* 1. *p.* 222. 7.—*Fn. Suec.* N° 151.—*Brun.* 135.—*Klein. Av. p.* 149. 1.

C 2

In

PLATE CCII.

In a former Volume of this Work* will be found a figure of the adult bird of this fpecies of Grebe; and which, on due comparifon, will appear to differ fo very confiderably from the young bird, that we are induced to include the latter in the prefent continuation. This difference is indeed fo very material, that it could fcarcely be conceived by a cafual obferver; or by any one not attentive to the progreffive alterations which take place in the plumage of certain birds from their early ftate till they arrive at full maturity.

The defcription of the Great Grebe in the volume before mentioned contains fome notice of thefe variations manifeft in the early growth of that fpecies: it is faid there, that " at firft they are perfectly downy and ftriped, efpecially down the neck, with black; after this, when about half grown, the ftripes on the neck are lefs diftinct, being rather mottled than ftriped, and the under parts, though white, are clouded with dufky; at this period, a fullnefs round the head is obferved: as the bird advances ftill further towards maturity, the brown and white appear clear and diftinct; the head becomes much tufted, and the horns are a little elongated. But we have great reafon to believe that the young bird does not obtain the full and perfect creft till the fecond year at leaft."

The length of the bird we have reprefented is eighteen inches and a quarter.

---

* Vol. III. Plate LXVIII.

PLATE

203

PLATE CCIII.

MELEAGRIS GALLOPAVO.

COMMON TURKEY.

GALLINÆ.

GENERIC CHARACTER.

Bill conic, incurvate : head covered with fpongy caruncles : chin with a longitudinal membranaceous caruncle : tail broad, expanfile : legs fpurred.

SPECIFIC CHARACTER
AND
SYNONYMS.

Front and chin carunculated ; breaft of the male tufted.

MELEAGRIS GALLOPAVO : capite caruncula, frontali gularique, maris pectore barbato. *Linn. Fn. Suec.* 198.— *Gmel. Linn. Syft. Nat.* 732. 99. 1.—*Lath. Ind. Orn.* 618. 1.

TURKEY : *Ray. Av. p.* 51. *n.* 3.
Albin. Av. 3. *p.* 33. *t.* 35.
Penn. Act. Angl. 72. 1. *p.* 67.

Dindon. *Buff. Hift. Nat. Oif.* 2. *p.* 132. *t.* 3.—*Pl. Enl. n.* 97.

It

PLATE CCIII.

It has ever been our wifh to render this Work as complete as the nature of the fubject would permit ; and under that impreffion we have never felt ourfelves reftricted to the exclufion of any bird which has poffeffed either intereft or beauty to recommend itfelf. In the earlier part of the Work fome varieties of the Peacock were introduced, and confeffedly with fatisfaction to our purchafers. The Turkey which we now introduce may not indeed poffefs an equal claim to our confideration ; it is, neverthelefs, an important bird, and much too interefting to be omitted.

The Turkey in a wild ftate was originally a native of the American woods, from whence it was introduced into Europe, and cultivated with uncommon fuccefs.

To a form confeffedly uncouth it unites a highly fingular and very varied appearance, and the feathers in various parts of the plumage are highly ornamental. Our figure is of the male bird in the ordinary ftate of plumage.

PLATE

204

PLATE CCIV.

LANIUS COLLURIO, *fem.*

RED-BACKED BUTCHER BIRD, *fem.*

GENERIC CHARACTER.

Bill ftraightifh, with a tooth on each mandible uear the end, naked at the bafe : tongue jagged at the end.

SPECIFIC CHARACTER
AND
SYNONYMS.

Tail fomewhat cuneated : back grey; four middle tail feathers uniform : bill lead colour.

LANIUS COLLURIO : cauda fubcuneiformi, dorfo grifeo, reĉtricibus
4 intermediis unicoloribus, roftro plumbeo.
Gmel. Syft. 1. *p.* 300. *Fn. Sv. N° 81. t. 2. f.* 81.
—*Linn. Syft.* 1. *p.* 136. 12.—*Lath. Gen. Syn.*
1. 167. 25.—*Ind. Orn.* 1. *p.* 69. 11.
RED-BACKED SHRIKE, LESSER BUTCHER BIRD, FLUSHER.
Albin. 11. *t.* 14.—*Will. Orn. p.* 88. 2. 89. 3.
Br. Zool. 11. *n.* 131.

The

PLATE CCIV.

The male of the Red-backed Butcher Bird has been reprefented in an early part of the prefent work : the female differs from the male in the general appearance, and the colours and marks upon the plumage as will be perceived from the annexed plate. In thefe particulars the female bears fome refemblance to the young of the male birds; but is larger than the adult male. The length eight inches.

It is fome years fince our figure of the male bird and that of the Wood Chat appeared; thefe have been oftentimes confidered by writers as varieties of the fame fpecies. We take this opportunity of ftating, that fuch an opinion is extremely erroneous, and can have arifen only from the want of due comparifon. We have been for fome years in poffeffion of a very fine fpecimen of the Wood Chat, and have feen a number of the Red-backed Shrike fince our former account was publifhed, the final refult of which is an unequivocal conviction in our own mind, that they are decidedly diftinct fpecies. And to this opinion, after fome unpleafant cavilling on the part of certain cotemporary writers, we have the fatisfaction to perceive the naturalifts of the prefent day concede.

PLATE

205

PLATE CCV.

TETRAO SCOTICUS, *mas.*

RED GROUS, RED GAME, MOORCOCK, *or* GORCOCK, *male.*

GALLINÆ.

GENERIC CHARACTER.

Near the eyes a fpot which is either naked, or papillous, or rarely covered with feathers.

SPECIFIC CHARACTER
AND
SYNONYMS.

Tranfverfely ftreaked with rufous and blackifh : fix outer tail feathers each fide blackifh.

TETRAO SCOTICUS: rufo et nigricante tranfverfim ftriatus, reftricibus 6 utrique exterioribus nigricantibus. *Lath. Ind, Orn.* 641. 15.

BONASA SCOTICA. *Brif.* 1. *p.* 199. 5. *t.* 22. *f.* 1.—*Id.* 8*vo.* 1. *p.* 55.—*Buff.* 2. *p.* 242.

ATTAGEN. *Brif.* 1. *p.* 209. 9.

Lagopus altera Plinii. *Ray. Syn. p.* 54. *A* 3 —*Will. Angl. p.* 177. *Albin.* 1. *t.* 23. 24.

VOL. IX.   D   RED

PLATE CCV.

RED GAME, MOORCOCK, GORCOCK. *Raii Syn. p. 54. A. 3.*
RED GROUS. *Br. Zool.* 1. *N°* 94. *t.* 43.—*Lath. Gen. Syn.* IV.
p. 746. 13.—*Id. Suppl. p.* 216.

The Red Grous inhabits the heaths of Wales and Scotland, the latter chiefly, for they have become rather fcarce for fome years paft upon the Cambrian heaths; and are found principally now among the more mountainous parts, as Merionethfhire and Caernarvonfhire.

The male is rather larger and more robuft than the female; its length about fifteen inches and a half. The plumage of the male a rich reddifh brown, fomewhat rufous, and gloffed with purple; and the whole very elegantly ftriated tranfverfely with black lines difpofed in fcallops, and nearly equidiftant. The female inclines to a yellower hue.

PLATE

206

PLATE CCVI.

COLUMBA DOMESTICA, *var.*

DOMESTIC PIGEON, *var.*

SILVER TUMBLER.

PASSERES.

GENERIC CHARACTER.

Bill ftraight, defcending towards the tip: noftrils oblong, half covered with a foft tumid membrane.

SPECIFIC CHARACTER
AND
SYNONYMS.

Cinereous: rump white, band on the wings and tip of the tail blackifh.

COLUMBA DOMESTICA: cinerea, uropygio albo, alarum fafcia, caudaque apice nigricante. *Linn. Fn. Suec. n.* 207.—*Scop. Ann.* 1. *n.* 177.—*Gmel. Linn. Syft. Nat.* 769. 2.—*Lath. Gen. Syn. Vol.* 2. *part.* 2.

Columba gyratrix, *var. Linn. Syft.* 1. *p.* 28, 29.

D 2

A very

PLATE CCVI.

A very elegant variety of the Tumbler Pigeon, named, from the delicate whitenefs of its plumage, the Silver Tumbler.

Thefe birds have been called from the early times of Ray and Petruver, the Tumbler Pigeon, alluding to the peculiarity of this variety tumbling repeatedly while flying in the air. This is effected by throwing themfelves backwards and falling again on their breaft and expanded wings.

Buffon enumerates many varieties of the Domeftic Pigeon, and fo alfo Linnæus and Latham, all which are diftinguifhed by fome peculiarity either of their plumage, habits, or manners of life. The Tumblers are a fingular kind, and fome of the varieties bear an enormous price: the Almond Tumbler, Dr. Latham fays, is fo highly valued, that the fum of eighty guineas has been given for a fingle bird.

PLATE

207

PLATE CCVII.

LANIUS EXCUBITOR, *fem.*

GREAT SHRIKE, *fem.*

ACCIPITRES.

GENERIC CHARACTER.

Bill ftraightifh, with a tooth on each mandible near the tip, the bafe naked: tongue jagged at the end.

SPECIFIC CHARACTER
AND
SYNONYMS.

Tail wedge-form, fides white: back hoary, wings black, with a white fpot.

LANIUS EXCUBITOR: cauda cuneiformi, lateribus alba, dorfo cano, alis nigris macula alba. *Linn. Syft.* 1. p. 135. 11.—*Fn. Suec.* Nº 80.—*Lath. Ind. Orn.* 67. 6.

The

PLATE CCVII.

The length of this bird is ten inches, the female rather larger and more robuſt than the male.

The male bird is figured in Plate LXXXVII. of this Work; the female differs from the male chiefly in the colour and markings on the breaſt, that part in the male being entirely white, in the female of a warm duſky tint with numerous tranſverſe ſemicircular pale brown lines or ſcallops diſpoſed in a pretty regular manner over the whole of the throat, breaſt, ſides, and belly.

It is ſome years ſince we deſcribed the male of this ſpecies: we then mentioned it as a rare bird, nor has later information leſſened the accuracy of this remark: within the ſpace of many years not more than three or four ſpecimens, and thoſe all males, are recorded to have appeared in England. The female is far more rare than the male.

As this bird has appeared on the eaſtern coaſts of Britain in the winter, it is conjeĉtured they have ſtraggled hither in their migrations from the north (where they occur in ſummer) to the more ſouthern regions, where they paſs the winter.

PLATE

208

PLATE CCVIII.

SCOLOPAX RUSTICOLA, *var.*

WOODCOCK, *Pied-white.*

GRALLÆ.

GENERIC CHARACTER.

Bill roundifh, obtufe, longer than the head: noftrils linear: face covered: feet four-toed, hind-toe confifting of many joints.

SPECIFIC CHARACTER
AND
SYNONYMS.

Bill ftraight, reddifh at the bafe: legs cinereous: thighs covered: head with a black band each fide.

SCOLOPAX RUSTICOLA: roftro re&o bafi rufefcente pedibus cinercis femoribus te&is, fafcia capitis nigra. *Gmel. Linn. Syft. Nat.* 660.—*Fn. Suec.* 170.

Numenius roftri apice lævi, capite linea utrinque nigra, re&ricibus nigris apice albis.—*Fn. Suec.* 1. *n.* 141.

Gallina ruftica. *Gefn. av.* 477.

Scolopax. *Bell. av.* 272. *Briff. av.* 5. *p.* 292. *n.* 1.

Bécaffe. *Buff. Hift. Nat.* 7. *p.* 462. *t.* 25.

WOODCOCK.

PLATE CCVIII.

Woodcock.   *Raj. av.p.* 104. *n.* 1.
               Will. Orn. p. 289. *t.* 53.
               Albin. av. i. *t.* 70.
               Lath. Gen. Syn. iii. 1. *p.* 129. *n.* 1.

The Woodcock in its ordinary ftate of plumage is a bird familiar to every one. The variety we have felected for reprefentation is very far from common : there are two or three other known varieties, in one of which an uniform cream or pale buff colour pervades the whole plumage ; and another is wholly white. Our prefent variety is of the pied kind partly buff, and partly pure white; the buff marked with fufcous.

Thefe varieties are about the fize of the Woodcock in its ufual ftate of plumage, the length being fifteen inches.

The Woodcock comes into England during the winter feafon, appearing firft about October, in November, and December or January. Never arriving in flights, but ftraggling individuals, or fome few in pairs, and chiefly in the evening or very hazy weather. They are obferved to penetrate inland as soon as they arrive.

Thefe birds inhabit northern countries, as Sweden, Ruffia, as far as Kamtfchatka ; breeding in the northern regions during fummer, and like many other birds of the migratory kind who pafs the fummer towards the polar climes, retire fouthward as the winter becomes fevere. In their migrations from the north, they vifit the more fouthern parts of Europe, and proceed as far as the temperate parts of Africa and Afia.

That

PLATE CCVIII.

That the Woodcook breeds in England has been authenticated upon the beft authority in many inftances : we have ourfelves both eggs and young produced in England. Thefe are, however, accidental circumftances, for it is clearly afcertained the Woodcock breeds more northward than the Britifh ifles. In warm countries inhabiting the plains only in winter, and retiring to the higher hills during fummer.

The flefh of the Woodcock is delicious, and in much requeft for the table of the epicure ; nor need it be added, that their entrails are never drawn : the prevailing opinion is, that this is unneceffary as they live by " fuction ;" they fubfift entirely on infects and the nutritious juices they extract from the various fnails, worms, and other fimilar food ; and the trail itfelf, as it is ufually termed, is ferved up at table with the bird.

The neft, one of which is in our poffeffion, confifts merely of a few withered leaves, and fibres laid loofely upon the bare turf or ground ; ufually fome fpot of earth felected clofe to the ftump of an old tree. The eggs four or five in number, the fize of a pigeon's egg, of a greyifh colour, with dufky blotches.

209

PLATE CCIX.

CORVUS FRUGILEGUS

ROOK.

Picæ.

GENERIC CHARACTER.

Bill convex, fharp-edged : noftrils covered with fetaceous recumbent feathers : tongue cartilaginous, bifid : feet formed for walking.

SPECIFIC CHARACTER

AND

SYNONYMS.

Black, front fomewhat cinereous : tail roundifh.

CORVUS FRUGILEGUS : ater, fronte cinerafcente, cauda fubrotundata. *Linn. Syft. I. p.* 156. 4.—*Fn. Suec.* Nº 85.—*Gmel. Syft. Nat. I. p.* 366.—*Lath• Ind. Orn.* 152. 5.

Cornix nigra frugilega. *Raii Syn. p.* 83. *A.* 3.—*Will. p.* 84. *t.* 18. —*Frifch. t.* 64.

Cornix frugilega. *Briff.* 11. *p.* 16. 3.—*Id. I. p.* 158.—*Klein av. p.* 59. 3.

Le Freux, ou la Frayonne. *Buff.* 111. *p.* 55. *Pl. Enl.* 484.

E 2 ROOK,

PLATE CCIX.

Rook. *Albin.* 11. *t.* 22.—*Will. (Angl.) p.* 123.—*Lath. Gen. Syn. I. p.* 372. 4.—*Id. Sup. p.* 76.

The Rook and Crow bear fuch a near resemblance to each other, that many fuperficial obfervers have been inclined to confider them as the fame bird. Naturalifts are however affured, that notwith-ftanding the two fpecies affimilate fo clofely in general appearance, they are fpecifically different.

The Rook is a trifle larger than the Crow, meafuring about twenty-one inches in length. But a better diftinction prevails in the ap-pearance of the fpace round the bafe of the bill and the noftrils, which appear bare of feathers; while the former part in the Crow is covered with feathers, and the noftrils with briftly hairs.

In their manners of life, the difference between the Crow and Rook is very material. The Crow feeds on carrion, which the Rook re-jects: the latter fubfifting wholly on feeds, worms, and infects.

The Rook is the known attendant upon the ploughman's labour, as upon the fowers, following at a little diftance behind in flocks to pick up the worms turned up by the ploughfhare, or the feeds caft in the furrows. They are neverthelefs, upon the whole, of infinite be-nefit to the hufbandman, as they deftroy myriads of thofe infect larvæ, which harbour in the earth, and gnaw the roots of corn, and other farinaceous plants.

Among

PLATE CCIX.

Among the more injurious kinds of larvæ, which harbour in fuch fituations, we may mention thofe of the beetle kind, particularly the chafers, and moft efpecially, as being more numerous and hurtful, the larvæ of the Melolontha vulgaris : thefe are in particular their favourite food.

In England the Rook remains throughout the whole year. In the more fouthern parts of Europe they are migratory. They affociate together in immense flocks, and ufually build upon the tops of high trees: their breeding places are denominated *rookeries;* there they congregate, and fit chattering, the males in turn relieving the females in feeding and guarding their young. They begin to build in March. After their breeding feafon they forfake the rookeries, and rooft elfe-where, but return to the rookeries again in Auguft, and after a while repair their nefts preparatory to the winter feafon.

A rook-pie is a favourite ruftic difh : it is only the young Rooks that are employed for this purpofe, and thefe are better ftripped of their fkins previoufly to being put into the pie.

The Rook is a general inhabitant of Europe, and the weftern parts of Siberia.

The plumage of the Rook appears particularly fplendid in the funfhine ; efpecially the head and neck, and alfo the upper part of the breaft and wings, which, in a ftrong light, are finely gloffed with blue and rich purple.

PLATE

PLATE CCX.

PARVO CRISTATUS, *fem. var.*

WHITE CRESTED PEA-HEN.

GALLINÆ.

GENERIC CHARACTER.

Bill convex, robuft : head covered with revolute feathers: noftrils large : feathers of the rump long: broad, expanfile, and covered with ocellar fpots.

SPECIFIC CHARACTER

AND

SYNONYMS.

Head with a compreffed creft; fpurs folitary. Female entirely white.

PAVO CRISTATUS γ: capite crifta compreffa calcaribus folitariis. *Linn. Syft. Nat. I. p.* 267. *I.—Fn. Suec.* N° 197.—*Gmel. Syft. Nat. I. p.* 729.—*Lath. Ind. Orn.* 616. *I.*

PAVO ALBUS. *Brif.* 1. *p.* 288.—*Id.* 8vo. *I. p.* 81.—*Frifch. t.* 120. —*Raii Syn. p.* 51. *A.* 2.—*Borowfk. Nat.* 11. *p.* 167.—*Ger. Orn.* 11. *p.* 74 *t.* 218.

LE

PLATE CCX.

Le Paon Blanc. *Buff.* 11. *p.* 323.
White Peacock. *Lath. Gen. Syn.* iv. *p.* 672.

About the fame fize as the female of the coloured Pea-hen: our figure is that of the female White Pea-hen, recorded by Dr. Latham as appertaining to the Leverian Mufeum.

This bird is entirely white, except the bill and legs, which are yellowifh.

PLATE

PLATE CCXI.

TETRAO TETRIX, *fem.*

BLACK GAME, *female.*

GALLINÆ.

GENERIC CHARACTER.

Near the eyes a fpot which is either naked, or papillous, or rarely covered with feathers.

SPECIFIC CHARACTER

AND

SYNONYMS.

Violet black ; tail forked : fecondary quill-feathers white towards the bafe : *male.* Red, varied with cinereous, and black tranfverfe ftriæ : *female.*

TETRAO TETRIX : nigro-violacea, cauda bifurca, remigibus fecundariis verfus bafin albis. *Lath. Ind. Orn.* 635.

TETRAO TETRIX : cauda bifurcata, remigibus fecundariis bafin
  verfus albis. *Linn. Fn. Suec.* 202.—*Scop.*
  Ann. I. n. 169.—*Brün. Orn. n.* 196.]

Tetrao ſ Urogallus minor. *Gefn. av.* 494.

VOL. IX.   F   Petit

PLATE CCXI.

Petit Tétras, ou Coq de bruyere à queue fourchue. *Buff. Hift. Nat. des Oif.* 2. *p.* 210. *t.* 6.—*Pl. Enl. n.* 172. 173.

BLACK COCK, BLACK GAME, or BLACK GROUS. *Ray. av. p.* 53. *n.* 2.—*Will. Orn. p.* 173. *t.* 31.—*Albin. av. I. t.* 22.—*Lath. Syn.* 11. 2. *p.* 733. *n.* 3.

No two fpecies of the fame genus can differ more confiderably from each other than the male and female of the Black Grous. The male is of a deep violet black; the female a warm brown, varied with cinereous and white, and marked throughout with black tranfverfe lines.

The female is fmaller than the male, the length of the latter being twenty-one inches, that of the female eighteen inches.

The Black Grous inhabits northern countries, and becomes gradually fcarce towards the fouth. They thrive well in the climate of the north part of Britain, but are fo much an object of requeft with the fportfman, that they can never become abundant except in extenfive domains where the brood can be protected.

PLATE

212

PLATE CCXII.

ANAS BICOLOR.

MOSCHOVY DUCK, *var?*

ANSERES.

GENERIC CHARACTER.

Bill convex obtufe, the edges divided into lamellate teeth : the tongue fringed, obtufe : three fore-toes connected, the hind one folitary.

SPECIFIC CHARACTER
AND
SYNONYMS.

Brown : head and nape greenifh black : neck, breaft, and quill-feathers white : legs fulvous.

ANAS BICOLOR : fufca, capite nuchaque nigro virefcente : collo pectore alarumque pennis albis ; pedibus fulvis. ANAS MOSCHATA *var.?*

Length twenty-five inches. Refembles fome varieties of the Mofchovy Duck, except in wanting the characteriftic mark of the naked papillous fkin round the eyes.

F 2   PLATE

213

PLATE CCXIII.

SCOLOPAX LAPPONICA, *fem.*

RED GODWIT, *female.*

GRALLÆ.

GENERIC CHARACTER.

Bill roundifh, obtufe, longer than the head: noftrils linear: face covered: feet four-toed, hind-toe confifting of many joints.

SPECIFIC CHARACTER

AND

SYNONYMS.

Bill yellowifh, legs black: body beneath rufous-ferruginous.

SCOLOPAX LAPPONICA: roftro flavefcente, pedibus nigris, fubtus tota rufo ferruginea. *Linn. Syft. I. p.* 246. 15. —*Fn. Suec. n.* 174.—*Gmel. Linn. Syft. Nat. I. p.* 667.—*Lath. Ind. Orn.* 718.

LIMOSA RUFA. *Brijf.* v. *p.* 281. 5. *t.* 25. *f. I.*—*Id.* 8*vo.* 11. *p.* 281.

La Barge Rouffe. *Buff.* VII. *p.* 504. *Pl. Enl.* 900.

RED GODWIT. *Edw. t.* 138.—*Br. Zool.* 11. *N°* 181. *t.* 67.— *Lath. Gen. Syn.* v. *p.* 142. 13.

The

PLATE CCXIII.

The extraordinary rarity of the Red Godwit may entirely juſtify the introduction of both ſexes of this very elegant and curious ſpecies in the preſent work.

It is recorded among the Britiſh ſpecies of the Snipe tribe, upon the authority of only a few inſtances in which it has been known to viſit England. One ſhot many years ſince, near Hull, is mentioned by Pennant; in the collection of birds, formed more than thirty years ago by Mr. Tunſtall, was a ſpecimen ſhot in Dorſetſhire; in the Muſeum of the great rival of Mr. Tunſtall, Sir Aſhton Lever, one or two varieties again occurred; and ſince the diſſolution of that Muſeum we have not ſeen it in any other collection.

The pair we have delineated were from the celebrated cabinet of the late Mr. Green, of Weſtminſter, whoſe collection was ſo rich as to contain Britiſh ſpecimens of both ſexes preſerved by himſelf, and which are now in our Muſeum.

Mr. Green diſſected both theſe birds, and was enabled to determine the ſex of each. According to his obſervations, the bird he aſcertained to be the female has the back and wing feathers uniformly cinereous; while in the male bird many of thoſe feathers are marked at the edges with rufous, ſome pretty deeply in, and the diſk of the feathers darker. This indeed ſeems to be the chief diſtinction in the general appearance of the two ſexes.

The length of this ſpecies is ſtated to be about eighteen inches, our ſpecimens do not exceed ſeventeen inches. In ſome birds the abdomen is ſaid to be red like the neck and breaſt; in both our ſpecimens the lower part of the belly is white.

This

6

PLATE CCXIII.

This fpecies is an inhabitant of America, where it breeds in the fens, retiring fouthward in winter. It is alfo found on the European continent, but very rarely except in the fouthern parts, and in the more temperate countries of Afia. They occur about the Cafpian fea in fpring; yet never appear fo far to the northward as Siberia.

PLATE

PLATE CCXIV.

ANAS ERYTHROPUS.

BERNACLE GOOSE.

ANSERES.

GENERIC CHARACTER.

Bill convex, obtufe, the edges divided into lamellate teeth : tongue fringed, obtufe : three fore-toes connected, the hind one folitary.

SPECIFIC CHARACTER
AND
SYNONYMS.

Cinereous, above undulated black and white; neck black : face and abdomen white.

ANAS ERYTHROPUS: cinerea fupra nigro alboque undulata, collo nigro, fafcie abdomineque albis. *Lath. Ind. Orn.* 843. 31.

ANAS ERYTHROPUS. *Faun. Suec.* N° 116. *(maf.) Frifch. t.* 189. *Linn. Syft. I. p.* 197. 11.—*Gmel. Syft. I. p.* 512.

BERNICLA. *Briff.* VI. *p.* 300. 14.— *Id.* 8vo. 11. *p.* 411.—*Raii Syn. p.* 137. *A.* 5.—*Will. Orn. p.* 274.

Anas Helfingen. *Olaff. Ifl.* 11. *t.* 33.

VOL. IX.   G   Anfer

3

PLATE CCXIV.

Anfer Brenta. *Klein Av. p.* 130. 8.—*Id. p.* 170. 12.
La Bernache. *Buff.* IX. *p.* 93. *f.* 5.—*Pl. Enl.* 855.
Canada Goofe. *Albin. I. t.* 92.
BERNACLE, or CLAKIS. *Br. Zool.* II. N° 269.
Will. Angl. p. 359.
Phil. Tranf. II. *p.* 853.
Gerard Herb. p. 1587.
Lath. Gen. Syn. VI. *p.* 466. 26.

The length of our bird is twenty-fix inches ; the bill fhort, black, with a pale or flefh-coloured fpot on each fide : head white, with a black fpot between the bafe of the bill and the eye : hind head and neck black : band on the fhoulders white : belly throughout pure white : back cinereous, the ends of the feathers black, with the tip white, forming a kind of black and white undulation : wings grey, the feathers, except the quill feathers, black towards the end, and terminating in a white tip : back and rump feathers black ; the fides with the tail coverts white : the tail black. The wings beneath are of a pale cinereous colour, as in the common Heron.

The confufion that prevails in the fynonyms of the Bernacle Goofe, Brent Goofe, and White-fronted Goofe, demands a very clofe attention.—The refult of this confufion is, that we find a Bernacle Goofe which is not a Bernacle Goofe, but the Linnæan Erythropus : the Brent Goofe, which is not the Bernacle, confounded under the trivial name of Bernicla ; and fome of the fynonyms of the true Bernacle, applied to the White-fronted Goofe, Anas Albifrons.

It

PLATE CCXIV.

It will tend materially to difpel this feeming confufion by bearing in mind, that the fpecies Erythropus of Linnæus is the true Bernacle Goofe, but which is better characterifed by the fpecific definition of Dr. Latham, as above quoted, than by the Linnæan character, " cinerea fronte alba." *Faun. Suec.* At the fame time recollecting, that notwithftanding the fimilarity of names, Anas Bernicla of Linnæus is the Brent and not the Bernacle Goofe.

This appears to be the Tree Goofe of Gerard's Herbal, in which the figure of the bird is rudely reprefented, and alfo the fhell (lepas antifera) in which it is affirmed with due gravity, that the birds are produced and hatched ; and after a certain period dropping out of the fhell, appear young Bernacle geefe ! In complete confirmation of this marvellous tale, we are affured, that the tails of the birds yet immatured, may be feen fticking out of the fhell ; alluding to the fingular feathered tentacula of the animal which inhabits the fhell* We fhould obferve after this, that it is not very unlikely the Bernacle and Brent Goofe might have been confounded ; or the honour of this Neptunian origin be afcribed to both. Gefner's Anfer Arboreus, which he calls alfo Branta f. Bernicla, feems indeed to be the Brent Goofe, called by Linnæus, as before ftated, Bernicla.

Anas Erythropus breeds in the northern regions of Europe, being found in Ruffia, Sweden, Norway, and Iceland, and it is believed

---

* A figure of this remarkable fhell may be feen in our work of Britifh Shells, Plate VII.

G 2

in

PLATE CCXIV.

in North America, though not commonly. During the winter feafon they are not uncommon in the northern parts of Britain, but become more rare towards the fouthward. In fevere winters it comes with other wild fowl to the London markets.

PLATE

215

PLATE CCXV.

ANAS STREPERA, *fem.*

GADWALL, *female.*

ANSERES

GENERIC CHARACTER.

Bill convex, obtufe, the edges divided into lamellate teeth: tongue fringed, obtufe: three fore-toes connected, the hind one folitary.

SPECIFIC CHARACTER

AND

SYNONYMS.

Wing-fpot rufous, black and white.

ANAS STREPERA: fpeculo alarum rufo nigro albo. *Linn. Fn. Suec.* 21.—*Gmel. Linn. Syft. Nat.* 520. 20.— *Brün Orn. n.* 91.—*Müll. Zool. Dan.* 118.— *Georgi it. p.* 166.

ANAS STREPERA. *Gefn. Av.* 121.—*Aldr. Orn.* 3. *p.* 234.—*Brif. Av.* 6. *p.* 338. *n.* 8. *t.* 33. *f. I.*

ANAS SUBULATA, *fem. S. G. Gmelin it. I. t. I.*

ANAS CINEREA, *maf. Gmel.* 2. 17.

Anas

PLATE CCXV.

Anas Platyrhynchos, roftro nigro pleno. *Aldr. Orn.* 3. *p.* 230.
t. 233.—*Raj. Av. p.* 145. *n.* 2.

Chipeau. *Buff. Hift. Oif.* 9. *p.* 187. *t.* 12.—*Pl. Enl. n.* 958.

GADWALL, or GRAY. *Raj. Av. I. p.* 145. *A.* 2.
Will. Orn. p. 374. *t.* 72.
Lath. Gen. Syn. 111. 2. *p.* 515. *n.* 61.

The male Gadwall will be found reprefented in Plate **CCXXV.** of the prefent work: the female is the fubject of the annexed Plate: the length is twenty-one inches.

PLATE

216

PLATE CCXVI.

TETRAO SCOTICUS, *fem.*

RED GROUS, RED GAME, MOORCOCK, *or* GORCOCK, *female*

GALLINÆ.

GENERIC CHARACTER.

Near the eyes a fpot, which is either naked, or papillous, or rarely covered with feathers.

SPECIFIC CHARACTER
AND
SYNONYMS.

Tranfverfely ftreaked with rufous and blackifh : fix outer tail feathers each fide blackifh.

TETRAO SCOTICUS : rufo et nigricante tranfverfim ftriatus rectricibus 6 utrinque exterioribus nigricantibus. *Lath.*
Ind. Orn. 641. 15, fem.

The

PLATE CCXVI.

The male of the Red Grous is delineated in Plate CCV. The female is a trifle ſmaller than the male bird: the colour more inclining to teſtaceous than dark red; the black tranſverſe lines, or ſcallops larger and placed more remotely; and the whole of the neck, breaſt, and back elegantly ſtudded with ſmall whitiſh ſpots.

PLATE

217

PLATE CCXVII.

TURDUS MERULA.

BLACKBIRD.

PASSERES.

GENERIC CHARACTER.

Bill ftraightifh: the upper mandible a little bending and notched near the point: noftrils naked, or half covered with a fmall membrane: mouth ciliated with a few briftles at the corners: tongue jagged.

SPECIFIC CHARACTER

AND

SYNONYMS.

Black, bill and eye-lids yellow.

TURDUS MERULA: ater, roftro palpebrifque fulvis. *Linn. Fn. Suec.* 220.—*Gmel. Linn. Syft. Nat.* 831. 22.— *Müll. Zool. n.* 29. 241.

MERULA. *Bell. Av.* 30. *b.*—*Gefn. Av.* 603.—*Aldr. Orn.* 2. *p.* 602, *t.* 604. 605.—*Briff. Av.* 2. *p.* 227. *t.* 10.

Merle. *Olin. ucc. p.* 29.

Merle. *Buff. Hift. Nat. des Oif.* 3. *p.* 330. *t.* 20.—*Pl. Enl. n.* 555.

VOL. IX. H Schwarze

PLATE CCXVII.

Schwarze Amfel. *Frifch. Av. t.* 29.—*Günth. Neft. U. Eyer. t.* 39.
BLACKBIRD. *Ray Syn. p.* 65. *I.—Will. Orn. p.* 190.—*Albin. I.*
pl. 37.—*Br. Zool. I.* N° 109. *pl.* 47.—*Lath.*
Gen. Syn. v. 2. *p. I. p.* 43.

———————————

The length of this bird is ten inches: the bill, infide of the mouth, and eye-lids in the male bird a fine yellow: plumage black, with the legs black brown. The female brown, inclining on the breaft and belly to rufous; the bill black.

The Blackbird is an inhabitant of Europe and Afia. Builds in thickets near inhabited places. Its neft is compofed of mofs, clay, hay, and dried twigs: the eggs four or five in number, of a blueifh green colour, with dark or dufky fpots.

Often kept in cages on account of the fweetnefs of its note: when tamed, it may be taught to whiftle tunes, or imitate the human voice.

PLATE

218

PLATE CCXVIII.

ANAS ADUNCA.

HOOK-BILLED DUCK.

ANSERES.

GENERIC CHARACTER.

Bill convex, obtufe, the edges divided into lamellate teeth : tongue fringed, obtufe: three fore-toes connected ; the hind one folitary.

SPECIFIC CHARACTER

AND

SYNONYMS.

Middle tail feathers recurvate: bill incurvate.

ANAS ADUNCA : verficolor, rectricibus intermediis (maris) recurvatis, roftro incurvato. *Gmel. Linn. Syft. Nat.* 538. 41.

Anas roftro incurvo. *Briff. Orn.* 6. *p.* 311. *n.* 2.

ANAS BOSCHAS ♀ : rectricibus intermediis (maris) recurvatis, roftro incurvato. *Lath. Ind. Orn.* 851.

HOOK-BILLED DUCK. *Albin.* 11. *t.* 96, 97.—*Id.* 111. *t.* 100.— *Will. Orn. (Angl.) p.* 381. *t.* 75.—*Lath. Gen. Syn.* VI. *p.* 495. D.

H 2

This

PLATE CCXVIII.

This bears a general refemblance to the Wild Duck, Anas Bofchas; and is confidered as a variety of that fpecies by Dr. Latham. Linnæus defcribes it as a diftinct fpecies under the name Adunca; and it cannot but be interefting to afcertain the bird Linnæus defcribes as a fpecies, whether it be really only a variety or not. We fhall therefore offer no apology for its introduction in this place.

The figure of the Hook-billed Duck in the annexed plate is taken from a very fine fpecimen in our Mufeum. The length is twenty-five inches: the form more flender than the Wild Duck in general: the neck white; the legs thicker, the fhanks rifing in a projection at the commencement above the foot, and again at the knee, and the bill incurvated in a moft fingular manner.

The fingular incurvation of the bill might eafily fuggeft an idea that it could be no other than an accidental deformity; but this cannot be the fact, fince the fame character prevails unerringly throughout the whole race.

PLATE

219

PLATE CCXIX.

SCOLOPAX LAPPONICA, *mas.*

RED GODWIT, *male.*

GRALLÆ.

GENERIC CHARACTER.

Bill roundifh, obtufe, longer than the head: noftrils linear: face covered: feet four-toed, hind-toe confifting of many joints.

SPECIFIC CHARACTER
AND
SYNONYMS.

Bill yellowifh, legs black; body beneath rufous-ferruginous.

SCOLOPAX LAPPONICA: roftro flavefcente, pedibus nigris, fubtus tota rufo-ferruginea. *Linn. Syft. I. p.* 246. 15.— *Fn. Suec. n.* 174.—*Gmel. Linn. Syft. Nat. I. p.* 667.—*Lath. Ind. Orn.* 718.

LIMOSA RUFA: *Briff.* v. *p.* 281 *e*.5. *t.* 25. *f. I.*—*Id.* 8*vo.* 11. *p.* 281.

La Barge rouffe. *Buff.* VII. *p.* 504.—*Pl. Enl.* 900.

RED GODWIT. *Edw. t.* 138.—*Br. Zool.* 11. N° 181. *t.* 67.— *Lath. Gen. Syn.* v. *p.* 142. 13.

This

PLATE CCXIX.

This figure reprefents the male bird, the length of which is feventeen inches.

The female is delineated in Plate CCXIII.

PLATE

PLATE CCXX.

MELEAGRIS GALLOPAVO, *var.* ε.

WHITE TURKEY.

GALLINÆ.

GENERIC CHARACTER.

Bill conic, incurvate : head covered with fpongy caruncles : chin with a longitudinal membranaceous caruncle : tail broad, expanfile : legs fpurred.

SPECIFIC CHARACTER
AND
SYNONYMS.

Front and chin carunculated; breaft of the male tufted.   Var. ε.
Gallopavo totus albus.

MELEAGRIS GALLOPAVO : capite caruncula, frontali gularique,
maris pe&ore barbato. *Gmel. Lin. Syft. Nat.*
732. 99. 1.
Var. ε. Gallopavo totus albus. *Gmel.* 732. 99. *I.*

Turkey

PLATE CCXX.

Turkey with the plumage entirely white ; longitudinal carunculated membrane of the throat red ; bill and legs yellowifh. Length three feet fix inches.

INDEX to VOL. IX.

ARRANGEMENT

ACCORDING TO THE

SYSTEM OF LINNÆUS.

ORDER I.

ACCIPITRES.

ORDER II.

PICÆ.

ORDER III.

ANSERES.

VoL. IX.   I   ORDER

INDEX.

ORDER IV.

GRALLÆ.

ORDER V.

GALLINÆ.

ORDER VI.

PASSERES.

VOL.

INDEX.

VOL. IX.

ARRANGEMENT

ACCORDING TO

LATHAM's SYNOPSIS of BIRDS.

I 2   ORDER

INDEX.

DIVISION II. WATER BIRDS.

ORDER VII. WITH CLOVEN FEET.

ORDER VIII. WITH PINNATED FEET.

ORDER

INDEX.

ORDER IX. WEB-FOOTED.

VOL.

INDEX.

VOL. IX.

ALPHABETICAL ARRANGEMENT.

ERRATUM. Vol. IX.

Plate CCX. Line 2.

For PARVO CRISTATUS, *read* PAVO CRISTATUS.

Printed by R. and R. Gilbert, St. John's Square, London.

THE

NATURAL HISTORY

OF

BRITISH BIRDS.

THE

NATURAL HISTORY

OF

BRITISH BIRDS;

OR, A

SELECTION OF THE MOST RARE, BEAUTIFUL, AND INTERESTING

BIRDS

WHICH INHABIT THIS COUNTRY:

THE DESCRIPTIONS FROM THE

SYSTEMA NATURÆ

OF

LINNÆUS:

WITH

·GENERAL OBSERVATIONS,

EITHER ORIGINAL, OR COLLECTED FROM THE LATEST
AND MOST ESTEEMED

ENGLISH ORNITHOLOGISTS;

AND EMBELLISHED WITH

FIGURES,

DRAWN, ENGRAVED, AND COLOURED FROM THE ORIGINAL SPECIMENS.

By E. DONOVAN.

VOL. X.

LONDON:

PRINTED FOR THE AUTHOR; AND FOR F. C. AND J. RIVINGTON, No. 62,
ST. PAUL'S CHURCH-YARD, AND No. 3, WATERLOO-PLACE,
PALL-MALL. 1819.

PLATE CCXXI.

ANAS SPONSA.

SUMMER DUCK, *female and young.*

ANSERES.

GENERIC CHARACTER.

Bill convex, obtufe, the edges divided into lamellate teeth : tongue fringed and obtufe : three fore-toes connected, the hind one folitary.

SPECIFIC CHARACTER
AND
SYNONYMS.

Creft pendent, double, varied with green, blue and white, *male.*

Lefs : body brown, beneath dirty white, varied with brown : and triangular white fpots : *female.*

ANAS SPONSA : crifta dependente duplici viridi, cæruleo alboque varia *mas.*

     Femina minor : corpus fufcum fubtus fordide album fufco varium, maculis triquetris pallidis. *Gmel. Syft. Nat.* 1. *p.* 539*.—*Lath. Ind. Orn.* *n.* 871. 97.

---

* *Femina* minor, pectoris maculis magis obfoletis *Gmel.* 65.

VOL. X.   B   Ana

PLATE CCXXI.

Anas Sponsa. *Linn. Syst.* 1. *p.* 207. 43.

Anas Aestiva. *Briff.* 6. *p.* 351. 11. *t.* 32. *f.* 2. *Id.* 8vo. 11. *p.* 455.

Beau Canard huppé. *Buff.* ix. *p.* 245.—*Pl. Enl.* 980, 981.

Summer Duck. *Catefb. Carol.* 1. *t.* 97.

Edw. t. 101.

Lath. Gen. Syn. 6. *p.* 546. 85.

The male of this very elegant fpecies of the Duck tribe appeared in the firft Plate of the Volume preceding the prefent. It was there obferved that the bird was introduced into the *Britifh Fauna* upon the authority of the Rev. Thomas Rackett, F.L.S. the fpecies having to his knowledge been found fome years ago in a wild ftate in Dorfet-fhire.

We now prefent figures of the female and the young, from a very pleafing group of this interefting web-footed family bred in England, and have thus completed the pictorial elucidation of this particular fpecies.

The female is rather fmaller than the male bird, and is alfo fome-what different in the tints of plumage, befides being deftitute of that flowing pendent creft fo very confpicuous in the male bird. Length about feventeen inches.

PLATE

222

PLATE CCXXII.

ANAS CANADENSIS.

CANADA GOOSE.

ANSERES.

GENERIC CHARACTER.

Bill convex, obtufe, the edges divided into lamellate teeth : tongue fringed and obtufe : three fore-toes connected, the hind-one folitary.

SPECIFIC CHARACTER

AND

SYNONYMS.

Cinereous : head and neck black : chin and throat white.

ANAS CANADENSIS : cinerea, capite colloque nigris, genis gulaque
        albis. *Linn. Syft. Nat.* 1. *p.* 198. 14.—*Gmel.*
        Linn. Syft. 514. 14.—*Phil. Tranf.* 62. *p.* 414.
        46.

Anfer canadenfis fylveftris. *Briff. av.* 6. *p.* 272. *n.* 4. *t.* 26.

L'Oie à cravate. *Buff. Hift. Nat. des. Oif.* 9. *p.* 82.

CANADA GOOSE. *Will. (Angl.) p.* 361. *t.* 70.
        Catefb. Carol. 1. *t.* 92.
        Sloane Jam. 2. *p.* 323. *n.* 6.
        Edw. Av. t. 151.
        Lath. Gen. Syn. 3. 2. *p.* 450. *n.* 14.

B 2   This

PLATE CCXXII.

This fpecies is larger than the common Goofe, its weight about nine or ten pounds, and the length forty fix inches.

It is chiefly in the northern parts of America that thefe birds abound, and especially in Canada: they extend during the fummer feafon as far as Greenland, to the northward, and fouthward in the winter to the Britifh Ifles. As a naturalized fpecies it is obferved ftill further to the fouth, being reared and bred freely in France as well as England. The flefh of the young birds is eaten, and their feathers are held in fome efteem. According to Dr. Latham, on the Great Canal at Verfailles hundreds are feen mixing with the fwans with the greateft cordiality, and the fame at Chantilly; and in England likewife, they are thought a great ornament to the pieces of water in gentlemen's pleafure grounds. They are very familiar.

About Hudfon's Bay this ufeful fpecies breeds in confiderable numbers, though the greater portion of them retire yet more northerly for the purpofes of incubation. Their firft appearance in the bay is from about the middle of April to the middle of May, when the inhabitants wait for them with expectation, being one of the chief articles of their food, and many years kill to the amount of three or four thoufand of them : thefe they falt and barrel for ufe. But thofe birds which they kill in their return from the north, which happens in Auguft, September, and October, they keep frefh for a winter ftore, in the fame manner as they preferve other wild fowl during the winter feafon ; that is by putting them, unplucked of their feathers, into large holes dug in the earth, which they flightly cover with mould, and clofe up the whole with ice and fnow ; and fuch is the feverity of the winter feafon, in that climate, that with this precaution merely, they may be eafily kept frefh for months.

We

PLATE CCXXII.

We may readily conceive that the capture of thefe birds is an object of the utmoft confequence to the Canadians, when we are informed that they diftinguifh the period of their firft arrival in fpring by the name of *Goofe month;* and that they purpofely prepare for that feafon rows of huts made of boughs, at the diftance of a mufket fhot from each other, in thofe fituations where the flights of geefe are expected to pafs. Here the Indians lie in ambufh, and as the geefe fly over they mimic their noife fo well as to entice the geefe within the reach of gunfhot, where each of the Indians being armed with two guns, fire both with all poffible expedition, and they are thus enabled to kill a confiderable number of them; fome good markfmen it is afferted have, in this manner, killed two hundred in one day. The Indians call them *Apiftifkifh*.

223

PLATE CCXXIII.

ANAS CLYPEATA, *fem.*

SHOVELER, *female.*

ANSERES.

GENERIC CHARACTER.

Bill convex, obtufe, the edges divided into lamellate teeth : tongue fringed, obtufe : three fore-toes connected, the hind one folitary.

SPECIFIC CHARACTER

AND

SYNONYMS.

End of the bill dilated, rounded, with an incurved nail.

ANAS CLYPEATA : roftri extremo dilato rotundato, ungue incurvo.
             Linn. Fn. Suec. 119.—*It. Goth.* 167.—*Gmel.*
             Linn. Syft. Nat. p. 518. 19.—*Lath. Ind. Orn.*
             856. 60.
ANAS CLYPEATA.   *Scop. Ann. I.* N° 70.
             Brun. N° 67, 68, 69.
             Borowfk. Nat. 111. *p.* 12. 5.
Anas platyrynchos altera.   *Raii Syn. p.* 143. *A.* 9.
             Will. p. 283.   *Mas.*

                                        Anas

PLATE CCXXIII.

Anas platyrynchos. *Raii Syn. p.* 144. 13.—*Will. p.* 283.—**xv.**—
  Id. 284. xvi. (*Femina.*)

Anas virefcens. *Marf. Dan. v. p.* 120. *t.* 58.

Avis latiroftra. *Klein. Av. p.* 134. 20.

Souchet. *Buff.* ix. *p.* 191.—*Pl. Enl.* 971, 972.

Loeffelente. *Bloch. Schr. der. Berl. Naturf. Fr.* 111. *p.* 373. 17.
  t. 7. *f.* 2.

Shoveler. (*Will. Angl.*) *p.* 370. 15. *male.*—*Id.* 371. 16, 17.
  female.
  Albin I. t. 97, 98.
  Catefb. Car. I. t. 96.
  Lath. Gen. Syn. 6. *p.* 509. 55.

The two fexes of the Shoveler Duck are materially diffimilar in appearance, and differ a little in point of fize: the head and neck of the
male bird is violet green, that of the female brown and fpeckled with
fufcous; and the breaft of the former white with the belly chefnut, and
the vent white; the female is alfo rather fmaller than the male bird.
The length of the male is twenty one inches, and the weight twenty
two ounces.

It appears that the Englifh name of Shoveler has been given to this
bird, in allufion to the very peculiar ftructure of the bill: this is
large, broad, and flattifh, and may be not inaptly compared, in its
general appearance, to the fpatulous portion of a fhovel, and to this
it may be added, that the bird employs it much in the manner of a
fcoop, and with fingular dexterity in catching its food. This confifts
                                                                      of

PLATE CCXXIII.

of the minor tribes of frefh water animals, as fifh and vermes, and infects of the aquatic kind, which latter it takes with facility as they glide over the furface of the water : the frefh water fhrimp in particular is a favourite article of food.

The Shoveler Duck, although an accurately known and well authenticated fpecies, is not confidered as a common bird. Willoughby mentions one that was found at Crowland, in Lincolnfhire; Dr. Latham once received a fpecimen from the London markets, and where alfo we have fometimes feen it. The fpecies is not fuppofed to breed in England.

Buffon fpeaks of their coming into France in February : and obferves that fome remain there during the fummer, and depart in September. It is faid to lay ten or twelve eggs, of a rufous colour upon a bed of rufhes, and in the fame places as the common Teal.

Vol. X.   C   PLATE

224

PLATE CCXXIV.

PODICEPS RUBRICOLLIS, *fem?*

RED THROATED GREBE, *female?*

ANSERES.

GENERIC CHARACTER.

Bill ftrong, flender, fharp pointed : noftrils linear : lore bare of feathers : tongue subbifid : body depreffed : no tail : legs four-toed : fhank compreffed, and befet with a double feries of ferrations behind ; toes lobate.

SPECIFIC CHARACTER

AND

SYNONYMS.

Somewhat crefted : fufcous : chin, cheeks, and region of the ears cinereous : neck beneath, and breaft ferruginous red, abdomen and fecondary quill feathers white.

PODICEPS RUBRICOLLIS : fubcriftatus fufcus, gula genis regioneque aurium cinerafcentibus, collo fubtus pectoreque ferrugineo-rubris, abdomine remigibufque fecundariis albis. *Lath. Ind. Orn. p.* 783. 6.

c 2 COLYMBUS

PLATE CCXXIV.

COLYMBUS RUBRICOLLIS: capite lævi, genis, gula, remigibus fecundariis et abdomine albis, jugulo ferrugineo.
Gmel. Syft. Nat, I. p. 592.

COLYMBUS SUBCRISTATUS. *Jacq. Vog. p.* 37. *t.* 18.
Gmel. Syft. I. p. 590.

COLYMBUS GRISEUS. *Faun Helvet.—Schæf. El. Orn. t.* 29.

Le Grebe à joues grifes. Jourgris. *Buff.* VIII. *p.* 241.—*Pl. Enl.*
931.

RED NECKED GREBE. *Lath. Syn. V. p.* 288. 7.
Id. Suppl. p. 69.
Arct. Zool. 11. *p.* 499. *C.*
Id. Sup. p. 69.

———

In an early part of the prefent work * we gave a plate of the
male fex of this very fcarce and fingular hird, from the fpecimen
originally in the collection of Dr. J. Latham, and which afterwards
came into our own poffeffion. The figure which we now prefent to
the attention of the reader is prefumed to be that of the female: there
is fome difference in the general appearance, which will be beft perceived on a comparifon of the two plates ; to which however it will
be proper to add that the prefent bird is rather fmaller than the
former.

This bird is of fuch confiderable rarity in every country where it
has been obferved, that fome erroneous conclufions have evidently
arifen refpecting it. There are writers who confider it as a variety

---

* Plate 6. Vol. I.

only

PLATE CCXXIV.

only of another fpecies, and others who have defcribed it as a non-
defcript, without having duly afcertained that it had been previoufly
defcribed. It is indeed obvious that the new fpecies *Subcriftatus* of
Jacquin, and which on the authority of that author has found a place
in the Gmelinian *Syftema Naturæ,* is no other than a variety of *Ru-*
bricollis (alfo defcribed in the laft-mentioned work.) It appears likewife to be confounded as a bird allied to Colymbus Urinator by one
writer; while another denominates it as a new fpecies *Grifeus*. And,
laftly, we may add that a variety of it, β of *Latham* feems to conftitute another diftinct fpecies in the work of *Gmelin,* under the fpecific name of *Parotis;* if this conclufion be correct, it is obvious the
varieties of Rubricollis form no lefs than three different fpecies in the
Gmelinian Syftem.

Length about feventeen inches. Found in Northern Europe as
far as the arctic regions.

PLATE

225

PLATE CCXXV.

ANAS STREPERA, *mas.*

GADWALL, *male.*

ANSERES.

GENERIC CHARACTER.

Bill convex, obtufe, the edges divided into lamellate teeth : tongue fringed, obtufe : three fore-toes connected, the hind-one folitary.

SPECIFIC CHARACTER

AND

SYNONYMS.

Wing-fpot rufous, black and white.

ANAS STREPERA : fpeculo alarum rufo nigro albo. *Linn. Fn. Suec.* 21.—*Gmel. Syft.* 520. 20.—*Lath. Ind. Orn.* 859, 69.
Borowfk. Nat. 111. *p.* 12. 6.
Klein. Av. p. 132. 6.
Cet. uc Sard. p. 325.

AN

PLATE CCXXV.

ANAS STREPERA. *Gefn. Av.* 121.

    Aldr. Orn. 3. p. 234.

    Brif. Av. 6. *p.* 339. 8. *t.* 33. *f.* 1,

    Klein. Av. p. 132. 6.

Schnarrente. *Frifch. Av. f.* 168.

GADWALL or GRAY, *Brit. Zool.* 2. *n.* 288,

    Arct. Zool. 2. *p.* 575. L.

    Ray. Av. 1. p. 145. *A.* 2.

    Will. Orn. p. 374. *t.* 72.

    Lath. Gen. Syn. 111. 2. *p.* 515. *n.* 61.

The female Gadwall is already figured in Plate CCXV. of this work; the prefent figure is that of the male bird.

The Gadwall is an elegant bird about the ordinary fize of the Wigeon, and is in particular diftinguifhed by the chafte variety of tints and markings on its plumage, that of the male bird efpecially. The female is rather more obfcure in colour, and is in particular deftitute of thofe elegant fcallops, or femi-circular dark lines which appear confpicuoufly elegant on the neck and breaft of the male bird.

This fpecies is found in England only during the winter months, when the feverity of the feafon compels them to retire fouthward; their habitation in the fummer being northward as far as Sweden, Ruffia, and Siberia: in the firft of which it is known to breed.

Like

PLATE CCXXV.

Like the reft of this tribe, the haunts of this bird are the fens and marſhes, where it reſides among the ruſhes. It is an excellent diver, and as it feeds only in the morning and evening it is not often ſhot. In ſome parts of England it is known by the local name of the Sea Pheaſant.

226

PLATE CCXXVI.

ANAS ALBEOLA.

SPIRIT, OR SPECTRE DUCK.

Anseres.

GENERIC CHARACTER.

Bill convex, obtufe, the edges divided into lamellate teeth : tongue fringed and obtufe : three fore-toes connected, the hind one folitary.

SPECIFIC CHARACTER

AND

SYNONYMS.

White : back and wings black : head and neck fhining, filky blue : hind head white.

ANAS ALBEOLA : alba, dorfo remigibufque nigris, capite colloque
        cærulefcente fericeo nitente, occipite albeola. *Lath.*
        Ind. Orn. 866. 86.

ANAS ALBEOLA : *Linn. Syft. I. p.* 199. 18.
        Phil. Tranf. LXII. *p.* 416. 18.
        Gmel. Syft. I. p. 517.

D 2

Sarcelle

PLATE CCXXVI.

Sarcelle blanche et noire, ou la Religieuse. *Buff.* ix. *p.* 284.—
Pl. Enl. 948.

Little Black and White Duck. *Edw. t.* 100·

Spirit Duck, Spectre Duck. *Arct. Zool.* 11. *N°.* 487.—*Lath.*
Gen. Syn. vi. 533. 75.

———————

Very rare in Britain.

Our specimen is rather smaller than the species is usually described, the length of our bird being only thirteen inches; while its ordinary length according to some authors is about sixteen inches. There can, however, be no distrust respecting the identity of the species, nor is there any doubt of our example being of the male sex.

The female is somewhat less than the male bird, and the prevailing colour of the plumage brown. The species is chiefly an inhabitant of the Arctic regions.

PLATE

227

PLATE CCXXVII.

TURDUS TORQUATUS, *fem.*

RING OUZEL, *or* ROCK OUZEL, *female.*

PASSERES.

GENERIC CHARACTER.

Bill, ſtraightiſh: the upper mandible a little bending and notched near the point: noſtrils naked, or half covered with a ſmall membrane: mouth ciliated with a few briſtles at the corners: tongue jagged.

SPECIFIC CHARACTER

AND

SYNONYMS.

Blackiſh, collar white, bill yellowiſh.

TURDUS TORQUATUS: nigricans, torque albo, roſtro flaveſcente.
> *Fn. Suec.* 221.

> *Scop. Ann.* 1. *p.* 198.

MERULA TORQUATA. *Geſn. Av.* 607.

> *Aldr. Orn.* 2. *p.* 620. *f.* 621. 622.

> *Briſſ. Av.* 2. *p.* 235. *n.* 12.

MERULA

PLATE CCXXVII.

MERULA CONGENER. *Raj. Av. p.* 67. *n.* 12.

Will. Orn. p. 195.

Merle à plaftron blanc. *Buff. Hift. Nat. des Oif.* 3. *p.* 340. *t.* 31.

—*Pl. Enl. n.* 516.

RING OUZEL or AMSEL. *Ray. Av. p.* 65. *n.* 2.

Will. Orn. p. 194.

Lath. Gen. Syn, 11. *I. p.* 46. *n.* 49.

———

Length eleven inches. The prefent figure is that of the female, which differs fo materially from the male, that we were induced to think it would be confidered as an acceptable addition to that of the male bird. The latter will be found in the Third Volume, Plate LXI.

Thefe birds are found chiefly in high and mountainous fituations.

PLATE

PLATE CCXXVIII.

ANAS CYGNUS.

WILD SWAN.

ANSERES.

GENERIC CHARACTER.

Bill convex, obtufe, the edges divided into lamellate teeth : tongue fringed, obtufe : three fore-toes connected, the hind one folitary.

SPECIFIC CHARACTER

AND

SYNONYMS.

Bill femicylindrical, black : cere yellow : body white.

ANAS CYGNUS: roftro femicylindrico atro, cera flava, corpore albo.
Linn. Fn. Suec. 107. — It. Weftgoth. 143.—
Scop. Ann. I. n. 66.—Kramer el. p. 338.—
Georg. it. p. 165.

CYGNUS FERUS. Briff. Av. 6. p. 292. n. 12. t. 18.—Bell. Av.
30. a.
Gefn. Av. 371. t. 372.
Aldr. Orn. 3. p. 10. t. 8.
Raj. Av. p. 136. A. 2.

Cygne

PLATE CCXXVIII.

Cygne fauvage. *Buff.* ix. *p.* 3.—*Pl. Enl.* 913.

WILD SWAN. *Will. (Angl.) p.* 356. *t.* 69.

Brit. Zool. 11. N° 264.

Phil. Tranf. LVI. *t.* x. *p.* 215. *f. I.* 2.

WHISTLING SWAN. *Arct. Zool.* 11. N° 469.—*Fl. Scot. I.* N°
204.—*Lath. Gen. Syn.* vi. *p.* 433.—*Id. Supp.*
p. 272.

The wild and tame fwan are confidered by the beft informed Ornithologifts as two diftinct fpecies : the former is found only in a ftate of uncultivated nature, having never been yet reduced to the bondage of domeftication, while the latter, foftered and protected by the hand of man, has been rendered fubfervient, if not ufeful, and thus repaid in an ample manner the pains beftowed upon its cultivation .

From a fimilarity of names, and the ideas we affociate in general to the appellatives of wild and tame, it might be readily concluded that the tame fwan muft be the domefticated offspring of the former ; but this is not the cafe : though nearly allied, they are obvioufly diffimilar, and offer characters that we can fcarcely hefitate to confider as fpecifically diftinct.

In the firft place, it will be obferved, that the wild fwan is fmaller than the tame kind : nor does this arife from the effect of domeftication merely, for both the fpecies are found in a ftate of wildnefs in the northern part of Europe, America, and Afia ; and in all thofe parts are known with fufficient accuracy to difpel doubt in this particular.

In

PLATE CCXXVIII.

In the conftruction of the bill there is a further difference, the bafe at the forehead being fmooth and entirely deftitute of the callofity or knob which we at once perceive upon that part in the tame or culti-vated fwan.

And befides this there is yet another very principal diftinction, the tame fwan being mute, or at leaft the found which it emits being nothing more than a hiffing noife, like that of the goofe; while the wild fwan, on the contrary, has a loud and piercing cry, which has been compared to that of a whiftle, and this it can exert with fuch effect, that it is affirmed it may be diftinguifhed when a flight of wild fwans is paffing overhead in the aerial regions, at fuch a prodigious elevation from the earth, as to be invifible to the naked eye.

It is from this laft mentioned circumftance that the wild fwan has obtained the more expreffive epithet of the whiftling fwan, and the other has been denominated the mute fwan.

The latter mentioned difference is alone fufficient to juftify the con-clufion of the two birds being fpecifically different. This diffimilarity in the voice arifes it has been long fince afcertained, from the anato-mical conformation of the wind-pipe; and which in the whiftling fwan is altogether fingular. The pipe enters the cheft at firft a little way only, and is then reflected into the form of a trumpet, after which it again enters, and then dividing into two branches proceeds to join the lungs. In the mute fwan the wind-pipe enters at once into the lungs, and hence there can be little doubt that the remarkable difference in the founds they emit arifes from this difference in the ftructure.

The whiftling fwan is about five feet in length, the mute fwan when

PLATE CCXXVIII.

full grown about fix feet. A material diftinction of the whiftling fwan confifts alfo in the colour of the bill as well as form; it is about three inches long, and from the bafe to the middle pale yellow, the other half black, and the legs are black, inclining fometimes to reddifh. Like the mute fwan the whiftling kind exhibits a tranfition of colour, from deep cinereous brown to pure white, as it proceeds through the different ftages of its growth. Our prefent figure is defigned to ex-hibit one of thefe tranfitions, the drawing being taken from a very fine fpecimen of the bird, and which had attained its full and perfect fize without having yet affumed the whitenefs of the older birds. Its ap-pearance when the plumage has become entirely white may be eafily conceived from that of the mute fwan, delineated at the conclufion of the prefent Volume *.

The wild or whiftling fwan is of the gregarious kind, affociating together in flocks of eight or ten in unmber; it is in fuch flocks that thefe birds ufually vifit the northern parts of Britain during the winter: unlefs, however, in feafons of particular feverity, they are never feen more foutherly with us than the Scottifh Ifles, and then very rarely in flocks of more than five or fix in number. In Iceland they affociate in larger flocks, and appear in ftill greater numbers towards the north-ward, where they breed, as in Lapland, the deferts of Tartary and Siberia as far as Kamtfchatka. A few breed in the Weftern Ifles. Von Troil informs us, that they alfo breed in Iceland, and that the greater part of the young brood ftay there the whole year, frequenting the lakes in fummer, and in the winter removing to the fea fhore. Their habits are the fame in America, the lakes to the fouthward abounding with them during the fummer, in the winter they appear upon the fea

---

* Plate CCXXIV.

coaft.

PLATE CCXXVIII.

coaft. It appears to be a vagarious bird, fpreading in fmall flocks about the Cafpian Sea, the Euxine Sea, Greece, and Egypt, though they are never known to pafs fouthward beyond the equator.

They appear about Hudfon's Bay in company with others of the goofe tribe, about the end of May : they lay four eggs, and hatch in July. The eggs as well as the young birds are efteemed very palatable food ; and the fkins dried are worn by the Indians with the feathers and down attached, the larger feathers being plucked to form the diadems of their chiefs, or to weave into cloaks and other articles of cloathing.

E 2   PLATE

PLATE CCXXIX.

LARUS CATARACTES.

SKUA GULL.

ANSERES.

GENERIC CHARACTER.

Bill ftraight, fharp edged, a little hooked at the top and without teeth lower mandible gibbous below the point: noftrils linear, broader on the fore part and placed in the middle of the bill.

SPECIFIC CHARACTER

AND

SYNONYMS.

Greyifh; quill and tail feathers white at the bafe; tail fomewhat equal.

LARUS CATARACTES: grifefcens, remigibufque bafi albis, canda fubæquali. *Gmel. Syft. Nat.* 603. 11.—*Linn. Syft. I. p.* 226. 11.
Lath. Ind. Orn. 818. 12.

Catarraĉtes et Catarraĉta, *Raii. Syn. p.* 128.

Catarraĉtes

PLATE CCXXIX.

Catarra&tes nofter. *A. 6.—Id.* 129. 7.

     Will. p. 265.*—Id. (Angl.) p.* 348. 349. *t.* 67.—

     Sibbald Scot. pars 2. *I.* 111. *p.* 20. *t.* 14. 2.

Cathara&ta Skua, *Brun. N°.* 125.*—Mull. N°.* 167.

Skua Hoyeri. *Cluf. exot.* 369.

Larus fufcus. *Brif.* 6. *p.* 165. 4.*—Id. 8vo.* 11. *p.* 405.

Le Goéland brun. *Buff.* viii. *p.* 408.

Brown Gull. *Albin.* 11. *t.* 85.

Catarra&tes or Cornifh Gannet. *Raj. Av. p.* 129. *n.* 7.*—Will. Orn.*

     p. 349. *t.* 67.

Skua Gull. *Brit. Zool.* 2. *n.* 243.*—Ar&t. Zool.* 2. *p.* 531. *A.—*

     Lath. Gen. Syn. vi. *p.* 385. 14.

The Skua Gull is one of the moft fierce and voracious fpecies of the Gull tribe: the general colour cinereous brown, varied with rufty, and fufcous longitudinal fpots: the bill and talons ftrong, and the whole afpe&t fingularly ferocious and gloomy.

This bird is of a large fize and very powerful: it feeds on fifh, and on the fmaller tribes of birds that frequent watery fituations where it lives and breeds. Length two feet.

It is an inhabitant of Europe, Afia, and America: in Britain the fpecies is confidered rare, being a local kind and confined chiefly to Cornwall, and the Hebrides.

PLATE

PLATE CCXXX.

ALCA ALLE, *fem.*

LITTLE AUK, *female.*

GENERIC CHARACTER.

Bill toothlefs, fhort, compreffed, convex, often tranfverfely furrowed; lower mandible gibbous near the bafe: noftrils linear: legs moftly three toed.

SPECIFIC CHARACTER

AND

SYNONYMS.

Bill fmooth, conic; beneath, and tips of the hind quill feathers white: legs black.

ALCA ALLE: roftro lævi conico, abdomine toto fubtus remigumque pofticarum apicibus albas, pedibus nigris. *Linn. Fn. Suec.* 142.—*Gmel. Syft. Nat.* 554. 5.— *Lath. Ind. Orn.* 795. 10.

Alca Alle. *Faun. Groenl.* N°. 54.

URIA

PLATE CCXXX.

Uria Minor. *Briff.* vi. *p.* 73. 2.—*Id. 8vo.* 11. *p.* 378.

Plautus Columbarius, *Klein. Av. p.* 146. *I.*

Mergulus melano leucus roftro acuto brevi, *Raii. Syn. p.* 125. *A.* 5.
Will. p. 261. *t.* 59.

Le petit Guillemot. *Buff.* ix. *p.* 354.—*Pl. Enl.* 917.

Small black and white Diver. *Will. (Angl.) p.* 343.

Greenland Dove. *Albin. I. t.* 85.

Little Auk. *Brit. Zool.* 2. *N°.* 233. *t.* 82.—*Arct. Zool.* 2. *t.*
429.—*Lath. Syn. V. p.* 327. 11.

This is the fmalleft fpecies of the Auk tribe found upon the Britifh coafts, and is alfo one of the leaft common. In point of fize it fcarcely exceeds the Ouzel; the length is nine inches. The prefent figure is that of the female ; which is very fcarce.

Alca Alle is an inhabitant of the icy regions of the north of Europe, and America, from whence it migrates to the fouth in the winter feafons; and in its return northward in the enfuing fummer a few remain and breed in the north of Britain.

Thefe birds have been found occafionally with the plumage wholly white.

PLATE

PLATE CCXXXI.

ANAS CYGNOIDES.

CHINESE GOOSE.

ANSERES.

GENERIC CHARACTER.

Bill convex, obtufe, the edges divided into lamellate teeth : tongue. fringed and obtufe : three fore-toes connected, the hind one folitary.

SPECIFIC CHARACTER

AND

SYNONYMS.

Bill femi-cylindrical; cere gibbous; eye-lids tumid.

ANAS CYGNOIDES: roftro femi cylindrico, cere gibbosa palpebris
tumidis. *Linn. Fn. Suec.* 108.
Gmel. Linn. Syft. Nat. 502. 2.
Anfer Guineenfis. *Briff.* VI. *p.* 280. 7.—*Id. 8vo.* 11. *p.* 435.
L'Oie de Guinée. *Buff.* IX. *p.* 72. *t.* 3.—*Pl. Enl.* 347.
Spanifh Goofe. *Albin. Av. I. t.* 91.
Swan Goofe. *Raj. Av. p.* 138. *n.* 8.
Will. Orn. p. 360. *t.* 71.

VOL. X.   F   Chinefe

PLATE CCXXXI.

Chinese Goose. *Arct. Zool.* 2. *p.* 571. E.

Lath. Gen. Syn. 111. 2. *p.* 447. *n.* 12.

This handsome species of Goose has been long since introduced, and naturalized with success in Britain. It does not appear to have been ascertained with any degree of accuracy whence the parent stock of this useful bird was derived, nor can any inference be drawn from the local names which various authors have assigned to it.

Linnæus calls it the Southern Goose: Brisson and Buffon the Goose of Guinea: Albin the Spanish Goose: Brown the West Indian, or Jamaica Goose; and Pennant the Chinese Goose.

That the species is found in every one of the above-named parts of the world we have little doubt, and this may offer some apology for the diversity of local names which the earlier writers have assigned to it. It is indeed to be confessed that this diversity of names had better been avoided, as it is scarcely possible but that some confusion may arise occasionally from this source, as to the identity of the species; and we must allow that it is always objectionable to indulge in the fantastic introduction of new names in the science of Natural History where others already applied are well known and sufficiently established to be understood. We believe upon the best information that the species abounds in a state of nature in the eastern parts of Siberia, where it frequents lakes and rivers, and that it occasionally migrates from thence to other countries of Asia, Africa, and Europe, and

hence

PLATE CCXXXI.

hence the fpecies might be perhaps diftinguifhed with more propriety by the appellation of the Siberian Goofe than any it bears at prefent.

In point of fize the prefent fpecies exceeds that of the ordinary Goofe, and approaches nearer to the Swan; the length being more than three feet: the annexed figure we truft in this cafe as in moft others, will be found at leaft fufficiently accurate to fuperfede the neceffity of entering into any minute detail refpecting the colours and markings by which the plumage is diftinguifhed.

F 2

PLATE

232

PLATE CCXXXII.

ANAS BERNICLA.

BRENT GOOSE,

ANSERES.

GENERIC CHARACTER.

Bill convex, obtuſe, the edges divided into lamellate teeth : tongue fringed and obtuſe : three fore-toes connected, the hind one ſolitary.

SPECIFIC CHARACTER

AND

SYNONYMS.

Fuſcous : head, neck, and breaſt black : collar white.

ANAS BERNICLA : fuſca, capite, collo pectoreque nigris, collari albo. *Linn. Fn. Suec.* 115.—*Gmel.* 513. 13.— *Scop. Ann. I.* N°. 84.—*Brun.* N°. 52.—*Friſch. t.* 156.—*Faun. Groenl.* N°. 41—.*Borowſk. Nat.* 111. *p.* 11. 3.—*Lath. Ind. Orn.* 844.

BRENTA, *Briſ.* vi. *p.* 304. 16. *t.* 31.—*Id.* 8*vo.* ii. *p.* 442.—*Raj. Syn. p.* 130. 8.—*Will. p.* 275. *t.* 69.—*Klein. Av. p.* 130. 8.

Le

PLATE CCXXXII.

Le Cravaut. *Buff.* IX. *p.* 87.—*Pl. Enl.* 342.

BRENT GOOSE, BRAND GOOSE. *Brit. Zool.* II. N°. 270.—
   Albin. I. *t.* 93.—*Will. (Angl.) p.* 360.
   Lath. Gen. Syn. VI. *p.* 467. 27.

Length, about twenty-one inches: the prevailing colour of the head, neck, and upper part black in the male, in the female brownifh, and in the younger birds, the collar of white fpots more or lefs confpicuous, and fometimes wholly wanting. The lower part of the breaft, fcapular and wing coverts afh colour, clouded with a darker fhade: tail dufky: legs black.

Thefe birds inhabit the north of Europe and America, where they frequent the iflands, and along the coafts, but are never obferved to fly inland: towards winter they proceed fouthward. They vifit the Scottifh ifles in great numbers: at Shetland are known by the name of HORRA *Geefe,* from being found in the Sound of Horra. In the winter time they occafionally occur in fome plenty about the fens and marfhes of Lincolnfhire, where they are brought to the London markets. The flefh is confidered palatable, unlefs when it partakes of a fifhy tafte, which happens when the birds have fubfifted for fome time upon the finny inhabitants of the water: at other times its ordinary food confifts of plants of various kinds, berries, and even grafs: worms, and fmall teftaceous animals which occur in the marfhes.

In the defcription of the Bernacle Goofe, Anas Erythropus (Plate CCXIV. of the prefent work,) we were led to notice the confufion

that

PLATE CCXXXII.

prevails among the Synonyms of the Bernacle Goose, the Brent Goose, and the White-fronted Goose, all three being occasionally confounded through the misapplication of the Synonyms, and each in its turn supposed to be the bird known among the credulous writers of the last age under the name of the Tree Goose. As the explanation there afforded will apply equally to the present bird as to Anas Erythropu , it may not be deemed impoper to introduce an extract from it in this place, referring at the same time for a more full detail to the description of Plate CCXIV. in the preceding volume. We there observed " The result of this confusion is, that, we find a Bernacle Goose which is not a Bernacle, but the Linnæan Erythropus: the Brent Goose, which is not the Bernacle, confounded under the trivial name of Bernicla ; and some of the Synonyms of the true Bernacle, applied to the White-fronted Goose, Anas Albifrons."

" It will tend materially to dispel this seeming confusion by bearing in mind, that the species Erythropus of Linnæus is the true Bernacle Goose, but which is better characterized by the specific definition of Dr. Latham, than by the Linnæan character, ' cinerea fronte alba.' *Faun. Suec.* At the same time reeollecting, that notwithstanding the similarity of names, Anas Bernicla of Linnæus is the Brent and not the Bernacle Goose." *Vide* Plate CCXIV.

PLATE

233

PLATE CCXXXIII.

FALCO APIVORUS, *var.*

HONEY BUZZARD, *var.*

ACCIPITRES.

GENERIC CHARACTER.

Bill hooked, the bafe covered with a cere: head covered with clofe fet feathers: tongue bifid.

* *Legs naked, and lefs.*

SPECIFIC CHARACTER

AND

SYNONYMS.

Cere black: legs half naked and yellow: head cinereous: tail banded with cinereous, tip white.

FALEO APIVORUS: cera nigra: pedibus feminudis flavis, capite cinereo, caudæ fafcia cinerea, apice albo. *Linn. Syft.* 1. *p.* 130.—*Faun. Suec.* N° 65.—*Gmel. Syft.* 1. *p.* 267.—*Raii Syn. p.* 16. 2.—*Will. p.* 39. *t.* 3.—*Lath. Ind. Orn.* 25. N° 52.

VOL. X.   G   Pojana.

PLATE CCXXXIII.

Pojana. *Linnan. p.* 84. *t.* 13. *f.* 75.

HONEY BUZZARD. *Br. Zool.* 1. *p.* 85. 6.—*Arct. Zool.* 11. *p.* 224. I.—*Albin.* 1. *t.* 2.—*Will. (Angl.) p.* 78. *t.* 3.— *Lath. Syn.* 1. *p.* 52.—*Id. Sup. p.* 14.

In the early part of this work we have already introduced a figure of the Honey Buzzard*, from a fpecimen originally in the collection of Dr. Latham; and the only one as that experienced Ornithologift informs us, with all his affiduity he was ever able to procure in a recent ftate; nor could he even afcertain the fex. And with refpect to the bird before us we have to acknowledge ourfelves under the fame uncertainty: it is introduced as an extraordinary variety, the plumage being throughout of a much darker hue than in the former bird.

Our prefent fubject is rather fmaller than the bird we have before delineated, and may prefent a tranfition in the growth of this particular fpecies; but this is uncertain, and we are rather inclined to think it an adult variety than a younger bird.

That thefe birds vary in appearance may be readily collected from authors; in the fpecimen figured by Albin, the tail was uniformly of one colour, without any bars, Linnæus defcribes it as having only one cinereous band, and the tip white, and the Britifh Zoology fpeaks of

---

* Plate XXXI. Vol. II.

three

PLATE CCXXXIII.

three dufky bars upon the tail; fometimes there are only two, and occcafionally the white at the tips is wanting.

Length from twenty inches to two feet: the fpecies feed on reptiles, mice, and other fmall animals, and is efpecially fond of bees, whence its name of Apivorus.

G 2   PLATE

234

PLATE CCXXXIV.

PLATALEA LEUCORODIA.

WHITE SPOONBILL.

GRALLÆ.

GENERIC CHARACTER.

Bill long, thin; the tip dilated, orbicular, flat: noftrils fmall, at the bafe of the bill: tongue fhort, pointed: feet four-toed, femi-palmate.

SPECIFIC CHARACTER

AND

SYNONYMS.

Body white: chin black: hind head fubcrefted.

PLATALEA LEUCORODIA: corpore albo: gula nigra, occipite fubcriftato. *Linn. Faun. Suec.* 160.—*Gmel. Syft. Nat.* 613. 1.—*Scop. Ann.* 1. N° 115.— *Brun.* N° 46.—*Sepp. Vog. t.* 88, 89.—*Klein. Av. p.* 126. *I.*—*Lath. Ind. Orn.* 667. *I.*

Platea, five Pelecanus. *Aldr.*—*Raii Syn. p.* 102. *I.*—*Will. p.* 212. *t.* 52.—*Briff.* v. *p.* 352. *I.*

La Spatule. *Buff.* VII. *p.* 448. *t.* 24.—*Pl. Enl.* 405.

Garza, o Becarivale. *Zinnan. Nov. p.* 111. *t.* 20. *f.* 99.

SPOON BILL, or Pelican. *Albin* 11. *t.* 66.

White

PLATE CCXXXIV.

WHITE SPOONBILL. *Brit. Zool. App. t. 9.—Arct. Zool.* 11. p. 441. *A.—Id. Sup. p. 66.—Lath. Gen. Syn.* v. p. 13. *I.—Pultney Catalogue, Dorfet. p.* 13.

A flock of thefe birds migrated into the marfhes near Yarmouth, in Norfolk, in the month of April, 1774, and upon the authority of this circumftance the fpecies firft obtained a place in the Britifh Fauna. They were obferved by Mr. Jofeph Sparfhall of Yarmouth, who tranfmitted a minute account of one of the birds to Mr. Pennant, the particulars of which are inferted in the Appendix to his Britifh Zoology.

Whether this fpecies had been previoufly afcertained as a Britifh bird feems rather doubtful. Ray informs us only that in his time they bred annually in a wood at Sevenhuys, not remote from Leyden, to which Mr. Pennant adds, that the wood is now deftroyed; and that thofe birds, with feveral others that formerly frequented the country, are at prefent become very rare. Albin give a figure of the bird which he faw in the poffeffion of Mrs. Legrand, but it is fufficiently plain from the tenor of his obfervations, he confidered it as a foreign bird. "In a certain grove (fays Albin) at a village, called Sevenhuys, not far from Leyden in Holland, they build and breed yearly in great numbers, on the tops of the high trees, where are alfo Herns and Night Ravens," &c. When the young ones are almoft fledged *, thofe

---

* This obfervation confirms the conjecture of Dr. Latham, who imagines the young birds (which are confidered as an article of food) are taken before they can fly, " for Willoughby," he obferves, " talks of their being fhaken out of the neft with a crook faftened to the end of a pole." *Vide Orn. p.* 289.

that

PLATE CCXXXIV.

that farm the grove, with hooks on the tops of long poles pull them down.——The bird is called by the Low Dutch, *Leplaer*, that is *Spoonbill*."

We have been informed that in very fevere winters this bird is fometimes obferved in Britain ; one of the beft authenticated inftances of this, occurred about twenty years ago, when a fingle bird which had been taken with other wild fowl in the marfhes, it is believed of Lincolnfhire, was brought to London and expofed for fale in Leadenhall Market. We are, indeed, affured that the example in our own poffeffion, the one from which the prefent figure is taken, was captured in the Hackney marfhes about eleven years ago, but we are unwilling to fpeak with too much confidence as our communicant is no more, and every means of afcertaining the particulars has faded with him. A very intelligent Naturalift, the late Dr. Pultney, in his " Catalogue of Dorfetfhire," has introduced the fpecies among the feathered tribes of that County. And, laftly, Mr. Montagu records the capture of two fpecimens within a few miles of Kingfbridge, Devonfhire, one was fhot in November, 1804, the other on the fixteenth of March, 1807, and it is remarked that both were killed with the common Heron.

The length of this bird is two feet eight or nine inches when at the full growth ; the bill large, long and flat, with the end fpatulous, or in the fhape of a fpoon ; whence its name of Spoonbill. The colour of the bill is various, being in fome birds black, in others brown, and upon being clofely viewed, appears rather of a yellow colour, varied and clouded with olive, and thickly fpeckled with darker, and the legs are varied in a manner fomewhat fimilar, though at the firft glance they

PLATE CCXXXIV.

they appear dark, and uniformly of one colour. The plumage is white throughout, or in the younger birds the quill feathers are tipped with black.

The fpecies is an inhabitant of various parts of Europe, being found as far north as Iceland, and to the fouth, according to Kolben, even to the Cape of Good Hope. They, preferring as it appears the more temperate climates, frequent the entrance of rivers about the fea coafts, where they build in lofty trees; the female lays three or four eggs the fize of thofe of a hen, and of a white colour fpeckled with red. They are faid to be very noify in the breeding feafon, and to feed on frogs, fnakes, and fifh, teftaceous animals, and plants: grafs, weeds, and the undigefted remains of the common ftickleback have been found in the ftomach on diffeftion.

The flefh is of a remarkably deep colour, but is reputed favoury, and without any fifhy tafte. The latter circumftance we fhould how-ever conclude, muft depend upon the nature of its food for fometime previous to its capture, for when it has been conftrained to fubfift for any confiderable period upon fifh it is fcarcely to be doubted that the flefh will imbibe that flavour.

PLATE

235

PLATE CCXXXV.

CUCULUS CANORUS, *jun.*

COMMON CUCKOW, *young.*

Picæ.

GENERIC CHARACTER.

Bill fmooth, a little curved: noftrils furrounded by a fmall rim: tongue fagittate or arrow fhaped, fhort, and pointed: feet formed for climbing.

SPECIFIC CHARACTER

AND

SYNONYMS.

Cinereous; beneath whitifh, tranfverfely ftreaked with brown: tail rounded, blackifh, and dotted with white.

Young. Body above fufcous, the margin of the feathers white: beneath banded with white and fufcous.

Cuculus Canorus: cinereus, fubtus albidus fufco tranfverfim ftriatus, cauda rotundata nigricante albo punctata. *Lath. Ind. Orn.* 207. *I.*

Vol. X.   H   Cuculus

PLATE CCXXXV.

Cuculus Canorus: cauda rotundata nigricante albo-punctata. *Linn.*
Faun. Suec. 96.—*Scop. Ann. I. p.* 44. *n.* 48.
Nozem, Nederl. Vogel. t. 61.
Il Cuculo, *Olin uc. t. p.* 38.
Coucou, *Buff.* vi. *p.* 305.—*Pl. Enl.* 811.

———————————

The plumage of the Cuckow in its early ftate of growth is fo very different from that of the adult bird, that we are entirely perfuaded no one unconverfant with the tranfitions it is known to undergo, would be eafily induced to believe it the fame bird. Neither can it efcape the obfervation of the accurate Ornithologift that fome errors have arifen on this fubject in the volumes of the early naturalifts before its hiftory was fully underftood. In this ftage of growth the common Cuckow is affuredly the red and rufous Cuckow of certain ancient writers; and even fome among the better authors of our own times are not wholly free from this or fimilar errors.

In the 29th Plate of the Second Volume of this work, the adult bird has appeared already; and the younger bird exhibiting the rufous variegated plumage, is the fubject chofen for our prefent Plate.

PLATE

236

PLATE CCXXXVI.

ANAS ANSER.

GREY-LAG GOOSE.

ANSERES.

GENERIC CHARACTER.

Bill convex, obtufe, the edges divided into lamellate teeth: the tongue fringed obtufe : three fore-toes conne&ted, the hind one folitary.

SPECIFIC CHARACTER

AND

SYNONYMS.

Bill femicylindrical, body above cinereous, beneath paler ; neck ſtriated.

ANAS ANSER : *(Ferus)* roſtro femicylindrico corpore fupra cinereo
fubtus pallidiore, collo ſtriato. *Fn. Suec.* 114.—
Linn. Syſt. I. p. 197. 9.—*Gmel. Linn. Syſt.*
510. 9.—*Lath. Ind. Orn.* 841. 26.—*Schaeff.*
El. t. 20.—*Raii. Syn. p.* 136. A. 4.—138. A. 3.
Anfer Sylveſtris, *Briſſ.* vi. *p.* 265. 2.—*Id. 8vo.* ii. *p.* 432.

H 2   Oie

PLATE CCXXXVI.

Oie-fauvage, *Buff.* ix. *p.* 30. *t.* 2.—*Pl. Enl.* 985.

Uæs Araki, *Forfk. Faun. Arab. p.* 3. N°. 6.

Oca Sylvatica, *Zinnan. Uov. p.* 104. *t.* 17. *f.* 91.

Wild Goofe, *Albin. I. t.* 90.—*Will. (Angl.) p.* 358.

Grey-lag Goofe, *Lath. Gen. Syn.* vi. *p.* 459 . 31.

The wild, or grey-lag Goofe is the undomefticated ftate of the tame or common Goofe of our poultry yards: a fpecies that inhabits the fens of England, and is fuppofed to remain with us during the whole year. It is at leaft well known that great numbers breed in the fens of Cambridge and Lincolnfhire, where they remain throughout the fummer, the feafon when all the birds that migrate northward are found in higher latitudes.

Thefe birds are tolerably prolific, producing feven or eight young at a brood; numbers are caught during the winter feafon in the decoys, many are killed for the fupply of the markets, and others put among the tame Geefe where they are eafily rendered tame, and become part of the domeftic ftock. In a wild ftate they affociate in flocks, and in their aerial flights are obferved to go forward in a ftraight line, or more frequently in a cuneated group or wedge, one point foremoft; the flock, whether in a ftraight line or wedge, being preceded by a leader.

The wild Goofe is an inhabitant of various parts of the world, but appears in other countries to be more of the migratory kind than with us: and is fometimes met with in flocks of three or four hundred. It occurs in Iceland, and on the continent from Lapland to the Cape of Good

PLATE CCXXXVI.

Good Hope. Kolben fpeaks of it as common in Arabia, Forfchal in Perfia, and Kæmpfer in China and Japan: it is alfo found in America, from Hudfon's Bay to South Carolina; and was obferved likewife by our navigators in fome of the iflands of the South Seas.

The wild Goofe is rather lefs than the tame Goofe, and weighs about ten pounds, while in a ftate of tamenefs it frequently attains to the weight of fifteen or twenty pounds, and often more: the length is about two feet nine inches, and the breadth five feet. The bird we have chofen for the annexed figure is rather above the ufual ftand-ard, meafuring from the tip of the bill to the end of the tail, twenty-nine inches and a half, and was confidered, in refpect to plumage, a very perfect bird.

PLATE

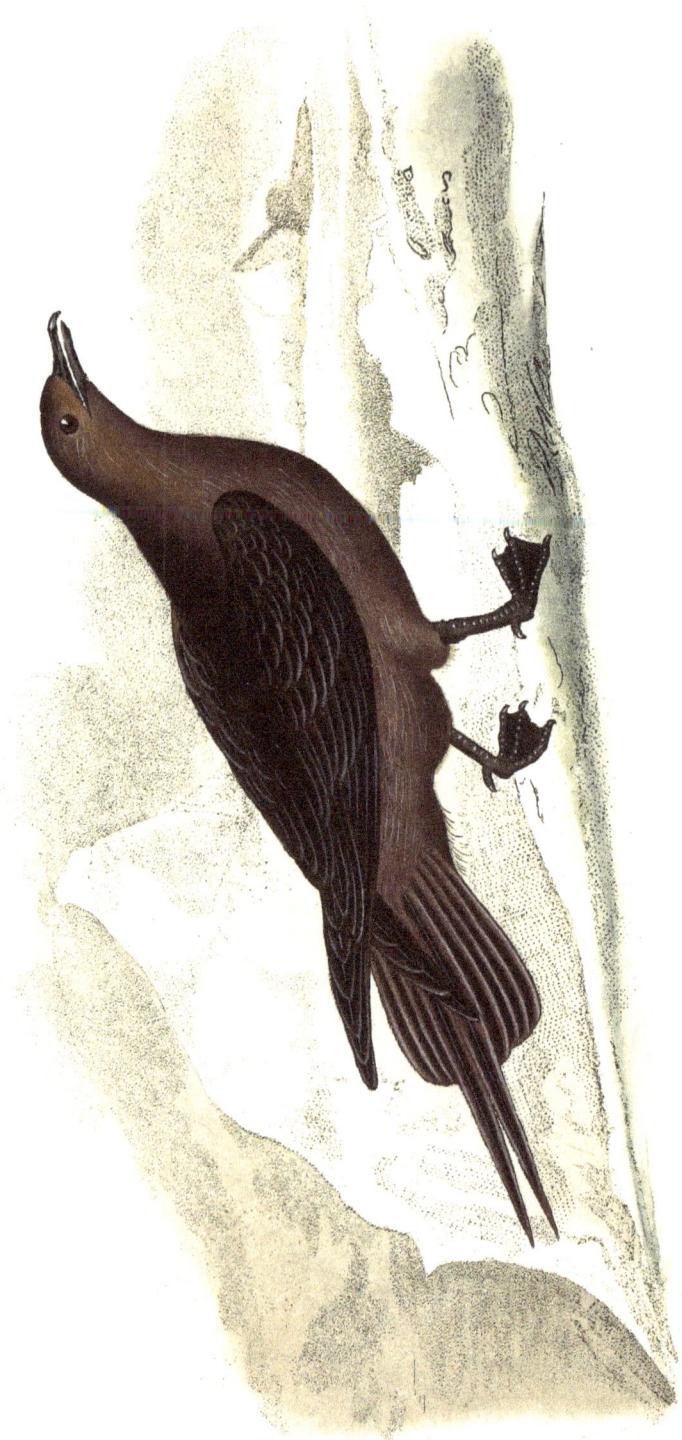

PLATE CCXXXVII.

LARUS PARASITICUS.

ARCTIC GULL.

ANSERES.

GENERIC CHARACTER.

Bill ftraight, fharp edged, a little hooked at the tip, and without teeth; lower mandible gibbous below the point: noftrils linear, broader on the fore part and placed in the middle of the bill.

SPECIFIC CHARACTER

AND

SYNONYMS.

Above black: collar, breaft, and abdomen, white: two middle tail feathers very long.

LARUS PARASITICUS: fupra niger, collo pectore et abdomine albis, rectricibus duabus intermediis longiffimis. *Lath. Ind. Orn.* 819. 15.

LARUUS PARASITICUS: rectricibus duabus intermediis longiffimis. *Linn. Faun. Suec.* 156.—*Gmel.* 601. 10.

Sterna

PLATE CCXXXVII.

Sterna re&ricibus maximis nigris. *It. Wgoth*. 182.—*Act Stockh*.
   1753. *p*. 291.

Cathara&a parafitica. *Brun. N.* 127, 128.—*Faun. Groen.* N° 68.

Larus fubfufcus major, &c. *Brown Jam*. 482.

Stercorarius longicaudus. *Brif.* vi. *p*. 155. 3. *(male)*.—*Stercorarius*
   Brif. vi. *p*. 150. *(female)*.

Sterna re&tricibus maximis nigris. *It. Wgoth*. 182.—*Act. Holm*.
   1753. *p*. 291.

L'Abbe à longue queue. *Buff.* viii. *p*. 445,—*Pl. Enl*. 762.

Arctic Birds. *Edw. t*. 148. 149.

Arctic Gull. *Br. Zool.* 11. N° 245. *t*. 87.—*Arƈt. Zool.* 11.
   n. 459.—*Lath. Gen. Syn.* vi. *p*. 389. 16. *t.* 99.

This is one of the rareft fpecies of the Gull tribe in Britain: it is
found to inhabit the Weftern Ifles. The male is diftinguifhed by the
length of the two middle tail feathers: thefe middle tail feathers in
the female are rather longeft, but by no means fo confpicuoufly different from the reft as in the male bird.

This is not the ftrongeft or moft powerful of the Gull tribe found
on our coaft, being much inferior to the Skua or the Wagel, and not
in point of fize exceeding fome others, but it is fierce and rapacious,
and does not hefitate to attack the other Gulls with impunity.
Too indolent to purfue and catch its own prey, it lies in wait upon
the rocks, watching with apparent unconcern thofe birds fwimming
and diving in the fhallows of the water, or turning up the fands in
queft of food, and when by thefe means, any one of them has obtained
a prize fuited to his inclination, he immediately darts down upon them

                 and

PLATE CCXXXVII.

and feizes it; or if the Gull attempts to fly, he follows in purfuit, and perfecutes him till he drops his prey, to efcape his fury.

The length of this bird is from twenty one inches to about two feet.

PLATE CCXXXVIII.

NUMIDA MELEAGRIS.

PINTADO, *or* GUINEA HEN.

GALLINÆ.

GENERIC CHARACTER.

Bill ſtrong, ſhort, the baſe covered with a carunculate cere receiving the noſtrils : head horned, with a compreſſed coloured callus : tail ſhort bending down : body ſpeckled.

SPECIFIC CHARACTER

AND

SYNONYMS.

Caruncles at the gape double : gular fold none.

NUMIDA MELEAGRIS : caruncula ad rictum gemina, plica gulari nulla. *Gmel. Linn. Syſt.* 744. *I.*

Phaſianus vertice calloſo, temporibus carunculatis. *Linn. Syſt. Nat.* x. *p.* 158.

Meleagris. *Briſſ. Av. I. p.* 176. *n. I. t.* 18.

Gallina vertice corneo. *Haſſelq. it.* 274.

I 2

Numida

PLATE CCXXXVIII.

NUMIDA MELEAGRIS nigra maculis rotundatis albis, remigibus
    extus albo tranſverſim ſtriatis, vertice corneo.
    Lath. Ind. Orn. 621. *I.*

Gallus et Gallina guineenſis. *Raii Syn. p.* 52. 8.—*Id. p.* 182. 17.
    —*Will. p.* 115. *t.* 26, 27.

Peintade. *Buff.* II. *p.* 163. *t.* 4.—*Pl. Enl.* 108.

GUINEA PINTADO. *Will. (Angl.) p.* 162.

      Lath. Gen. Syn. IV. *p.* 685. *I.*—*Id. Sup. p.* 204.

Theſe birds, originally from Africa, are now naturalized, and become abundant throughout civilized Europe : and in America, as well
as the adjacent iſlands. It is highly prolific, and the fleſh, that of the
younger birds eſpecially, in much eſteem.

The ordinary ſize of this bird exceeds that of the common Cock,
the length twenty-three inches : the female is diſtinguiſhed from the
male by having the wattles of a ſmaller ſize, and of a red colour,
while in the male they incline to blue.

The female lays a number of eggs in the ſeaſon : they are ſmaller
than that of a hen, and of a more rounded form, the colour reddiſh
white obſcurely freckled with darker, and may be hatched under the
Common Hen. It is obſerved of theſe birds, by Dr. Latham, that
they are " fond of having a large range, but if there is much ſhelter,
the hen will often ſecrete a neſt and appear on a ſudden with more
than twenty young ones at her heels." And he further remarks, that
" the ſpecies is very clamorous the day through, having a creaking
                  harſh

PLATE CCXXXVIII.

harſh kind of note, ſomewhat like a door turning on its ruſty hinges, or an ungreaſed axle-tree ; and when at rooſt is often ſo eaſily diſturbed as to hinder the reſt of the family the whole night through, from its noiſe."

There are varieties of this bird, the breaſt being ſometimes white ; and ſometimes the whole plumage white.

PLATE

PLATE CCXXXIX.

COLUMBA DOMESTICA *var.* C.
TURBITA.

TURBIT PIGEON.

PASERES.

GENERIC CHARACTER.

Bill ftraight, defcending towards the tip: noftrils oblong, half covered with a foft tumid membrane.

SPECIFIC CHARACTER

AND

SYNONYMS.

Colours various, generally cinereous, rump white: band on the wing, and tip of the tail black.

Var λ *feathers of the breaft recurvate.*

COLUMBA DOMESTICA, cinerea uropygio albo, alarum fafcia, caudaeque apice nigricante. *Lin. Fn. Suec. n.* 207.
λ C. pennis in pectore recurvatis. *Linn. Syft. Nat.* XII. *I. p.* 280.—*Gmel. Syft. Nat.* 769. 2.

COLUMBA

PLATE CCXXXIX.

COLUMBA DOMESTICA: minor verſicolor, dorſo inferiore albo.—
λ C. pennis in pectore recuvis.   *Lath. Ind. Orn.*
592. 2.
Briſſ. Av. I. p. 75. F.
Friſch. Av. t. 147.
Pigeon à Cravate.   *Buff. Hiſt. Nat. des. oiſ.* 2. 513. *t.* 23.

This is a pleaſing and very curious variety of the domeſtic **Pigeon**, the colour of the plumage generally cinereous, and white : tail at the end, and bars on the wing black : eyelids tumid, bill and legs red ; head and neck white, and the ſides of the breaſt a fine gloſſy green, paſſing backwards in a ſomewhat ſemilunated form, and uniting below the nape of the neck.

But the chief character by which this variety is diſtinguiſhed is the remarkable projecting ruff of feathers on the breaſt, which deſcends from below the throat in a longitudinal direction : theſe feathers open or divide in a perpendicular line, forming two diſtinct tufts, one of which bends outwards to the right, and the other to the left, or in the more ſimple language of ſcience, theſe feathers are recurvate. Our ſpecimen is about the ſame ſize as the common variety of the domeſtic Pigeon.

PLATE

PLATE CCXL.

FRINGILLA VIRENS.

GREENFINCH.

GENERIC CHARACTER.

Bill conic, ſtraight and pointed.

SPECIFIC CHARACTER

AND

SYNONYMS.

Olive, beneath fleſh colour: wing coverts white, in the middle black : wings and tail black.

FRINGILLA VIRENS: olivacea, ſubtus incarnata tetricibus alarum albis, medio nigris, remigibus rectricibuſque nigris.

FRINGILLA CŒLEBS. β fem. Linn. Syſt. I. p. 318. 3.—Gmel. Syſt. I. p. 901.

Fringilla alis et cauda nigris. Le Pinçon à Ailes et queue noires, Briſſ. 111. p. 153. A.—Id. 3vo. I p. 348.— Buff. IV. p. 121. I.

VOL. X.   K   Chaffinch

PLATE CCXL.

Chaffinch var. A. *Lath. Gen. Syn.* 258. 10.

Fringilla cœlebs β. An feminæ varietas? *Lath. Ind. Orn.* 437. 12.

We are not entirely convinced that we are proceeding with fufficient caution in feparating the Greenfinch from the Chaffinch, as two dif-tinct fpecies. Such a feparation is contrary to the opinions of very able writers upon the fubject of Ornithology, but notwithftanding this we are much inclined to believe that future obfervations may juftify the accuracy of our conclufion, or at leaft afford a prefumptive evidence in its favour.

Every practical Ornithologift will admit of this diftinction between the two birds, and be prepared to point out the difference that pre-vails between the two fpecies, and the fexes of each; and hence arifes an obvious difficulty in endeavouring to determine whence this differ-ence of opinion between the practical and fcientific Ornithologift has originated. The naturalift affirms that the Greenfinch is a variety of the female Chaffinch, but in reply to this, the practical obferver points out the two fexes of each kind. It requires, therefore, more than ordinary caution in attempting to combine the two opinions, and we muft finally conclude that the Greenfinch is a mere variety of the Chaffinch, or reject the opinions of the fcientific naturalift as not fully authorized by the facts of nature.

That the varieties of the Chaffinch are numerous, as indeed are thofe of many common birds will be conceded, but when as it appears

that

PLATE CCXL.

that we are enabled to determine from the external afpect of the birds, as well as from diffection that both the fexes of the Greenfinch are diftinctly known, we can fcarcely hefitate to think they muft be fpecifically different.

We have been long fince in poffeffion of what we confider as the two fexes of each of thefe birds, the Chaffinch and the Greenfinch, and have no idea in our mind that they can be the fame. We are aware that this was alfo the perfuafion of the late Mr. Green, a very intelligent obferver of the fmaller tribes of our common Englifh birds, and who was inclined and able to beftow more attention upon this curious fubject of inquiry than moft other collectors; —he was fully fatisfied they are fpecifically diftinct: we have the two fexes of both birds very beautifully fet up by his hands.

It is very well known that the Greenbird and the Chaffinch occafionally affociate: it is alfo known that they migrate feparately in flocks. The Greenfinch, for example, retires from Sweden and Holland in autumn, while the Chaffinch remains, and paffes the winter alone, and is again vifited by their fuppofed mates in fpring. With us in Britain, both the Chaffinch and the Greenfinch remain throughout the year, and yet fometimes flocks of the Greenfinch are feen without a fingle Chaffinch, and again the latter obferved in abundance without any intermixture of the former, precifely as was before obferved of their migrations upon the Continent. With thefe fuggeftions and facts before us, we can fcarcely avoid believing the Greenfinch and Chaffinch to be fpecifically diftinct: we are not inclined to fpeak with too implicit confidence, fince it muft affuredly remain for

K 2   future

PLATE CCXL.

future obſervation to determine the point with any poſitive degree of certainty *.

The Greenfinch is one of the moſt abundant ſpecies of the ſmaller tribes of birds in this country.

---

* Dr. Latham appears to be the only author inclined to our opinion, if we may collect this from the doubtful manner in which he ſpeaks of our preſent bird under the ſpecifical ſynonyms of Fringilla cœlebs. " An feminæ varietas ?" *Ind. Orn.*—" *A variety of the female Chaffinch* ?"

PLATE

241

PLATE CCLXI.

PROCELLARIA PUFFINUS.

SHEARWATER PETREL.

ANSERES.

GENERIC CHARACTER.

Bill toothlefs, a little compreffed, hooked at the point ; mandibles equal : noftrils cylindrical, tubular, truncated, lying on the bafe of the bill : feet palmated : the back toe pointing downwards, feffile fharp and fpur-like.

SPECIFIC CHARACTER

AND

SYNONYMS.

Body above black : beneath white : legs rufous.

PROCELLARIA PUFFINUS: corpore fupra nigro fubtus albo, pedibus rufis. *Brünn. Orn. n.* 119.—*Fabr. Fn. Groenl. n.* 56.—*Gmel. Syft. Nat.* 566. 6.—*Lath. Ind. Orn.* 824. 11.

Avis diomedea, Shearwater, *Raii Syn. p.* 133. *I. et. A.* 2.—*Will. p.* 251. *Id. (Angl.) p.* 332. 334.

Puffinus Anglorum, *Raii Syn.* 134. *A.* 4. *Will. p.* 252.

Le

PLATE CCXLI.

Le Puffin, *Buff.* ix. *p.* 321.—*Pl. Enl.* 962.

Manks Puffin, *Edw. t.* 379.—*Will. (Angl.) p.* 333.

Shearwater Petrel, *Brit. Zool.* 11. N° 258.—*Id. fol.* 146. *t.*
M.—*Arct. Zool.* 11. N° 462.—*Flor. Scot. I.*
N° 198.—*Lath. Syn.* vi. *p.* 406. 11.—*Id. Sup.*
p. 269.

The Shearwater Petrel is about fixteen inches in length, the prevailing colour of the plumage black above, beneath white.

As a Britifh bird the fpecies is almoft entirely confined to the northern fea coafts, particularly to the Calf of Man and the Orknies. As a northern bird the fpecies is known to extend as far as Denmark, Iceland, and Greenland; and it has been befides obferved in the Arctic regions, and in the fouthern feas. Kalm fays it is every where common in the Atlantic, from our channel to the coaft of America.

Except in the breeding feafon thefe birds are chiefly obferved out at fea, and not unfrequently at a confiderable diftance from the land. They frequent the fhores in fpring, about February, March, and April, but only for a fhort time at intervals. During their ftay on fhore, like many other of the fea-birds, they take poffeffion of fome rabbit burrows which are either before deferted, or the inmates of which they expel, and there bring up their young. The female Shearwater lays but one egg, which is blunt at each end : the young are taken in the beginning of Auguft in great numbers, killed, falted, and barrelled, and are eaten boiled with potatoes. 'Some are pickled like the young

of

PLATE CCXLI.

of the common Puffin. The fkins prepared with the feathers on, and alfo the feathers without the fkin, are applied to various ufeful purpofes.

It is obferved of thefe birds, that after the young are hatched, the adult birds are abfent during the day time purfuing their ufual habits of fifhing, but return regularly every evening to feed their young.

PLATE

242

PLATE CCXLII.

FALCO NISUS.

SPARROW HAWK.

ACCIPITRES.

GENERIC CHARACTER.

Bill hooked, the bafe covered with a cere: head covered with clofe fet feathers: tongue bifid.

SPECIFIC CHARACTER
AND
SYNONYMS.

Cere green, legs yellow ; abdomen white undulated with grey: tail banded with blackifh.

FALCO NISUS: cere viridi, pedibus flavis, abdomine albo-grifeo undulato, cauda fafciis nigricantibus. *Linn. Faun. Suec.* 68.—*Scop. Ann. I. p.* 17.—*Gmel. Linn. Syft. Nat.* 280. 31.

Accipiter. *Brifs. Orn. p.* 89. *n. I.*

Accipiter fringillarius. *Gefn. Av.* 51.

Aldr. Orn. I. p. 344. *t.* 346, 347.

VOL. X.   L   *Bell.*

PLATE CCXLII.

Bell. Av. 19. 6.

Ray Av. 18.—*Will. p.* 51. *t.* 5.—*Klein. Av.
p.* 53. 23.

Nifus ftriatus fagittatus. *Frifch. t.* 90, 91, 92.

Sperber. *Gunth. Neft. u. Eyer. t.* 6.

Epervier. *Buff. 1. p.* 225. *t.* 11.—*Pl. Enl.* 412. 467.

SPARROW HAWK. *Alb. I. t.* 5.—*Id.* 111. *t.* 4.

Will. (Angl.) p. 86.

Br. Zool. I. N° 62.

Arct. Zool. 11. *p.* 226. IV.

Lath. Gen. Syn. I. p. 99. 85.—*Id. Supp. p.* 26.

A widely diffufed fpecies, being a general inhabitant of Europe, Afia, and Africa, and we have little doubt of America alfo. The length of the male bird is twelve inches, that of the female fourteen or fifteen inches.

This is a fierce and active kind of Hawk, but not very powerful, the bill being fmall and the legs unlike moft of the Hawk tribe, long and flender. It is trained like the reft for the fports of Hawking, but its attacks are confined to the fmaller tribes of birds, and as the trivial appellation implies, of the fparrow chiefly.

Among the recorded varieties of this fpecies one β is fpotted (Accipiter maculatus of *Brifs,*) and another γ lacteus is entirely of a milky white.

PLATE

243

PLATE CCXLIII.

ALCA IMPENNIS.

GREAT AUK.

ANSERES.

GENERIC CHARACTER.

Bill toothlefs, fhort, compreffed, convex, often tranfverfely furrowed; lower mandible gibbous near the bafe: noftrils linear; legs generally three-toed.

SPECIFIC CHARACTER

AND

SYNONYMS.

Bill compreffed, furrowed: an oval fpot each fide before the eyes.

ALCA IMPENNIS: roftro compreffo ancipiti fulcato, macula ovata utrinque ante oculos. *Linn. Faun. Suec. n.* 140.—*Brünn. Orn. n.* 105.—*Fabr. Fn. Groen n.* 52.—*Pall. Spic.* 5. *p.* 2.—*Gmel. Linn. Syft. Nat.* 550. 3.—*Lath. Ind. Orn.* 791. *I.*

Alca major, *Briff. Av.* 6. *p.* 85. *n. I. t.* 7.—*Id. 8vo.* 11. *p.* 382.

L 2   Penguin

PLATE CCXLIII.

Penguin. *Raii. Syn. p.* 118.

          Will. p. 242. *t.* 65.—*Id.* (*Angl.*) *p.* 322. *t.* 65.

Le grand Pingoin. *Buff.* IX. *p.* 393. *t.* 29.—*Pl. Enl.* 367.

GREAT AUK. *Brit. Zool.* 11. *p.* 229. *t.* 81.—*Arct. Zool.* 11.

        N°. 424.—*Lath. Syn.* v. *p.* 311. *I.*

This is the largeſt ſpecies of that ſingular tribe of aquatic birds which inhabit our rocky coaſts, and are known by the names of Auks; and by our fiſhermen under the more local appellation of "Sea Parrots." The ſize exceeds that of the common Goofe; the length about three feet.

Our figure of this ſcarce and intereſting bird is copied from the well known ſpecimen, originally in the collection of Sir Aſhton Lever, and which we obtained by purchaſe, for our own Muſeum, at a price not very inconſiderable *, that example being at the time alluded to the only one we believe known. Since that period a few of theſe birds have been killed in the Orknies, the exact particulars of which are not, however, within our knowledge. There was formerly a preſerved ſpecimen in the collection of that eminent cotemporary collector of Sir Aſhton Lever, the late Mr. Tunſtall, the fate of which it is no longer poſſible to aſcertain, as that collection was diſperſed by public ſale, long previous to the diſſolution of the Leverian collection; but which, it is believed, was ſuffered to decay through want of care.

---

* Ten Guineas at the public hammer.

Dr. Latham

PLATE CCXLIII.

Dr. Latham fpeaks of this fpecimen: it appeared to him to be a young bird, the oval fpot between the bill and the eye being fpeckled black and white, and the bill itfelf marked with only a few furrows.

The Great Auk is never feen more foutherly in Britain than the Scottifh iflands ; upon fome of which it is known to breed, though in very fparing numbers. Dr. Latham obferves that this bird is fometimes feen on the Ifle of St. Kilda appearing there the beginning of May and retiring in June. It lays one large egg clofe to the fea mark, fix inches long, white irregularly marked with purplifh lines, and blotched at the larger end with black or ferruginous fpots ; and it is faid that if the egg be taken away the bird will not lay a fecond. It is fuppofed thefe birds are hatched late in the feafon, as the young in Auguft are only covered with a grey down.

When thefe birds leave the Scottifh coaft, they retire northward to Norway, the Ferroe ifles, Iceland, Greenland, and Newfoundland. They appear to be common in the Arctic regions, where the natives are faid to ufe their fkins for garments, and otherwife apply them to purpofes of utility and ornament.

The old birds are rarely feen on land: they fubfift principally upon fifh, in queft of which they ufually go out fome diftance to fea, and being expert divers, fhew much facility in the capture of their prey.

PLATE

244

PLATE CCXLIV.

ANAS OLOR.

MUTE SWAN.

ANSERES.

GENERIC CHARACTER.

Bill convex, obtufe, the edges divided into lamellate teeth : tongue fringed, obtufe : three fore-toes conne&ed, the hind one folitary.

SPECIFIC CHARACTER

AND

SYNONYMS.

Bill red, with a black flefhy tubercle at the bafe : body white.

ANAS OLOR: roftro rubro, bafi tuberculo carnofo nigro, corpore albo. *Lath. Ind. Orn.* 334. 2.

ANAS OLOR: roftro femi-cylindrico atro, cere nigra, corpore albo· *Gmel. Syft. Nat.* 501. 47.

Cygnus Manfuetus. *Raii. Av. p.* 136. A. *I.*

Anas Cygnus (Manfuetus) *Linn. Syft. I. p.* 194.—*Faun. Suec.* Nº. 107. β.

Le Cygne. *Buff.* IX. *p.* 3. *t. I.—Pl. Enl.* 913.

Tame

PLATE CCXLIV.

Tame Swan, *Albin.* III. *t.* 96.—*Br. Zool.* II. *N°.* 265. *t.* 60.

MUTE SWAN, *Arct. Zool.* II. *N°.* 470.

Lath. Syn. VI. *p.* 456. 2.

The Mute Swan occurs in a wild ftate in Siberia and Ruffia, whence it migrates foutherly in the winter; in a ftate of domeftication it is very generally diffufed over Europe and Afia.—In England the Swan is under the protection of penal laws and ftatutes, and hence there are few of our principal rivers entirely deftitute of this noble bird,

" The pride of filver ftreams."

The Swan was formerly held in high efteem for the table, and the Cygnets or young Swans are yet in fome requeft in England. In Ruffia the older birds are in as great repute for the luxurious repaft as they were formerly in England, where no great feaft or public entertainment was thought complete without at leaft one of them upon the table. The flefh of the Swan is of a deep red, and of a rich and full flavour, without any ill tafte even in the adult ftate : we have eaten of it and confider it not amifs.

Thefe birds feed on fifh, aquatic infects, plants of various kinds that grow in the water, as the nymphea, &c. and alfo grafs. Though nothing can furpafs the graceful dignity of its motions when failing upon the fcarcely rippled furface of the lake, it walks with an awkward gait, and appears to peculiar difadvantage when on the land.

Swans

PLATE CCXLIV.

Swans build their nefts in the high grafs near the water, laying from fix to eight eggs which are of a large fize and whitifh colour, and are faid to be depofited in fucceffion, one every other day : the females cover the eggs for the fpace of two months before they hatch. They are feen in all their ftages of growth upon the river Thames, where they are efteemed royal property, and are under the protection of the corporation of the city of London. At certain appointed periods the companies are rowed up in the city barges in great ftate to infpect the brood, a ceremony that has obtained the name of *Swan-hopping*. It need be fcarcely added, that ftealing the eggs of the Swan is felony.

When it has attained its full growth the Swan is about five feet in length, or rather more, the weight twenty-five pounds : the young birds are brown, and do not affume the perfect whitenefs of their plumage till they have become of pretty confiderable fize.

INDEX to VOL. X.

ARRANGEMENT

ACCORDING TO THE

SYSTEM OF LINNÆUS.

ORDER I.

ACCIPITRES.

ORDER II.

PICÆ.

ORDER III.

ANSERES.

Anas

M 2

INDEX.

ORDER IV.

GRALLÆ.

ORDER V.

GALLINÆ.

ORDER

INDEX.

ORDER VI.

PASSERES.

VOL.

INDEX.

VOL. X.

ARRANGEMENT

ACCORDING TO

LATHAM's SYNOPSIS of BIRDS.

DIVISION I. LAND BIRDS.

ORDER I. RAPACIOUS.

ORDER II. PIES.

ORDER III. PASSERINE.

ORDER IV. COLUMBINE.

ORDER

INDEX.

INDEX.

VOL. X.   N

I N D E X.

V O L. X.

ALPHABETICAL ARRANGEMENT.

Parafiticus

I N D E X.

GENERAL

SYSTEMATIC ARRANGEMENT

OF

SELECT BRITISH BIRDS,

ACCORDING TO LINNÆUS.

ORDER I.

ACCIPITRES.

Eagle, Falcon, Owl, and Shrike or Butcher Bird.

EAGLE.

					Plate
Falco Ossifragus.	Sea Eagle	-	-	-	105
Haliætus.	Osprey	-	-	-	70
Milvus.	Kite	-	-	-	47

BUZZARD.

				Plate
Falco Apivorus.	Honey Buzzard	-	-	30
Apivorus, var.	Honey Buzzard variety		-	233

FALCON.

				Plate
Falco Peregrinus.	Peregrine Falcon	-	-	53

VOL. X.   O   *Falco*

CONTENTS.

ORDER II.

PICÆ.

Crow, Roller, Oriole, Cuckow, Wryneck, Woodpecker, King's-fisher Nuthatch, Hoopoe, Creeper.

CROW.

CONTENTS.

CROW.

O 2   *Picus*

CONTENTS.

ORDER III.

ANSERES.

Swan, Goofe, Duck, Merganſer, Auk, Petrel, Pelican, Diver,
Grebe, Gull, Tern.

SWAN.

CONTENTS.

SWAN.

GOOSE.

DUCK.

Anas

CONTENTS.

PELICAN.

CONTENTS.

PELICAN.

DIVER.

GREBE.

GULL.

CONTENTS.

ORDER IV.

GRALLÆ.

Spoonbill, Heron, Ibis, Snipe, Sandpiper, Phalarope, Plover, Avofet, Oyfter-catcher, Coot, Rail.

SPOONBILL.

STORK, HERON.

IBIS

CONTENTS.

IBIS.

SNIPE.

SANDPIPER.

VOL. X.   P   *Tringa*

CONTENTS.

ORDER

CONTENTS.

ORDER V.

GALLINÆ.

Peacock, Turkey, Pheafant, Pintado, Grous.

P 2

Tetrao

CONTENTS.

ORDER VI.

Passeres.

Pigeon, Lark, Stare, Thrush, Chatterer, Grosbeak, Bunting, Finch, Fly-catcher, Warbler, Wagtail, Titmouse, Swallow, Goat Sucker.

Pigeon.

Lark.

Stare.

Ouzel

CONTENTS.

OUZEL THRUSH.

CHATTERER.

GROSBEAK.

BUNTING.

FINCH

CONTENTS.

FINCH.

FLY-CATCHER.

WARBLER.

Motacilla

CONTENTS.

Printed by R. Gilbert, St. John's Square, London.